Ethnobotany of Jalgaon District,

Maharashtra

Ethnobotany of Jalgaon District, Maharashtra

Shubhangi Pawar
M.Sc., M.Phil., Ph.D.
Reader in Botany,
Pratap College, Amalner – 425 401
District Jalgaon (M.S.)

D.A. Patil
M.Sc., Ph.D.
Reader in Botany,
Post-Graduate Department of Botany,
S.S.V.P.S's L.K. Dr. P.R. Ghogrey Science College,
Dhule – 424 005 (M.S.)

2008
DAYA PUBLISHING HOUSE
Delhi - 110 035

ISBN10 81-7035-515-X
ISBN13 978-81-7035-515-1
EAN 9788170355151

Published by : **Daya Publishing House**
1123/74, Deva Ram Park
Tri Nagar, Delhi - 110 035
Phone: 27383999
Fax: (011) 23260116
e-mail : dayabooks@vsnl.com
website : www.dayabooks.com

Showroom : 4760-61/23, Ansari Road, Darya Ganj,
New Delhi - 110 002
Phone: 23245578, 23244987

Laser Typesetting : **Classic Computer Services**
Delhi - 110 035

Printed at : **Chawla Offset Printers**
Delhi - 110 052

PRINTED IN INDIA

Dedicated to

Late Reverned Sane Guruji

for his Love, Dedication and Service for the Cause of Humanity and Nation

Preface

Ethnobotanical knowledge is the wisdom developed by mankind over many generations for proper utilization of the ambient plant-wealth. The man has always remained successful on Planet Earth due to his ability to exploit the natural resources in his surrounding. Nay, his civilization, economic thoughts and systems are interwoven inherently with such resources. Human societies all over the world began to dwell in organized patterns like hamlets, villages and then in urban areas. Nevertheless majority of them still remained almost static over many centuries. Thus, such primitive or semi-primitive human communities are the 'living fossils' reflecting unevolved to less-evolved cultural heritage even in modern times.

Attempts to record plants and their medicinal significance in India were initiated by the Portuguese. The first book, 'Colloquios dos simples e drogas e cousas medicinals da India', was authored by Garcia da Orta from Goa in 1563. Later, it was followed by the publication of 'Tracado de las drogas medicinas de las Indias Orientalis' authored by Christobal Acosta in 1578. However, the first major work on the floristics and Ethnobotany of Malabar was undertaken by the Dutch which resulted in a treatise in Latin, 'Hortus Malabaricus', published in classic 12 volumes during 1676-1693. It was compiled by Hendrik Adriaan Van Rheede, the then Dutch Governor of Cochin. (*cf.* Manilal, 1999). William Roxburgh (1832) during his floristic inventories recorded indigenous medicinal herbs from south India. George Watt (1889-1896, 1908) contributed significantly to the traditional plant-based uses in India. His work truly reflected 'Ethnobotany' and 'Indigenous Knowledge' in our country. Works carried out by the botanists particularly Dr. E.K.Janaki Ammal and Dr.S.K.Jain and their persuations nurtured the subject of Ethnobotany in India.

In recent times, the factors such as the accelerated loss of indigenous plantlore, consideration of plantlore as a part of national heritage, research on natural products, discoveries of new drugs from plants and opening of alternative approaches to healing are responsible for explosion of interest in ethnobotany. Moreover, going back to nature and herbalism encompassing both healthcare and beauty-care products caught on worldover in the last few decades. World Health Organization is also promoting the concept of 'traditional medicines' for developing countries. The new patent policy under GATT further inspired these trends throughout the world.

The literature survey warranted a dire necessity for in-depth ethnobotanical inventory in the district. Our pains-taking efforts in the region are shaped in the present book-form. We are deeply in debt to Professor Dr.R.M. Pai, Ex-Head, P.G. Department of Botany, Dr.B.A. Marathwada University, Aurangabad (Maharashtra), for his constant inspirations throughout the period of work. Many tribal and rural people kindly gave us their time, benefits of their knowledge and secrets, besides ever-time hearty welcome during field work. We would sincerely like to thank them all. We are grateful to forest officials, authorities of Botanical Survey of India, Western Circle, Pune; Agharkar Research Institute (M.A.C.S.), Pune; colleagues, friends and respected Principals of our colleges for help, facilities and inspiring words. Our endeavour will be fruitful if the readers find this book useful.

Dr. D.A. Patil

Dr. (Mrs.) Shubhangi Pawar

Contents

ETHNOBOTANICAL ENUMERATIONS

SECTION-I: TRIBALS

SECTION-II: NON-TRIBALS

SECTION-III: VERNACULAR PLANT NAMES

APPENDICES

Introduction

Ethnobotany encompasses the total, natural and traditional man-plant relationships. It reveals the role played in the culture of people, their social behaviour, intellectual achievements and survival. It recognizes the important role of the ambient vegetation in the economic life of people. It was earlier defined as the inter-relationship between the primitive humans and plants. This concept has undergone changes and presently it is no more restricted to tapping down data of plants used by aboriginal people. Apart from botany and ethnology, this science includes areas of folklore, mythology, sociology, ecology, forestry, agriculture, medicinal sciences, phytochemistry, pharmacology, economics, linguistics and literature. It has by nature emerged as an interdisciplinary science. The scope of ethnobotany has now greatly enlarged, both in terms of its theoretical contributions to an understanding of man-plant relationships, as well as for the practical applications of the biological knowledge of aboriginal people in medicine, health, agriculture and industry.

The ancient man gained an insight into the knowledge of selective uses of different plants or their parts or products. However, their methodology is still a subject of many investigations. The curing properties of certain plant species and their parts might have been discovered by ancients. By trial and errors, they would have acquired knowledge of deciphering which plant could be used as food and which could be avoided. This knowledge is passed on from generation to generation orally. However, in the last two centuries, this important knowledge has been engendered and slowly becoming extinct because of distraction of forests and rapid acculturation effecting the tribal cultures. This pressure and the renewed interest in natural drugs, foods and other necessities of mankind made ethnobotany an organized science.

The people of Jalgaon district, a part of Khandesh region of the state of Maharashtra, have their own culture, customs, rites, taboos, totems, myths, proverbs, folk songs, foods etc. The tribals such as Bhils, Pawaras, Tadvis and Vanjaras dwell in inaccessible or less accessible forest pockets of Satpura mountains. Moreover, the rural folks constitute the major segment of population of the district. They still depend largely on nature for essential requirements of their day-to-day life. An intimate relationship still exists between them and the surrounding plant wealth. Undoubtedly, the role of indigenous culture of this region is reflected in its plantlore. It appeared that their indigenous knowledge about their ambient vegetation is facing danger of being lost to oblivion by different forces of pressures.

Review of literature suggests that this district has not been fairly investigated ethnobotanically. Presently, ethnobotany has become an important discipline of research and development in resource management, conservation of biodiversity and socio-economic development of a region. It can inform on the biotic interference in this region of the state. Apart from the miscellaneous uses, tribal or folk medicines are still their secrets. The role and extent of utilities of plants have yet not been studied judiciously. The results of present investigations would help blend resources into development. It was, therefore, thought worthwhile and at the suggestion of senior author Dr.D.A.Patil, to undertake a detailed ethnobotanical investigation in the district.

The man-plant relationship is broadly categorized into two *viz.*, concrete and abstract. The former includes mainly the material use, whereas the latter considers the importance of faith in good or bad powers of plants, festivals and games, worships, sacred plants, taboos etc. The results of studies of either of these are incorporated in this contribution.

GEOGRAPHICAL POSITION

Jalgaon district lies between 20° and 21° north latitude and 74°55′ and 76°28′ longitude. On its location in the upper Tapi basin, it forms a distinct topographical unit separated from neighboring Madhya Pradesh state by Satpura and from the south by Satmala hilly ranges. An arc of extended ranges of Sahyadris stretches in the easterly direction. From east to west the river Tapi passes through the district with a deep cut-up bed and black soil fertile plains along its bank. The district is bordered in the north by the boundaries of Madhya Pradesh and, in the south, it is continuous with the periphery of Aurangabad district. Limits of Madhya Pradesh again together with Buldhana district of Maharashtra are in its east. Nasik and Dhule districts of Maharashtra constitute its western borders (Map-II).

Jalgaon is one of the thirty one districts of Maharashtra, situated at the northern most border of the state (Map-I). It is a part of erstwhile region of Bombay Presidency. 'Khandesh' was divided in 1906 into two districts *viz.*, 'West Khandesh' and 'East Khandesh', which were, later on, renamed after the district headquarters as Dhule (or Dhulia) and Jalgaon districts respectively.

Jalgaon district stretches nearly 128 km along the Tapi river and varies in breadth from 112 to 144 km. Jalgaon forms an upland basin, one of the most northerly sections of the Deccan plateau. The district at present comprises thirteen administrative tehsils *viz.*, Jalgaon, Bhusawal, Raver, Yawal, Muktainagar (Edlabad), Chopda, Erandol, Amalner, Pachora, Jamner, Parola, Bhadgaon and Chalisgaon. The district headquarter is located at Jalgaon. The total forest area in the district is 72685.27 hectares. It is distributed among 305 villages accounting for 20.20 per cent of the villages in the district.

TOPOGRAPHY

Jalgaon district includes varied topographical features and landscapes, consisting of wild hills and forests, stretches of barren plain, low rolling rocky hills and densely gullied topography near major river banks. Regionally, from east to west, parallel with the Tapi, are well marked belt of the country in the center, rich Tapi valley, in the north, the high and wild Satpura, and in the south and south-west, bare ridges and rich well-watered valleys flanked by the Ajanta range.

The Tapi bank is high and bare. North of the Tapi, the whole length of the rich alluvial plain is bounded by the steep southern face of Satpura–a belt of mountain land from 32 to 35 km broad. The northern boundary of the district is marked by the longitudinal depression of the Aner river and its eastern counterpart the Mamat tributary of the Suki river. These two longitudinal valleys separate the southern ranges of the Satpura from their northern terrain.

Within Jalgaon limits are three chief hill ranges, the Satpura to the north, the Hatti in the south-east, and the Ajanta or Satmala in the south. The Satpuras, a broad belt of mountain land, stretching in a wall-like line along the north bank of Tapi, rise from the first range of hills, ridge behind ridge, to the central crest about 675 m high and then slope gently to river Narmada.

GEOLOGY AND SOILS

The physiography of the district is made up of high hill ranges on the north, alluvium in the center and low hill ranges to the south of Tapi. On the north, the hill ranges stretch east-west and form part of Satpuras, the highest peak being about 1175 m. The Aner river flows east-west and forms the northern boundary of the district.

Among important tributaries of the Tapi are the Purna on the east and the Girna on the south-west. Besides the Tapi and its main tributaries, numerous streams originating from the hilly ranges find their way to the main rivers. Deccan traps cover almost the whole of this district, except a few strips of alluvium-covered land on both sides of the major streams. These trap rocks are the result of outpouring of enormous lava, which spread over vast areas of western, central and southern India at the end of mesozoic era. They came through long, narrow fissures and cracks in the earth-crust and spread out as nearly horizontal sheets. They are called 'plateau basalt'.

The trap rocks are abundant in labradorite and enstatite-augite with varying proportions of interstitial glass which on alteration gives rise to secondary minerals like palagonite, chlorophaete, iddingsite etc. Sometimes, porphyritic basalt is seen showing phenocrysts of felspars and glassy matters. Magnetite occurs as

Map I: Showing the State of Maharasthra in India

MAHARASHTRA

minute discrete grains amidst other minerals as well as in the glassy ground mass. In few cases, olivine is also present. The intertrappean beds represented by clays, sandstones and limestones formed in depressions during quiescent period between the successive eruptions of lava flows have not been reported so far from the district. But they are likely to exist at places.

The trap weather with characteristic spheroidal exploitation gives rise to large rounded boulders on the cutcrops and are very common throughout the district. The trap soils produced by erosion and weathering is deep brown to rich red or black (regur). These black soils are very rich in plant nutrients. They are sticky when wet and on drying, produce conspicuous cracks. Another alteration products are laterite formed in tropical climate under alternate dry and wet seasons. It is porous, pitted, clay-like rock of variegated colours and has a limonitic crust on the exposed surface. It is soft

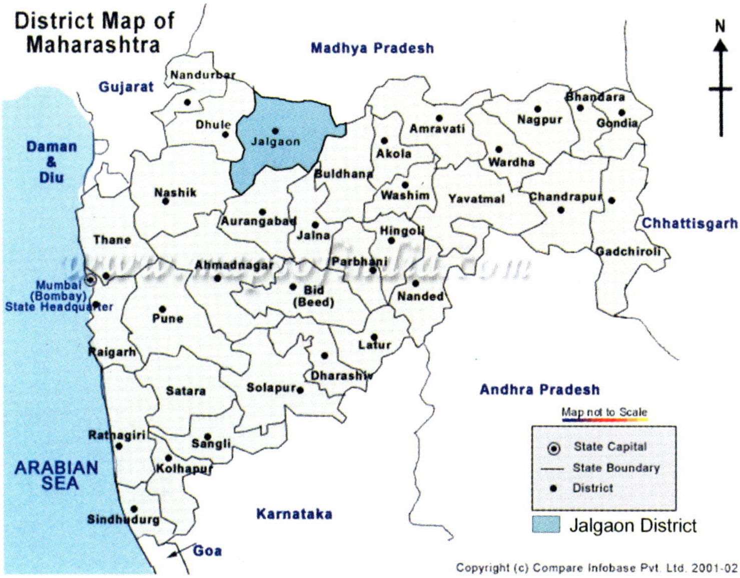

Map II: Showing Jalgaon District in Maharasthra

when freshly dug but on exposure to air it quickly dehydrates and become quite hard.

Sporadic patches of laterite are recorded at places capping the hills in southern parts of the district. The river Tapi flows from east to west throughout the district forming a strip of alluvium-covered land on both sides. Though alluvium layers are much thicker and wider on the west, they thin out on the east and trap rocks are exposed near Bhusawal. These alluvium layers are composed of reddish and brownish clays, with intercalation of gravel and 'kankar'. To the east and south-west are the valleys of the rivers Purna and Girna respectively.

Jalgaon district is composed entirely of trap rocks which are generally barren of any economically useful and important minerals. Secondary minerals like agate, jasper, chalcedony and rock crystals, which fill up the cavities in vesicular basalts are sometimes used as semiprecious stones. However, trap is used extensively for building purposes as road metal, railway ballast and in cement concrete. Kandar, the small nodular calcareous matter, is abundant in all black soils. On burning, the material yields almost hydraulic lime, which is locally used in making plaster and in colour washing.

The soils are essentially derived from the underlying basalt, though older alluvium forms deep cover all along broad Tapi valley. The fertility of these soils depends largely on their position vis-à-vis relief and their nearness to the stream courses. The central belt of the wide Tapi valley, about half of the area, consists either of a black alluvial clay, highly retentive of moisture, or of a loam overlying a stratum of yellowish clay of good depth. Along the bank of rivers, the soil is mixed or overlaid with lime nodules. At some places, the surface is entirely washed away. Occasionally remnant patches or strips or rich alluvial deposits are seen.

Hot springs are recorded in the district. The Unapdeo hot spring lies about four km north-west of Adavad in Chopda taluka and about half km from the grit spur of Satapuras. The rocks are trap and the ground is hard and black near the spring. It has temperature of 140°C. The water is testless. It has peculiar smell. It is believed to cure skin diseases. Another similar hot spring is in the same range in Chopda taluka–Sunapdeo and Najhardeo. The water is slightly sulphurous.

GENERAL VEGETATION

The forests of Jalgaon district are of the tropical, dry deciduous type wherein the *Tectona grandis*, *Hardwickia binata*, *Boswellia serrata* are the predominant species. The vegetation varies with the changes in altitude, aspect and rainfall. As these factors are variable, various subtypes of forests are met with in the region. In the eastern region of the district, in general, the dominant forest trees are *Hardwickia binata* Roxb. and Boswellia serrata Roxb. However, these are being eliminated due to illicit cuttings. In the eastern part, the Pal forest is dominated by *Hardwickia binata* Roxb. Manudevi forest is of *Tectona grandis* L. type, whereas Padmalaya forest is dominated by *Boswellia serrata* Roxb. The different subtypes of forests may be considered into: (I) *Tectona grandis* L. type (II) *Hardwickia binata* Roxb. type (III) *Boswellia serrata* Roxb. type and (IV) Scrub type.

Tectona grandis L. Type

Tectona grandis L. chiefly occurs in the eastern as well as western part of the district. This type is generally found in reserved as well as protected forest areas. The best forest of this subtype are seen at Manudevi (Yawal taluka), Eklagna (Erandol taluka), Satrasen and Unapdev areas (Chopda taluka). The other trees generally observed are *Anogeissus latifolia, Acacia ferruginea, Boswellia serrata, Garuga pinnata, Diospyros melanoxylon, Dalbergia paniculata, D.lanceolaria, D.latifolia, Haldina cordifolia, Mitragyna parvifolia, Madhuca longifolia* var. *latifolia, Schleichera oleosa, Schrebera swietenioides, Terminalia crenulata, Soymida febrifuga, Bridelia retusa, Butea monosperma, Bauhinia racemosa, Cassia fistula, Wrightia tinctoria, W.arborea, Holarrhena pubescens, Cordia dichotoma, C.macleodii, Salmalia malabarica, Cochlospermum gossypium, Toona ciliata* etc.

Hardwickia binata Roxb. Type

This type of forest occurs in eastern as well as western part of the district. The chief localities of these forests are in eastern part at Pal (Raver taluka) and in western part at Lasur (Chopda taluka). The other common trees in these forests are *Hardwickia binata, Boswellia serrata, Tectona grandis, Garuga pinnata, Mitragyna parvifolia, Sterculia urens, Terminalia crenulata, Schrebera swietenoides, Moringa concanensis, Diospyros melanoxylon, Cochlospermum gossypium, Cassia fistula, Holoptelea integrifolia* etc. The other common plants in these forests along nallas and in the plains are *Helicteres isora, Securingea leucopyrus, S.virosa, Grewia flavescens, G.villosa, Rhus sinuata, Dichrostachys cinerea, Mimosa hamata, Acacia polycantha, Maytenus emarginata, Lantana camara, Carrisa congesta* etc.

Boswellia serrata Roxb. Type

This subtype is chiefly found in Padmalaya forest (Erandol taluka). *Boswellia serrata* is very common in this area. The other common trees found in these forests are *Holarrhena pubescens, Wrightia arborea, W.tinctoria, Madhuca longifolia* var. *latifolia, Anogeissus latifolia, Acacia polycantha, A.leucophloea, Diospyros melanoxylon, Holoptela integrifolia, Ougeinia oojenensis, Cordia dichotoma, C.macleodii, C.gharaf, Cassine albens, Dalbergia paniculata, D.lanceolaria, Pongamia pinnata, Cassia fistula etc. Common shrubs in these forests are Mimosa hamata, Grewia flavescens, Maytenus emerginata, Euphorbia neriifolia, Dichrostachys cinerea, Rhus sinuata, Securingea virosa, Carrisa congesta* etc.

Scrub Forests

In this type, the area is covered with open grassland with few scattered shrubs and trees. These occur in the plain areas near Padmalaya, Eklagna and Pal. A large number of rivulets and streams drain in these areas. The common tree and shrub species in scrub forests are *Acacia polycantha, Mimosa hamata, Dolichandrone falcata, D.atrovirens, Balanites aegyptiaca, Azadirachta indica, Butea monosperma, Securingea leucopyros, S. virosa, Flacourtia indica, Lantana camara, Maytenus emarginata, Acacia leucophoea, A.nilotica, Dichrostachys cinerea, Ziziphus caracutta, Z.glaberrima, Capparis sepiarla, Euphorbia neriifolia, Opuntia* spp. etc.

CLIMATE

The climate in this district is generally dry except the monsoon period. The year may be divided into four seasons. The cold season from December to February is followed by the hot season form March to May. June to September is the monsoon season and October to November form the post monsoon season.

The climatic factors that appear to be of greatest importance are (a) Temperature, (b) Rainfall, (c) Humidity, (d) Wind.

Temperature

December is the coldest month with the mean daily minimum temperature 11.9°C and mean daily maximum 29.8°C. Cold waves which pass over northern India sometimes affect the district and minimum temperatures may drop down to within two degrees of the freezing point of water and slight frost may occur. Temperature begin to rise steadily from about the beginning of March. May is the hottest month of the year. The mean daily maximum temperatures riches 42.5°C. The highest maximum temperatures recorded at Jalgaon is 47.8°C. Temperature drops appreciably with the onset of monsoon by the end of the first week of June. The monsoon period is generally pleasant. With the withdrawal of the monsoon by the end of September day temperatures rise a little in October and both day and night temperatures begin to drop rapidly by November. In general, in the Satpuras the heat is somewhat tempered by the forest. The Tapi valley is the hottest part of the district.

Rainfall

The average rainfall in the district is 740.7 mm. About 81 per cent annual rainfall is received during monsoon months June to September, July being the month of highest rainfall. The central part of the district comprising Jalgaon, Bhusawal, Jamner and Pachora Tehsils get a little more rain than in the rest of district. The details of the rainfall since 1993 to 2001 are given in Table-I. The rainfall is heavier in the hilly ranges of northern part of the district. The heaviest rainfall during these years was in 1993. It was 1049.3 mm.

Humidity

Except during the monsoon the air is generally dry, particularly in the afternoon. Summer is the driest part of the year.

Winds

In summer, winds are generally light in the mornings and blow from direction between south west and north west. In the afternoon they strengthen and blow to a northerly direction. In the monsoon month winds are stronger and predominantly south-westerly. During the post monsoon and winter months the winds are light.

Table 1: Showing Average Annual Maximum and Minimum Temperature and Rainfall from 1993 to 2001

Sl.No.	*Year*	*Temperature (°C)*		*Rainfall (mm)*
		Maximum	*Minimum*	
1.	1993	43.3	09.9	1549.3
2.	1994	43.1	10.8	705.9
3.	1995	41.6	10.0	668.4
4.	1996	42.5	10.9	626.1
5.	1997	42.5	10.9	1023.8
6.	1998	44.1	8.3	583.4
7.	1999	42.3	10.4	555.0
8.	2000	42.3	10.2	632.59
9.	2001	42.2	10.1	554.54

HISTORY

In the following, the origin, social set-up, customs, religions, occupation etc. of the people of Jalgaon district in general and that of tribal people in particular have been briefly discussed. The investigator feels that such observations are necessary for studies like ethnobotany since rural and other ethnic communities are the primary sources of ethnobotanical information. This will substantially explain and prove the needs of the uses of plants or their products in the region under study.

Jalgaon district is the eastern part of the Khandesh region of the state of Maharashtra. The early name of the Khandesh was Rsika. Under the Yadavs this region constituted part of the Seunadesa during 10-12th century. It later became known as Khandesh under Muslim rule. It was divided in 1906 into 'East Khandesh' and 'West Khandesh' under British rule, the district headquarters were Jalgaon and Dhule (Dhulia) respectively. These districts are renamed as Jalgaon and Dhule districts after their headquarters in 1960.

The earliest formal reference of Rsika is found in the Ramayana where search of Seeta was ordered. Until a few years ago, the prehistory of Jalgaon district was shrouded with obscurity. Recently the Tapi and Girna valleys have been brought to light in reference to material culture and the history of man as a tool using animal since his nomadic state.

Paleolithic tools made from Deccan trap and quartzite are known from the banks of the Tapi, a region now part of Dhule district. Achulian, hand axes and cleavers are known to occur in the open river bed of Changdev. Microlithic tools are also recorded at many places in the Tapi and Girna river systems. They are made up to agute, chalcedony, jasper and crystal. In the Jalgaon district these have been known from Changdev and Patan (on Patna) near Chalisgaon (Plate 16).

Excavations at Bahal and Tekwada villages about 18 miles off north of Chalisgaon on the right and left banks of Girna respectively, have revealed painted black-on-red pottery of Chalcolithic period (1st Millenium B.C.-500 B.C.). Hatched diamonds, criss-cross and wavy lines, ladders etc. are painted on their necks and rims. These excavations also revealed shells and semi-precious stone beads. Copper was scarcely used for artistic work. Excavations at Bahal also produced glass-wares attributed to 4th-3rd century B.C. roughly corresponding to the period of Ashoka, the Mauryan ruler. Punch marked coins are also obtained at this site. Inscriptions at Pitalkhora caves, near north of Chalisgaon and cave number 10th-12th of Ajanta show prevalence of Buddism in Jalgaon district. Chalolithic site is also discovered near Vakad near Pahur. Ruling of Satvahanas is indicated by the caves of Pitalkhora during 2nd B.C.to 1st century A.D. Then Abhiras or Ahirs are of considereable importance in the history of Jalgaon. Towards the close of 13th century, this region was under the reign of Vakatus, Sendrakas, Rashtrakutas and lastly the Yadavas. Since 1295 upto 1760 it was under domain of Muslims. After Yadavas, the name of this region was changed to 'Khandesh' to suit the name of 'Khan' given to the Faruqi king Malik II by Ahamad I of Gujarat. Khandesh was under the regime of Peshwas (Maratha period). From 1760 till 1818. Afterwards, it was under the British domain till independence of India.

THE PEOPLE AND THEIR CULTURE

Jalgaon is located in the upper Tapi basin and comprises its Tehsils *viz.*, Jalgaon, Bhusawal, Raver, Muktainagar (Edlabad), Yawal, Chopda, Erandol, Amalner, Parola, Pachora, Jamner, Bhadgaon and Chalisgaon. Jalgaon is the district headquarter. It has 17 towns and 1510 villages. According to 1991 census, the population of the district is 3,87,634. The aborigines inhabit the hilly forested areas of Satpura mountain. While a few of them also have settled in the plains. The forests are mostly found in Chopda, Yawal, Raver, Edlabad and Chalisgaon Tahsills, while in others small disturbed forest pockets are observed.

The district has predominantly Hindu population. Every household is found to contain some small corner where the family images are kept in Devharas and worshipped. The public places of worship and prayer are the temples of Lord Mahadeo, Vishnu and Ganapati. Popular forms of Vishnu are Vithoba and Rama temples. There are other temples like Maruti, Dattatraya, Goddess Bhavani and those of local deities. They also worship gods and goddesses such as Bhairav, Bahiroba, Jakhai, Kalimata, Janai, Kelkai, Mukai, Nawalai, Phiranga, Satvai, Takai, Vaghoba, Khandoba, Mhasoba and Vetal. These are represented mostly by carved stone images. Maruti and Mhasoba are besmeared with Shendur (red ochre) and oil. Every village has Maruti (Hanuman) temples on the boundary of village and sometimes inside. On Saturday it is worshipped and strings of flowers especially of Ruchkin or Rui [*Calotropis gigantea* (L.) R.Br.] are used to garland Maruti. The most feared spirit by the rural folk is Mhasoba. Most of the deities are promised of a sacrifice of a good fowl or coconut. People in the district on some occasions also visit and worship the gods/goddesses in the state of Maharashtra and even outside.

Inhabitants of this district observe usually two rites *viz.*, (i) Rotpuja, (ii) Chakrapuja. The former is a worship in honour of Kajubai. It is performed on the first sunday after Nagpanchami in 'Shravan' month. A strict fast is observed on the worship by all members of the family. The latter is in honour of Ranubai. After worship, dinner is served and it is strictly observed that no remnant of food is wasted or given away. The surplus food is buried in a pit near the house. 'Atithipuja' is another practice in Khandesh. No head of the family would take food without a guest, waiting for a guest while throwing Kakabali (a portion of food for crows and dogs) was/is enjoyed by the house hold.

There are two chief styles of houses in rural areas locally called (i) Dhaba type (ii) Chhapper type. The former type lasts many ears if head of family repairs. The walls are of entirely clay or of stones and lime or cement. The flat roof rests on strong teakwood (*Tectona grandis* L.) beams running wall to wall. Over the beams is laid a layer of strong branches of trees and a coating of leaves of sugarcane (*Saccharum officinarum* L.) or stems of cotton stalks (species of Gossypium), pigeon pea [*Cajanus cajans* (L.) Millsp] The latter type has clay walls or partition walls of cotton stalks or rarely branches and other plant material. Of late, the use of burnt bricks, cement and concrete is largely used to construct 'pakka' house. The staple food in the district consists of jawar, bajra, wheat, rice and few pulses. The edible fruits, vegetables and other plant products are as those mentioned elsewhere in the thesis. Animal food or products of animals are also part of the diet of some communities. Use of spices and condiments as in other parts of Maharashtra are observed in this region. All classes of people have habit of drinking tea in morning and afternoons.

The articles of Jalgaon people are not materially different from those in other parts of Maharashtra, except the dressing pattern of tribal communities. In case of men wears, Dhoti and Sadra (Kurta) fashion now prevails mostly in rural and mostly in peasant's families. The old, barabandi, kopri, angarkha and dagla are now not in vogue anywhere. The old fashion of headdress consisting of turban, rumal, patka and safa has been replaced in the last few decades by white caps. But it has also ceased as necessity. A full nine meter saree and a bodice constitute the dress of grown up women in rural as well as urban areas. The length of saree in modern times has been cut short to six meters only. The dresses of boys and girls now widely differ. Ladies prefer golden and silver ornaments as in other parts of Maharashtra, except those of the aborigines.

Marriages and the funerals are similar to Hindus elsewhere in the state and do not differ much.

ETHNOLOGY OF TRIBES

Bhil

Bhils are the largest tribe of western India and the third largest in the country. They are concentrated in some parts of Rajasthan, Gujarat, Madhya Pradesh and Khandesh region of Maharashtra. Bhils are the oldest indigenous tribe of India. It is thought that the word 'Bhil' denotes the ancient community expert in using bow (and arrow). The words 'Bhil, Beel or Billa, all mean bow. They have been called in Sanskrit texts 'Vanputras' (sons of forest) and Nishadas. A tale in puranas refers to the origin of Nishadas from Lord Mahadeo. One of his children was born from a beautiful woman who lived in forest. He was ugly and ill tempered. He was cursed of Lord Mahadeo for killing his sacred bull. Lord Mahadeo turned him out into the forest, the descendants of whom are known as Nishadas or Bhills. They are known for their bravery, carefree disposition and excellence at archery. They are spread all over Satpura mountain in Khandesh region and have also settled down in the plains (Plate 12).

A typical Khandesh Bhil is dark, well built, stunted, active and hardy with high cheek bones, wide nostrils and in some cases coarse features. An inquiry in this regard revealed that they belong to different groups of sub-castes such as 'Garasiya Bhils', Dungari Bhils and Bhilta. Their dialect Bhill is a mixture of Marathi, Gujarati and even Hindi (Hindustani) especially in Khandesh region.

Bhils mostly reside in the hilly areas of Satpura mountain. Usually they occupy the tops of isolated hills. Most of them still live in thatched huts or houses. The roofs and partition walls are made up of bamboo or stems of Nirgundi (*Vitex negundo* L.), Karvi (*Carvia callosa* Nees, leaves of [*Butea monosperma* (Lam.) Taub.], teak (*Tectona grandis* L.*f*.) and few other forest plants. Walls of huts and houses are usually of unburned bricks, stones and mud. Their abodes are generally divided into two parts – the front is varandah and the inner part comprising one or two small rooms. Bhills settled in the plains are generally concentrated on one side of a village. In such case, their place of residence is referred as Bhilati. In the plains, the roof is usually made up of cultivated plants like jawar (*Sorghum bicolor* Moench.) and leaves of sugarcane (*Saccharum officinarum* L.). Species of *Gossypium* etc. Each household has glasses of metal or earthware, earthen, wooden or metal platters, few cooking utensils, a hand mill, a stone slab with roller, a one or two cots and few guilds, fishing nets, a bow and arrows, an axe or sometimes a sword. A cow, a few fowls and one or two dogs are the only live stock generally reared by them.

Main food of Bhils are jawar [*Sorghum bicolor* (L.) Moench.] maize (*Zea mays* L.), onion (*Allium cepa* L.), chilli (*Capsicum annum* L.), apart from produce of wild plants. Very few of them practise agriculture. Now-a-days, they are usually watchmen (Jaglya or Rakholya) on fields of others farmers. They also like flesh of goats and fishes. They are mostly addicted to wine drinking. Wine selling also partly helps their sustenance. The Bhil dress is simple. A male wears traditionally a turban (but replaced by a cap in modern times), langoti (loin cloth) and 'Sadra or Kurchi' as shirts. The women wear coarse tattered sarees and 'Choli' (blouse). Girls wear skert or 'Lehenga' and 'Polka'. Men like ear-rings. Ladies are fond of ornaments like 'Zabal' or 'Bali' in ears, 'Chapalya' of coins of rupees and chains around neck, 'Kada' on arms, bangles on hand wrist, 'Lachhya' (Chains) around foot ankle and 'Nathani' as nose rings (Plate 12).

Bhils differ in their religious beliefs and practices from place to place. Some worship only the tiger god called 'Vagdev'. Most of them are devotees on Mata, Lord Mahadeo, Hanuman and goddess Durga, apart from the local deities like Khandoba, Kanhoba, Bahiroba, Mhasoba, and Shitalmata. They worship the spirits of their ancestors and believe in witchcraft and omen. Their witchcrafts are called Badvas. Self-made musical instruments are *e.g.* Dhol, Dholki, Pawri, Flute etc. Folk songs and folk dances of various kinds performed at different ceremonies are the other ethnic characteristics of Bhils which need mention.

They have no temples of their own. They ride sacred images in opens or open sheds. Stones are consecrated and set on a mud terrace. They hold Banaras sacred and visit other shrines in places like Nasik and Jejuri. Their chief festivals are Holi and Dasara when goddess Durga is appeased by them by sacrificing goats.

The marriage ceremonies are very much akin to the other classes of Hindus. A marriage proposal is made from the bridegrooms side. Then a formal proposal is made by the father of a boy. The girl is brought out and seated among the guests. A packet of sweetmeats is offered to the girl on behalf of father of the boy. An astrologer fixes a day of betrothal. A saree, bodice-piece and sweetmeats are offered at the girls house. A final announcement is made. The father of boy gives her dowry called 'ghun' in Bhili language. After feast the evening is spent in music and dance. Next day the father of a boy fixes the day of marriage. Widow marriage is settled in them. A married woman can get divorce and remarry the man of her choice if the proposed husband is ready to pay the first money which he had spent at the time of first marriage of the woman.

On the point of death of a Bhil, his relatives distribute money among the poors in his name. After death more or less similar method is practiced of carrying the corpse by four bier bearers to the burial place. The body is laid in the grave. Cooked rice and bread are placed in the mouth of deceased. It is covered with earth. The funeral party then bathes and accompanies the chief mourner to his house. A fire is lit in front of his house. Some women's hair are burnt in it. Each mother takes neem leaves (*Azadirachta indica* A.Juss) and throws on fire. Open hand palms passed through the smoke is rubbed on face. Thus one is purified and returns to his home. On the third day, a neem twig steeped in cows urine is rubbed on right shoulders of bier bearers to wash the oil or milk applied on them.

Pawara

Pawara is a distinct tribe and inhabit especially in Satpura mountain in Khandesh region of Maharashtra. They are also found in the adjacent state of Gujarat. They are actually 'Rathwa Palkwara' but generally referred simply as 'Pawara'. 'Barela Pawara' is another sub-caste in this region. Erroneously, Pawaras are sometimes considered as a sub-caste of Bhils and hence referred as 'Pawara Bhils'. However, Pawaras are clearly distinctive in respect to their festivals, customs, rites, sacraments, beliefs, gods and goddesses, dialect and even the physical features. Their origin is also different from those of Bhils. They are originally resident of Pawagad in Gujarat and hence called 'Pawara' and are Rajput in origin. They migrated before 500 to 700 years and some of them settled in Satpura mountains. Pawagad is also mentioned in their folk tales and folk songs. During Muslim rule in India, some of them left their residence in Rajasthan. Famines and diseases in those days are also thought other reasons for their migration.

Pawaras are fairly coloured, fairly tall and good stature, eyes being brown and nose straight. Their dialect is Pawari. It has the impact of Gujarati, Rajasthani and Hindi language. In Khandesh region, Pawari has received some impact of dialects like Ahirani, Bhili or Bhilori and Marathi pronounciation of some words differ from place to place. Pawaras have their own songs, tales, proverbs, idioms etc. Their dialect is fairly rich. Pawara's dressings and ornaments are traditional and quite different from other tribal people in the region. Male wear dati (Dhoti) which extend below up to knees only. Those who live interior of forests wear Khutti (waist cloth). They wear Topi (Cap) or Pagadi (Turban) as head-dress and Daglaw (Shirt). If some body visit their houses, they immediately put on shirts and wrap towels around waist. Girls wear Nati (Saree), Falka and Dagli (Skert). Ladies wear Nati (nine meteral saree) and Dagni (blouse). Old ladies use Kachav (Choli or blouse) and Ghagara. Gents and ladies both like to wear silver ornaments. Males use silver ornaments like Hakul (Chain) in ears, Hakuw (Chain around waist), Ukarta (Chain around neck) and Wawa or Pewa (Rings) round wrist. Ladies wear silver ornaments like Bewatua or Batye, Hawala and Bahut on upper part of arms; Dewa, Silva, Gokhara or Gakhara on wrist; Kangetla and Kidnya-ridu in ears, Ahadi-Hakul, Tabali (neckless of silver coins); Pingna Tuda, Wala, Wakia and Sawad around legs, Anguthi mundi and Kilya on fingers of hands and legs; Pansale and Nathani in nose etc. (Plates 1–5 and 9).

Maize (*Zea mays* L.) and jowar [*Sorghum bicolor* (L.) Moench.], bajra [*Pennisetum americanum* (L.) K.Schum.] contribute their staple food. Rice is also used especially on festival days. Pulses form part of their diet. Wild plants are also added seasonally depending upon their availability and necessity. They enjoy non-vegeterian meals occasionally. They are very fond of Mahuwa wine.

Pawaras consider marriage as an essential part of their life. Unmarried persons are hated and in their opinion such persons are not welcomed in 'Swarg' (heaven). They do marry with paternal and maternal uncle's daughters. They have maintained 'Kula' system and strictly observe prohibiting marriages within Kulas. In interior region Pawaras follow polygamy and is thought prestigious. Different trends while marring are prevalent in Pawara community *e.g.* Jabri vivah (marriages by force), Gharghushi vivah (by entering the house of a boy). Devara-Babhi vivah (marriage with brother's wife after death of one's brother), widow marriage, remarriage, arranged marriage and love marriage. In every case, it is the father of a girl who receives dowry (Dej) which is decided by the wise people of their community.

Marriage ceremonies fall under three types *viz.*, (*i*) Aytula viyav, (*ii*) Churyo viyav, (*iii*) Gudekudan viyav. These generally depend upon the financial conditions of the two parties. During Ayatula viyav, turmeric paste is not applied to body of a bride and bridgegroom. Simple useful cloths are worn by them. The family deity called Girhun is worshipped at the beginning of marriage ceremony. Ceremony of Churyo viyav is completed at night. About 5 to 6 persons from girl's side gather at boy's house. The family deity is worshipped. Simple feast is served at night. The ceremony is declared completed after it. Total programme is completed during night period before dawn without knowing to anybody. Hence this form of marriage is called Churyo viyav. The third form of ceremony that is Gudekudan is also known by other names like 'var viyav', 'motali chhak' and 'Avid of Halad viyav'. Turmeric paste is applied to the boy and girl to be married. The marriage ceremony may last for four days. On every day good feast is served. Only prestigious and rich family can observe this type of ceremony.

Interestingly, the two weekly bazar days before 'Holi' festival are very important for selecting life partners. A weekly bazar of fortnight before the 'Holi' is called 'Gulalya Bazar', on this day Pawaras from all villages meet each other, express good wishes and enquire about each other. On such day, young girls and boys enjoy the bazar day, observe and try to select their life partners. If the boy and girl are in love or like each other, the boy apply gulal (red powder) to the girl of his like. They ran away and get married. To select and marry in this way in thought by Pawaras very prestigious and feel proud of it. Another weekly bazar, just a week ahead of 'Holi' is known as 'Bhongarya Bazar', Pawaras gather at the place of village of Bhongarya Bazar with drums. They come in groups of each hamlet and each group is headed by their police-patil. Police-patil wears shirt, coat and 'Dhoti'. Another person hold umbrella on his head. All dance rhythmically holding bows, arrows and sticks in hand. There is also a competition of drum biting. Meanwhile, the young girls and boys meet each other and run away intending their marriage.

Pawaras have animistic faith and are deeply steeped in superstitions. They worship gods and goddesses. Smooth stones are smeared by red lead and are placed in open on raised platform of earth and stones or under the shade of trees. No temples are constructed. They establish their family deity in house which is made up of teak wood of 2.5 to 3 feet length. The goddesses Uli (or Huui, Hawavi) is worshipped at the time of Holi festival. Indiraja (God Indra) is most dear god of Pawaras. Likewise, they worship Bapdev (Vaghdev) so that they are protected from diseases. Sitamata or Dhartimata (who nurses them), Govidev (Gavdev) (who protect their village), Kovder (who give light and show path), Atthi (who gives warm), Hudovdev (his worship is thought of good beginning of various ceremonies), Karhan or Kansari (Goddess of food grains), Khandi-raja (worship of bullock of god), Virdati (Creator of universe), Rani Kajol (Goddess of happiness), Ochhawat (Hanuman) and many such other, *e.g.* Bilatdev, Lakdiv Rana, Kumbhaj Rana, Choho Rana, Sohy Rana, Chand hurej (moon and sun) Kharivahulo (Goddess of wind), Bijlakuwar (Goddess of lightness), Dudodarya (God of sea), Menghjo Taro (Polar star) etc..

Festivals are very dear to Pawaras. The most commonly celebrated festivals are Olpujavu or Oputtar (worship agricultural instruments), Milapi pujan (worship of forest), Duda-mukay (worship of maize), after which pawaras allowed new food grains Kakada nawy (worship of cucumber and vegetable), Naway (worship of rice after which Pawaras are allowed to consume it as food). Indal (means god Indra), Diwali, Pola (festivals of bullocks) and few others.

Pawaras live in large hut-like houses which range in size from (approx.) 75x50 ft to 100x50 ft. These are constructed using teakwood. Each pillar is large and long carved out of single trunk of teak tree.

The walls are mostly made up of bamboo splits or stems of Karvi (*Carvia callosa* Nees). The roof is covered by slopes of tiles. The walls are also thatched by using local plant materials. Huts are without or with inner partition walls. These is a single door as a main entrance. In one part or in a adjacent part of huts, the domestic animals are kept. The hut is two tiered. In the upper part, mostly the food grains, agricultural implements, musical instruments, fishing nets, weapons etc. are kept.

The household things generally observed are Chambu (drinking pot), Guda masha (four legged wooden structure for keeping beds and shawls, Dusar (water vessel), Kuthi (small food-grain container), Mothi or Muwa (large food grain container), Gatli (Food grain grinder), Khatulu (Cot), Khanjano (stone mortor), Cudu (strong pestle), Muhav (wooden pestle), Ukhavavu (mortor) and utensil like Kalachi (large spoon), Chatu (large wooden spoon), Sansanbhi (pincer), Thawi (plate) etc.

The funeral rite in Pawaras is called maylanpan. The method of carrying corpse is similar to those of Hindus.

Vanjara

Vanjaras of Jalgaon district are said to be denizen of Rajasthan. They were primarily trading nomads in search of trade of food grains, spices and salt etc. This community is divided into four sub-castes *viz.*, (i) Charan, (ii) Mathuria, (iii) Labhan and (iv) Dhadi. Sometimes, Charan are also called Gor or Govar in the state of Maharashtra. The word Vanjara also signifies them as moving about in forests. The Vanjaras of this district were primarily the traders of bullocks. Their temporary settlements were called 'Tanda'. Now they have settled more or less permanently in all talukas of Jalgaon district. However, they have concentrated more in Chalisgaon, Raver and Jamner talukas. They are generally referred as 'Gor' of Govar Vanjaras in this region.

Gor Vanjaras have stout body of 5 to 5.5 ft height. Straight nose, larger eyes and black hair are amongst their prominent physical features. Their ancestors or old persons were/are found growing curled hair on head and turban too. However, this pattern is being obscured in modern times. Males usually wear shirts, Dhoti, Pagote (Turban) or a cap on head. The dresses of females are colourful and very artistic. They were Phetya (Parkar), Kachali (Blouse) and Phadki

(Odhani). The Phetya is composed of different parts called 'Chit', 'Lawan', 'Lepo' and 'Chid'. The Kachli is hand-woven and composed of different parts such as 'Khadpa', 'Dariyaachi' and 'Peti'. They stitch their own clothes. Vanjara women are very fond of wearing ornaments and fashionable articles of different kinds. They fix a 2-3 ft long stick on the head upright called 'Chundo'. 'Topali' or 'Ladi-ghugri' tied in hair, Bindi on forehead and 'Tarota' in ears are very characteristic. Other ornaments of females are 'Balya' (made of white plastic), 'Kangani' (of silver) both in hands; 'Kalka' and 'Bichawa' (of silver) both in legs; and the coins of Indian rupees, costing one or two, are strung and wear around the neck. Males wear 'Kalda' of silver in hands and legs and also a ring in fingers (Plate 13).

Different classes of Vanjaras have their own goddesses (called Kuldaiwat) *e.g.* Tulaja-Bhawani of Rathod, Kankali of Jadhav, Motimata of Chavan and some others worship Dhulmata. Male goats are offered usually to their goddesses to fulfill their wishes.

They celebrate the festivals like 'Diwali', 'Dasara', 'Holi' and 'Tij' or 'Dhamoli' in the month of 'Shravana' (Plate 14). The newly born child, especially a male child, wears white clothes only after his first 'Holi' festival. On the second day of Holi, he is advised to pray 'Holi' and he can wear white clothes. They celebrate and enjoy a programme called 'Dhund' on this day. A long wood is bitten by sticks. While doing so, they recite or pronounce some 'Shlokas'. They have also another interesting function. A pole is fixed in the foreground of a house. It is brother's wife, called Bodi, obstructs him from removing it from the place of fixation. Male Vanjara tries, time and again, to pull out the fixed pole. While pulling out it, he is bitten by sticks by his bodi. Other ceremonies of Hindus in the region are also celebrated by Vanjara people.

Marriages are performed only at the houses of bridegroom. They do not marry with a daughter of their maternal uncle. The ladies sing songs at night a day before marriage. This programme of singing at night is called 'Vadav'. A hot needle is touched on left arm of a bridegroom. This is called 'Vadai'. They dance at the time of marriage.

Generally, two days before the marriage ceremony, the boy and girl are rubbed with turmeric paste. They sit side by side on low wooden stools on marriage day. The hems of their garments are tied in the accompaniment of music. Priests repeat auspicious verses and women on both sides sing songs and sprinkle handfuls of jawar [*Sorghum bicolor* (L.) Monech] on heads of couple. On the second day, the couple is bathed together. The women stand round them and sing songs. After this, fathers of the boy and girl interchange presents of turbans and waist-cloth. On the third day, the priest is pelted with onions (*Allium cepa* L.) and shells.

Widow marriage is allowed but as far as possible, a woman should not leave her family. Husband's younger brother has claim over her for wife-hood. The fact is noted in caste council and no ceremony is necessary.

Vanjaras have their own dialect known as Vanjari. In Maharashtra, they speak the state language Marathi and also Ahirani in Jalgaon district, a local dialect of Khandesh region.

Their huts are made up of wood, grasses and leaves with slanting roofs on both sides and have narrow entrance. As they are leading a settled life in this region, they also construct permanent houses like other rural people. Houses are built using stones, bricks, mud, wood etc.

Due to modernization and attainment of a settled life in this region, they prefer farming and animal husbandry. They are hard toiling and also work on daily wages for agricultural work.

Tadvi

Tadvi is a small community inhabiting in and at the foot hills of Satpura mountains. It is considered that during the period of Mughal emperor Aurangajeb, some Bhils are converted to Islam. This tribe developed a mix culture under influence of muslims and the previous culture as a Hindu. Their language is also 'Ahirani', a local language of Hindus near Satpuda. It is Urdu mixed type. Tadvi are heighted and dark in colour. Men wear 'Dhoti', shirt and 'Topi', while ladies wear nine meter saree with Gujarathi style 'padar'. Ladies prefer silver ornaments like thick bangles called 'patalya', 'tode', black garland as 'Mangalsutra' and 'wakya'. Tadvi leave in huts of mud called 'Kudachi zopadi'.

Tadvi celebrate 'Gudhipadva', 'Sarva-pitri Amawasya', Nag-panchami, Dipawali, Moharum, Ramjan and Bakri-id. These people are non-vegetarian but they never eat beaf of cow just like other Hindus. In the houses of Tadvi, Khandoba, Bahiroba and Pir are worshipped.

Tadvi never weds with muslims. Brahmin or a priest is essential to determine the date of engagement (Muhurta). After engagement, feast of sweet rice is arranged for the relatives. In presence of 'panch', a leader of society, relatives of a bride offer 'Shela' and 'Pagadi' to the bridegroom, while the relatives of bridegroom offer saree to the bride.

At the time of marriage ceremony, black garland as 'Mangalsutra' is offered to bride and both visit to the mosque. Kazi (Muslim priest) reads the 'Lalama' and then marriage ceremony takes place as per the rites of Islam. Like other tribes, in Tadvi, there is also a council of village elders called panchayat which looks after the social life of people. All cases of civil or criminal nature are heard and disposed off. Fines are imposed as a punishment for all sorts of offences. If the fine is not paid, the criminal is discarded from the tribe.

FESTIVALS IN JALGAON DISTRICT

The festivals in a region bring life and gaiety in public life and also dispel the prevailing monotony of life. Most of the festivals are in honour of some deities. Festivals, which are observed in other part of the state of Maharashtra, are also observed in Jalgaon district. The ceremony or major ones are: Gudhi-padva, Ram-navami, Hanuman-jayanti, Gauri-pujan, Akshay-tritiya, Vat-savitri pournima, Nag-panchami, Narali-pournima, Janmastami, Ganesh-chaturthi, Navratra and Dasara, Diwali, Tulashi-vivah (Khopdi), Makar Sankrant, Mahashivratra, Holi or Shimga and few others. Some of these are celebrated by using entire plants of their parts or plant products. The subject matter is, however, explained in part II of the book.

GAMES AND RECREATIONS IN JALGAON DISTRICT

Games in the district fall under two categories, major games and minor games. Hututu (Kabaddi), Langadi and Kho-kho are usually played in educational institutions, while 'Vitidandu' and 'Atyapatya' are also fairly common in the rural parts of the district. Wrestling is also common but mostly noted at the time of pilgrims. Minor games are many *e.g.* sivasivi, kolanti-udi, ghoda-ghoda, agagadi, palkhi, andhali-kosimbir, lapandav, badabadi or raparapi (or abadhobi), toba, vagh-bakri, surparambi, playing of sagargoti (*i.e.* seeds of *Caesalpinia bonduc* (L.) Roxb.), songatya and phugadi. The phugadies are of different types such as donda-phugadi, bus phugadi and bhui-phugadi. Some other variants of phugadi *viz.*, Zimma, Kombda and Pinga are also noted in this region. Kite flying and swinging are usually observed at the time of Sankranti and Akhaji festivals respectively.

The traditional forms of religious and devotional sermons like purana, kirtana, pravachana and bhajana provide entertainment. Gondhalies are a community by themselves and devotees of the goddess Ambabai in whose honour they perform 'gondhal' on some auspicious occasions like marriage. Tamasha performance has a great popularity in the region.

CROPS IN JALGAON DISTRICT

Jalgaon being predominantly agricultural district of the state, a large variety of plants are cultivated. These are as follows.

Food Plants: Cereals, Millets and Pulses are Major Crops

Cereals and Millets

Bajara [*Pennisetum americanum* (L.) K. Schum.], Jowar (*Sorghum bicolor* Monech), Nagli [*Eleusine coracana* (L.) Gaertn.] Rice (*Oryza sativa* L.), Maka (*Zea mays* L.).

Pulses

Tur (*Cajanus cajan* (L.) Millsp), Harbara [*Cicer arietinum* (L.) Millsp], Udid [*Vigna mungo* (L.) Hepper.], Mug (*Vigna radiata* L.), Kulith [*Macrotyloma uniflorum* (Lamk.) Verdc]. Math [*Vigna aconitifolia* (*Jacq.*) Morechal], Chawali (*Vigna unguiculata* (L.) Walp. subsp. *sublobata* (Roxb.) Verdc.

Condiments and Spices

Lasun (*Allium sativum* L.), Kanda (*Allium cepa* L.), Mirchi (*Capsicum annum* L.), Kothimbir (*Coriandum sativum* L.)

Sugarcane

Us (*Saccharum officinarum* L.) It is one of the important crop of the district.

Fibre Yielding Plants

Kapus (*Gossipium arboreum* L., *G.herbaceum* L.), Ketaki (*Agave vera-cruz* Mill.), Ambadi (*Hibiscus cannabinus* L.).

Oil Yielding Plants

Bhui-mug (*Arachis hypogea* L.), Til (*Sesamum orientale* L.), Erandi (*Ricinus communis* L.), Kali Til [*Guizotia abyssinica* (L.*f.*) Cass] Suryaphool (*Helianthus annuus* L.) etc.0

Edible Fruits

Papaya (*Carica paypaya* L.), Mosambi [*Citrus sinensis* (L.) Osbeck.), Amba (*Mangifera indica* L.), Peru (*Psidium guayava* L.), Bor (*Ziziphus mauritiana* Lamk.), Dalimb (*Punica granatum* L.), Keli (*Musa paradisiaca* L.) etc.

Vegetables

Bhendi [*Abelmoschus esculentus* (L.), Moench.], Pankobi and Phoolkobi (*Brassica oleracea* var. capitata and *B. oleracea* L. var. *botrytis*) Khira (*Cucumis sativus* L.), Gajar (*Daucus carota* L. var.*sativa* DC.), Dodke [*Luffa acutangula* (L.) Roxb.], Gilke [*Luffa cylindrica* (L.) Roem], Tamate (*Lycopersicon esculentum* Mill.), Mula (*Raphanus sativus* L.), Wange (*Solanum melongena* L.), Methi (*Trigonella foenum-graecum* L.), Palak (*Spinacia oleracea* L.), Tandulja (*Amaranthus roxburghianus* Herski.), Alu [*Colcoasia esculenta* (L.) Sch.], Dhemse (*Citrullus fistulosus* Stocks.), Kashiphal [Cucurbita moschata (Duch.) Poir.], Bhopala (*Cucurbita maxima* Duch.), Dhudhi Bhopala [Lagenaria siceraria (Molina) Standley] etc.

PLANT BASED INDUSTRIES

Jalgaon is predominantly a agricultural district. The general pattern of non-agricultural employment leans considerably on village and cottage industries. These can be categorized into two types *viz.*, (i) Agricultural based industries, (ii) Forest based industries. The former type vitalizes the products of agriculture as raw materials. The primary processing of crops to convert them into suitable forms for human consumption are dehusking of paddy and pulses, crushing of oil seeds, and flour milling or for use industries like cotton ginning, pressing, spinning and weaving. The major by-product of sugar manufacture is molasses and bagasses. Distilleries are also run for utilizing molasses. The latter type of industries utilizes the forest product like timber for wood-working industries *e.g.* saw mills, furniture, packing cases, bullock carts, packing cases, bidi-wrapping, bamboo working etc. Naturally, these industries are patterned to suit the needs, although partly of people of the district.

PREVIOUS ETHNOBOTANICAL EXPLORATIONS IN THE STATE

The Indian region is amongst the richest ethnobotanical treasures in the world awaiting full documentation, utilization and conservation. The forests are species-rich and provide a valuable pharmacopoeia and gene pool to improve crops. Man has a long history in our country. It was one of the cradles of agriculture in the world and wide spread in it by 2000 B.C. India with diverse flora, coupled with large number of aboriginal tribes inhabiting ecologically different regions, offers immense scope for ethnobotany.

The resurgence of ethnobotanical investigation in our country is recent, only about 40 years old. Ethnobotany was initiated by Dr.E.K.Janki Ammal (1897-1984). She studied subsistence of food plants of certain aborigines of south India. She opened up an "Ethnobotanical Section" of Botanical Survey of India at Allahabad in 1960. Dr.S.K.Jain searched intensively among the tribals of central India. He triggered ethnobotanical activity in India by impressing on the scientists of different disciplines. During the last two and half decades, work has been initiated at NBRI Lucknow. HBPGR Delhi, CCRAS and CCRUM. It has now become a part of curricula of Indian universities.

As far as state of Maharashtra is concerned, Late Dr.V.D.Vartak, his associates and students extended ethnobotanical observations in the western part of Maharashtra. Their mission is still continued in this region. The data on ethnobotany in Maharashtra is largely available through their efforts (*cf.* Ghate, 1998; Ghate and Vartak, 1996; Ghate, Vartak and Kumbhojkar, 1990; Gadgil and Vartak, 1973, 1974, 1976, 1981; Gaikwad, 1988; Godbole, 1934; Godbole; 1994, 1996; Gunjatkar and Vartak, 1982, 1983; Kulkarni and Kumbhojkar, 1992a,b,c, 1993, 1996, 1999, 2000; Kulkarni, Kumbhojkar and Khedkar, 1998; Kumar, Kulkarni and Singh, 1980; Kulkarni, Kumbhojkar and Nipunge, 1990; Kumbhojkar, Kulkarni and

Upadhye, 1991; Kumbhojkar, Upadhye and Kulkarni, 1996; Kumbhojkar and Kulkarni, 1998; Niiengaonkar, Vartak and Chitre, 1985; Nipunge, Kumbhojkar and Vartak, 1998; Nipunge, Vartak and Kumbhojkar, 1992; Nipunge, Kulkarni and Vartak, 1993; Upadhye and Kumbhojkar, 1992, 1996, 1998; Upadhye, Kumbhojkar and Vartak, 1986, 1988; Upadhye, Kumbhojkar and Kulkarni, 1997a-b, 1999; Vartak, 1957, 1959a,b,c, 1980, 1981a,b, 1982a,b; Vartak and Gadgil, 1980, 1981, 1990; Vartak and Kulkarni, 1987; Vartak and Mandavgane, 1981; Vartak, Kumbhojkar and Ghate, 1985; Vartak, Kumbhojkar and Nipunge, 1987; Vartak and Ghate, 1990, 1994).

Some others, as a part of their floristic studies, also added information from this region (*cf.* Almeida and Almeida, 1981; Anonymous, 1982; Caius and Radha, 1939; Jayanand Tosh, 1981; 1996; Jayanand Tosh and Vartak, 1959; Ryan, 1903; Tribhuvan and Peter, 1992; Vedprakash and Mehrotra, 1987). Scientists of Botanical Survey of India, Western Zone, also added few publications as a part of their floristic study in the state. (*cf.* Anand Kumar, Kulkarni and Singh, 1981; Ansari and Rao, 1973; Ansari and Nayar, 1978; Ethoven, 1930; Janardhan, 1961, 1963; Kamble and Pradhan, 1980; Lisboa, 1887; Mudaliar, Kulkarni and Sharma, 1987; Rao 1960, 1964; Santapau, 1956; Shah, 1984; Shah, Yadav and Badrinath, 1983; Sharma and Malhotra, 1984). Few research papers have also been added over years (*cf.* Chandras, 1966; Datar and Vartak, 1975; Deokule and Magdum, 1992; Kirtikar, 1906; Sensarma, 1984; Shekhawat and Sunil Anand, 1984; Tarafder, 1984; Tekale, 2002).

Vidarbha region of the state of Maharashtra has made a headway in this line (*cf.* Badhe and Sharma, 1981; Badhe and Pandey, 1988; Badhe, Sharma and Pandey, 1997; Bhogaonkar and Devarkar, 2001a-b; Chute and Tiwari, 1999; Das and Misra, 1988; Deshmukh and Dhore, 1994; Deshmukh, Muratkar and Rothe, 1999; Giri, 1994; Malhotra and Moorthy, 1973; Mirashi and Salpekar, 1975).

From Konkan region of the state are added few publications (*cf.* Ghate and Sane 1994; Kothari and Moorthy, 1987, 1996; Kulkarni, 1968).

Kolhapur district of Maharashtra also has been surveyed ethnobotanically (*cf.* Khandekar and Badve, 1968; Mahajan and Divan, 1968; Majgaonkar and Phadke, 1999; Sharma and Kulkarni, 1980; Upadhye, Kumbhojkar and Vartak, 1986).

Recently Marathwada region of the state has received some attention of botanists (*cf.* Bhutkar, 2002; Gupta, Hussain and Imam, 1997; Korekar, and Papdiwal, 2002; Naser Rafiuddin and Vaikos, 2002; Rathor, Zare, Kagne and Chavan, 2002; Rohidas and Bankar, 2002; Wangikar, 2002).

Khandesh region, especially Dhule and Nandurbar Districts have been investigated through the efforts of Borse, Bhamare and Patil (1990), Bhamare (1995, 1998a,b), Bhamare and Patil (2005), Yadav and Bhamare (1989), Tayade and Patil (2003a,b, 2005, 2006). As far as Nasik districts are concerned, Patil and Patil (2000a,b,c, 2001, 2002, 2005a,b,c, 2006a,b,c,d), Lakshminarasimhan and Sharma, (1990); Sharma and Lakshminarasimhan (1986) published few research papers.

Pawar and Patil published recently some ethnobotanical information on Jalgaon district (2000a,b, 2001, 2003, 2004, 2005a,b, 2006a,b,c,d), Karnik (1966) added information on some medicinal plants from this region. Pawar, Patil and Patil (2004, 2005, 2006) and Patil, Pawar and Patil (2002, 2004, 2006) have made contribution on this line from north Maharashtra region. Patil (1998a,b, 2002), Patil and Patil (2007) also added information on Khandesh Region of Maharashtra.

The survey of pertinent literature warranted that Jalgaon district has largely remained unstudied ethnobotanically. The present efforts undertaken, therefore, appear amply justified. These efforts have generated vast indigenous knowledge and will be very much relevant to the conservation of biodiversity and sustainable use of plant resource of this district.

METHODOLOGY

Field work forms a important tool in ethnobotanical studies. Intensive and extensive ethnobotanical surveys were conducted in different parts of the district during 1997–2004. Information has been collected regarding plant part used, purpose of use, methods of processing and preparation of drug, its dosage and administration. Apart from the rural folks, tribals such as Bhils, Pawaras, Vanjaras, Tadvis etc. inhabit in this district. The ranges of Satpura mountain extend on northern boundary of the district. The offshoots of western ghats (Sahydri mountain) also extend on western part of this district. The district has dry deciduous type of forests. Besides common tribal and rural people, particularly their senior men and women, Bhagats, Buwas, Maharaj and heads of tribal communities were interviewed. These interviews were centered on the uses of plants in their surroundings. Routine enquiries, observations, folk gatherings, pilgrims, ceremonies added and aided a considerable portion in the methods of this study (Plates 3–5, 9, 14, 16, 17, 20–22). The forest guards and rangers also helped to contact some people of interest in the district. Different forest pockets, talukas, villages, hamlets (padas) were visited in different seasons to collect information from nearly all corners of the district. Mostly the knowledgeable informants were taken to the field and then collection of plants for voucher specimens, uses of plants etc. as revealed by them were recorded. To determine the authenticity of data accrued during fieldwork, repeated verifications from different informants and in different areas were done. The reliable information was thus obtained. In order to make these information useful to general public in future, special attention was also paid to record local names of plants. Even the reasons for such naming of plant/s was/were particularly enquired. Tribals and rural people sometimes explained very interesting origin of these names. Festivals, pilgrims and temples were visited on some special occasions to note down the plant part used for different worships. This was also done enquiring famous Pujaris or Brahmins in the different cities and villages. For proverbs, elder ladies and gentlemen were initiated through discussions. Their actual wording, meaning and occasion/s on which they are said or used, were enquired. Many common people in everyday life helped on this line. Folk songs were similarly studied. In case of songs, riddles, puzzles and proverbs, apart from English, they are also given in Devnagari script in parenthesis. They are then followed separately by their respective theme.

During ethnobotanical field studies plant specimens, their parts or products were gathered. Each collection, place of collection was noted in field diaries. The traditional method of pressing, drying, preserving and mounting was followed that of Jain and Rao (1977). The identification was done using district, state, regional and national floras. The family of the plants was deciphered by using

Patil's key to the angiospermic families of North Maharashtra (1992). The flora such as (a) The flora of British India Vol. I-VII by Hooker, J.D. (1872-1897), (b) The flora of the presidency of Bombay Vol. I and II by Cooke, T. (1958), (c) Flora of Maharashtra state: Monocotyledons by Sharma B.D. *et.al* (1996) (e) Flora of Marathwada Vol.I and II by Naik V.N. *et.al* (1998). (f) Flora of Dhule and Nandurbar district (2003) by Patil D.A. (2000), Forest Flora of Jalgaon District by Kshirsagar, S.R. and D.A.Patil (2007) and few other floras published from the state of Maharashtra were of particular help in identification of species. The herbarium specimens and the botanicals have been housed in the Herbarium, Department of Botany, Pratap College, Amalner (District Jalgaon), Maharashtra.

Few photographs of plants or their parts and articles collected during field trips have been provided to authenticate the information. For ethnological information, it was obviously easier to observe, note and ask the necessary information with regard to caste, tribe, ceremonies, rites, customs, dressings, ornaments, their abodes and physical features too. Few interpreters for language/dialects were of particular help throughout the course of this study. Few craftsmans, artists and folk or tribal singers were also consulted to better the information. This study focused wild as well cultivated plants used by tribal, rural and urban and semi-rural areas of the district. The urban population was also of help on some occasions to give information. Many house wives in urban societies and particularly the old ladies or grandmothers were of particular help in case of time tested house-hold remedies.

The aboriginal or folk uses of these plant species presently included are compared with those of: (i) Wealth of India (Anonymous, 1948-1976), (ii) Dictionary of Economic Products of India (Watt, 1889-1899), (iii) The Useful Plants of India (Ambasta, 1986), (iv) Dictionary of Indian Folk Medicine and Ethnobotany (Jain, 1991). It is given under 'Notes From Classics'. This is adopted to bring the information gathered from this district given under 'Ethnobotanical Notes' in a clearer focus from cross-cultural point of view or also to point out as new information. Recent literature is also consulted. Critical notes have, therefore, been added. The cross-cultural ethnobotanical study is aimed at assessing the creditability of folk claims of Jalgaon district and the rest part of our country.

Ethnobotanical information presented in part I of the book given under classical literature are terse in the following sequence: Root, stem, bark, leaf, flower, fruit, seed and then plant product like latex, gum etc. In case of ethnobotanical information from Jalgaon district, the recent botanical name, family name, local name. The plant families in Part I of the book are arranged according to Bentham and Hooker's system with some accepted splitting of the families. The genera and their species are arranged alphabetically. However, it is not followed in Part-II of the book to suit the data obtained.

Ethnobotanical Enumerations

Part I
CONCRETE RELATIONSHIPS

Family: Ranunculaceae

Clematis heynei M.A.Rao

Local Name

Ranjai

Ethnobotanical Notes

Medicine

1. Juice of roots, about 10 ml, is drunk with a glass of cow milk to cure haematuria. It is advised daily at night for few days.
2. Extract of leaves, about a spoon, mixed with black pepper (*Piper nigrum* L.) and some sugar, stops vomiting.

Notes from Classics

Leaves are used as a remedy in leprosy and also to cure fever.

(Watt, 1889-1893)

Plants are used in leprosy.

(Anonymous, 1948- 1976)

Plants are used in leprosy. Stem yield fibres.

(Ambasta, 1986)

Plants are used for boils and itches.

(Jain, 1991)

Critical Notes

Root juice is effective to stop vomiting. This is a first report.

Family: Annonaceae

Annona squamosa L.

Local Name

Seetaphal

Ethnobotanical Notes

Medicine

1. Seeds are crushed in urine of cow and applied on head to prevent hair fall.
2. Seed powder is applied on scalp to kill lices.
3. Seed powder, about 1 gm, mixed in water is taken orally for abortion. It is administered till effect.

Veterinary Medicine

1. On the wound of domestic animals seed powder is applied for healing.

Notes from Classics

Roots are purgative. It is used in acute dysentery. Leaf paste is applied on abscess, for suppuration of tumour and prolapse anni in kids. Dried unripe fruit powder causes abortion. Ripe fruits with salt are applied to malignant tumour. Fruits are good for digestion.

(Watt, 1889-1893)

Roots are drastic purgative. Leaves are used as poultice for suppuration of skin eruptions.

(Anonymous, 1948-1976)

Seeds are abortifacient and yield fatty oil.

(Ambasta, 1986)

Roots are used against abdominal pains and dysentery. Stem bark is used for diarrhoea. It is also a tonic. Leaves are applied on cuts, to remove dandruff and to expel guinea worms. Vapours of leaves are inhaled for hysteria. They are used for fits and ulcer. Thalamus is used by women for antifertility. Fruits are used for diarrhoea, dysentery and tumour. Seeds are used to expel guinea worms, as abortifacient and to kill lices. The plant is used for cold and fever.

(Jain, 1991)

Critical Notes

Seeds constitute a useful medicine to prevent hair fall in case of human beings. They are also beneficial in healing wounds of domestic animals. Both medicinal uses did not find place in the published literature. The plant species is exotic, native of tropical America cultivated in our country for its fruits. It is although under cultivation, it has also naturalized on waste places.

Polyalthia longifolia (Sonner) Thw.

Local Name

Ashok

Ethnobotanical Notes

Medicine

1. To avoid infection during menstrual bleeding especially of a woman who has crossed the age of forty, leaf extract about spoonful mixed with ghee is administered for a week.
2. A spoonful of decoction of leaves is given thrice a day for a month to cure back-ache after delivery.
3. Infusion of dried flowers, about half cup, is given once a day to control sugar level in diabetes.
4. Powder of wood is applied locally on piles till cure.

Notes from Classics

Whole plant is used as febrifuge.

(Watt, 1889-1893)

Bark is febrifuge.

(Anonymous, 1948-1976)

Bark is febrifuge. Wood is used for making drums, carriages, shafts, packing cases, pencils, match boxes etc. Fruits are eaten during scarcity.

(Ambasta, 1991)

Critical Notes

Leaves to avoid infection during menstrual bleeding, and to cure back-ache, flowers for diabetes and wood for piles are interesting and new to science of ethnobotany. It is native of tropical Asia, usually planted as an avenue.

Family: Menispermaceae

Cissampelos pareira L. var.*hirsuta* (DC.) Forman

Local Name

Pahadwel

Ethnobotanical Notes

Medicine

1. Juice of leaves is dropped in ears for deafness.
2. Paste of leaves is applied on hands and legs for swellings.
3. Paste of leaves is also applied on joints for rheumatism.
4. Juice of leaves about half cup is given twice a day for gastric troubles.

Notes from Classics

Dried roots and bark are used as mild tonic and diuretic in acute and chronic cystitis. Root decoction is used for ulcer of kidney, in jaundice, pains in stomach, dyspepsia, diarrhoea, dropsy and cough. It cures prolapse uteri. It is applied externally on snake bite and scorpion sting. Leaves are applied on abscess.

(Watt, 1889-1899)

Leaves are externally applied for sores and itch.

(Anonymous, 1948-1976)

Roots are diuretic, antiperiodic and purgative. Roots are useful to cure dyspepsia, dropsy and urinary troubles. Plants yield strong fibers.

(Ambasta, 1986)

Roots are good remedy on antifertility, useful in the treatment of body-ache, catarrhal infection of bladder, cold and colic pains. They are useful in diarrhoea, dysentery, dysuria and fever. Roots cure stomach ulcer, gastric and urine complaints, leucoderma and leprosy wounds. These also give relief in head-ache. Leaves are applied on abscesses, boils, burns, pimples, wounds, sores and skin diseases. Leaves are useful for cold, fever, eye complaints and heart burns. They cure jaundice, inflammation, small-pox and stomach-ache. Fibers are obtained from stem. Stem pieces cure head-ache.

(Jain, 1991)

Critical Notes

The medicinal uses of leaves against deafness, rheumatism and gastric complaints are nor documented yet.

Cocculus hirsutus (L.) Diels

Local Name

Vasanvel

Ethnobotanical Notes

Medicine

1. 10 gm seed powder with honey is consumed twice a day for seven days after menstrual period, it helps to reduce fertility in women.
2. Leaf paste is applied around eyes to treat conjunctivitis.
3. Equal amount of powder of leaves and cow ghee is mixed. It is taken orally, one spoonful every day, for one month to cure piles.
4. Leaf extract, about half glass, is taken orally at morning to strengthen body.

Miscellaneous

1. Leaves are used as fodder for goats.
2. Stem is used as cordage to fie fodder bundles.

Notes from Classics

Roots are heating, laxative, sudorific, alterative and cure rheumatic and venereal pains. Roots are rubbed with Bonduc nuts in water and administered as cure belly-ache and in bilious dyspepsia. Roots are used as tonic and given in fever. Goats and cows cherish the plants.

(Watt, 1989-96)

Roots are used for rheumatism, dyspepsia and stomach-ache in children. Leaf juice mixed in water is used as refrigerant. Mucilage is also applied on eczema, prurigo and impetigo. Plant is reputed medicine for rheumatism, stomach-ache, gonorrhoea, dyspepsia and externally applied on eczema. Plant is laxative, alterative and demulcent.

(Anonymous, 1948-1976)

Ambasta (1986) documented uses similar as in Anonymous (*loc.cit.*).

Roots are useful in gout fever, rheumatism and venereal complaints. Leaves are used to prevent bed-wetting of children. Leaves control conjunctiva, fever, gonorrhoea. Leaf paste if applied externally, it cures eczema, irritation of skin, cuts, sores and wounds. Leaves give relief in rheumatism, urine diseases and venereal complaints. Leaves and stems are used as vegetable.

(Jain, 1991)

Critical Notes

Leaves are being communicated for the first time useful to treat conjunctivitis and piles. They are also beneficial to reduce fertility in women and to strengthen body. It is common on hedges along roads and also a common weed of cultivated fields.

Family: Papaveraceae

Argemone mexicana L.

Local Name

Piwla Dhotra, Vilyati Dhotra

Ethnobotanical Notes

Medicine

1. A cup of root decoction is useful as a tonic by women, especially during pre and post-natal care.
2. Crushed roots are mixed in cow urine. After two or three hours, a cup of this urine is administered to check vomiting.

3. Root extract is used as aphrodisiac. Two spoonful of this mixture is advised daily for a fortnight.
4. Roots are also considered useful in stomach-ache. Two spoons of extract is advised once a day till cure.
5. A glass of leaf juice is homogenized with black peper. (*Piper nigrum* L.) A spoonful of this mixture is advised daily for rheumatism.
6. Leaves are applied locally to relieve pains.
7. Seed extract, about two spoons for a week, is taken orally to reduce body heat.
8. Decoction of the entire plant, one spoonful twice a day, is taken orally to cure cough.
9. Decoction of the entire plant is advised also to increase urination. About 25 ml of decoction is taken in the morning daily for seven days.
10. Juice of plant is mixed with equal quantity of whey and ginger (*Zinziber officinale* Rosc.). About 20 ml of this mixture is advised after meals for fever.
11. A cup of decoction of entire plant is drunk daily at night for gonorrhoea.
12. Half cup of a fresh juice of this plant with a pinch of root powder of *Hemidesmus indicus* (L.) R.Br. is mixed in whey. The whole mixture is advised twice a day for urination.
13. Two to three drops of latex is dropped in eyes regularly for five days or more to cure cataract.

Notes from Classics

Latex is used for dropsy, jaundice and cutaneous affections. It is diuretic and relives blisters. Latex is taken internally for scabies. Latex in ghee is consumed to cure gonorrhoea. Seed and seed oil is beneficial to control asthma. Seed oil exercises soothing influence if applied externally in head-ache and herptic eruptions. It is purgative and expectorant.

(Watt, 1886-1899)

Latex is applied on scabies and opthalmia. Seed oil is used for skin disease. In small amount, it is used as cathartic.

(Anonymous, 1948-1976)

Mixed with dried oil, seed oil is used in plant industry and soap making.

(Ambasta, 1986)

Roots are used for dog-bite, eye complaints, inflammation, mouth wash and piles. Leaves are useful for the treatment of boils, ulcer, dropsy, jaundice, eye complaints, leprosy and skin diseases. They are efficacious on scabies and syphilis. Leaves are also used for expelling round worms. Seeds give good relief in asthma, cough, caries, gum trouble, head-ache, indigestion and skin disease.

(Jain, 1991)

Critical Notes

Roots are found useful during pre and post-natural care, to check vomiting, as aphrodisiac and stomach-ache are interesting and have not been so documented in the past literature. Leaves are also observed by the present investigator useful to relieve body pains and against rheumatism. These are likewise not recorded earlier. Decoction of entire plant is beneficial for urination. This is a new finding. Plant juice for urination and latex for cataract are also not recorded.

Plant is exotic, native of Mexico. Although the plant is exotic noxius weed of waste lands and cropped lands. It is useful for many purposes.

Family: Brassicaceae

Brassica juncea (L.) Czern. et Coss.

Local Name

Mohri, Rai

Ethnobotanical Notes

Medicine

1. Seeds are crushed in water and the paste is applied on lymph glands.
2. Seed powder is mixed in sesamum oil (*Sesamum orientale* L.) and warmed. This paste is applied on knees for arthritis.

Notes from Classics

Seed paste is used as poultice. Also, it is externally applied in neuralgia and rheumatic affections. Seed paste is externally applied on abdomen to improve digestion.

(Watt, 1886-1896)

Leaves are eaten raw or pickled.

(Ambasta, 1986)

Seeds are consumed for boils and ring worms. Seeds are used as condiment. Leaves are used as vegetable.

(Jain, 1991)

Critical Notes

Medicinal importance of seeds for lymph glands and arthritis are being reported for the first time. It is a crop plant.

Raphanus sativus L.

Local Name

Mula

Ethnobotanical Notes

Medicine

1. Extract of root about half glass is drunk at morning to cure piles. It is advised for 15-20 days.
2. Powder of root is applied inside the mouth to cure stomatitis.
3. Spoonful of root extract is drunk with honey twice a day for three days to cure cough.
4. Spoonful of root extract mixed in honey and spoonful amount of ginger (*Zingiber officinale* Rosc.) thrice a day for three days to cure fever.

Notes from Classics

Roots are used for in urinary and syphilic complaints. Roots are stimulant, diuretic and stomatic, also useful in piles and gastrodynic pain. Juice of root is used in dysuria, strangury and in case of ischuria. Juice is drunk to improve appetite before meal and also to increase digestive power. Seeds are diuretic, laxative and lithotriptic, also have emmenogogue properties. They are demulcent, used in gonorrhoea.

(Watt, 1889-1899)

Fruits are refreshing, depurative, used in liver and gall bladder troubles, neuralgia and head-ache. Roots are used in urinary complaints, piles, gastrodynia. Juice of fresh leaves is used as a diuretic and laxtive. Seeds are peptic, expectorant, diuretic and carminative.

(Anonymous, 1948-1976)

Roots are diuretic, used for piles and gastrodynia. Roots are eaten as a salad or cooked as a vegetable. Roots are appetizing. Juice of fresh leaves is diuretic and laxative. Seeds are expectorant, diuretic and carminitive. Seeds yield non drying oil suitable for soap. It is edible, used as illuminant. Seed cake used as a manure. Fruits are eaten raw or cooked.

(Ambasta, 1986)

Leaves are employed for acidity. Seeds are applied on rashes, ring-worm and skin eruptions. Leaves, fruits and roots are used as vegetable.

(Jain, 1991)

Critical Notes

Roots are useful in treating stomatitis and fever. These uses are not recorded yet. This species is cultivated for its edible leaves and roots. It is exotic, native or Europe.

Family: Capparidaceae

Cadaba fruticosa (L.) Druce

Local Name

Rangat-chopda

Ethnobotanical Notes

Medicine

1. Powder of leaves mixed in coconut oil is applied on chronic bleeding wounds for quick healing.

Notes from Classics

Roots are leaves are anthelmintic, deobstruent and are prescribed as decoction in uterine obstructions. Leaves are also used as poultice for sores.

(Anonymous, 1948-1976)

Ambasta (1986) recorded similar uses as those in Anonymous (*loc.cit.*).

Leaves are used in gastro-intestinal disorders and urine complaints. It is vermicidal

(Jain, 1991)

Critical Notes

Leaves are useful as a remedy to recover bleeding wounds early. This use is not found in classic literature.

Capparis zeylanica L.

Local Name

Jakhambel, Wagati

Ethnobotanical Notes

Medicine

1. Roots and bark paste are applied together on rheumatic swelling of joints.

Miscellaneous

1. Food: Pericarp of unripe fruits is used a vegetable.
2. Fodder: Leaves are served especially to goats as fodder by the goatherds.

Notes from Classics

Green fruits are pickled.

(Watt, 1889-1893).

Leaves are applied as counter irritant and as a cataplasm in boils, swelling and piles. Root bark is bitter, sedative, stomachic, cholagogue, antihidrotic, used in cholera. Fruits are pickled.

(Anonymous, 1948-1976)

Root bark is sedative, stomachic and used in Cholera. Fruits are pickled.

(Ambasta, 1986)

Root and bark are anthelmintic. Their paste is applied on sores, rheumatic joints and swelling, also useful in cholera. Plant is used for blisters, colic, neuralgia, pleurosis and pneumonia. Fruits are also used as vegetable.

(Jain, 1991)

Critical Notes

The unripe fruits are used as vegetable by tribal people. Natural dispersal is thereby affected. It is largely served as goatfeed. These forces of biotic interference appear to have depleted the population of this species. It is also to be noted that it finds place in different systems of medicine.

Capparis decidua (Forsk.) Edgew.

Local Name

Kerdu

Ethnobotanical Notes

Miscellaneous

Food

1. Fruits are used as vegetable and also for pickled.
2. Flower buds are used for vegetable.
3. Spiny leafless dried branches are used for fencing as mechanical barrier.

Notes from Classics

Flower buds are cooked when fresh as pot herb. Fruit is largely consumed, ripe fruits are made into pickles.

(Watt, 1889-1899)

Small round fruits and flower buds are edible, and commonly pickled. Wood is used for making tool handles, boat knees and cart wheels.

(Anonymous, 1948-1976)

Flower buds are used as pot herb. Fruits are edible and pickled. Wood is used for knees of boats, axles of cart wheels and tool handles.

(Ambasta, 1986)

Fruits and buds are edible. Fruits are pickled.

(Jain, 1991)

Critical Notes

Food value of fruits and flowers is already documented. These should be assessed on chemical ground if not evaluated earlier.

Cleome gynandra L.

Local Name

Pandhri-tilwan

Ethnobotanical Notes

Medicine

1. Paste of leaves is applied on the joints for arthritis.
2. Few drops of juice of leaves are dropped in ears for ear-ache.

Notes from Classics

Decoction of root is mild febrifuge. Leaves are eaten as pot-herb and in flavouring sauces. They are also pickled. Leaves are rubefacient, vesicant and used as counter-irritant in head-ache, neuralgia, rheumatism and other local pains. Leaves are applied as poultice on boils to prevent formation of pus. Juice of leaves with oil is dropped in ears for ear-ache. Leaves are taken internally for certain bilious disorder. Seeds are anthelmintic, rubefacient, given internally for expulsion of round worms. Infusion of seeds is given to treat cough. Paste of seeds is applied to kill maggots on sore, to kill lice and used as fish-poison

(Anonymous, 1948-1976)

Ambasta (1986) reports similar uses as those in Anonymous (*loc.cit.*).

Leaves are used in boils, head-ache and ear complaints. Leaves applied on knees for rheumatic pains. They cure pains in stiff neck, and also used as vermifuge. Seeds are used in cough, head-ache, and help to expel round worms. Seed paste is applied externally on skin for skin diseases. Plant is used to treat asthma, eye disease, fever and wounds. Plant is used as antidote to snake bite.

(Jain, 1991)

Critical Notes

Leaves are beneficial ear-ache. These uses have remained unnoticed yet.

Cleome viscosa L.

Local Name

Piwli hilwan, Bhera, Kanphuti

Ethnobotanical Notes

Medicine

1. Juice of leaf, three to four drops, is dropped in ears to cure malarial fever.
2. Few drops of leaf juice are dropped in ears in case of ear-ache and to prohibit pus formation.
3. Juice of leaf is also useful for tooth-ache, if the juice is dropped in an ear to apposite side of the aching tooth.

Notes from Classics

Leaves are rubefacient, vesicant and sudorific. Juice of leaf is mixed in ghee for the treatment of inflammation of middle ear. Leaf paste is applied on wounds of ulcer. Poultice of leaves is efficacious as counter-irritant to chronic pains of joints. Seeds are rubefacient, anthemintic, sometimes used as condiments.

(Anonymous, 1946-1976)

Ambasta (1986) recorded similar uses as in Anonymous (*loc.cit.*).

Infusion of leaves is applied over boils, wounds, also used for ear-ache and inflammation of ears. Seeds are anthelmintic, useful in convulations, diarrhoea and skin disease. Seeds check pus formation in case of boils. Stem is useful as vegetable.

(Jain, 1991)

Critical Notes

Leaf juice is medicinally useful for aching tooth. It has not been recorded outside Jalgaon district.

Maerua oblongifolia (Forsk.) A. Rich.

Local Name

Motiwel

Ethnobotanical Notes

Medicine

1. Leaf juice mixed in turmeric powder (*Curcuma longa* L.) is applied at night for cracks on foot sole.

Notes from Classics

Roots are alterative tonic and stimulant. Unripe fruits are eaten after boiling.

(Anonymous, 1948-1976)

Ambasta (1986) documented similar properties as those in Anonymous (*loc. cit.*).

Critical Notes

Use of leaves to treat cracks on foot sole is not tapped in past.

Family: Flacourtiaceae

Flacourtia indica (Burm *f.*) Merr.

Local Name

Par-bor

Ethnobotanical Notes

Miscellaneous

1. Ripe fruits are edible.

2. Leaves are used for goat feed.

Notes from Classics

Fruits and leaves are eaten. Fruits are eaten either raw or cooked. Leaves are used as cattle fodder.

(Watt, 1889-1899)

Fruits are edible, excellent for jam and jellies. Branches and leaves are lopped for fodder.Wood is used to turning agricultural implements.

(Anonymous, 1946-1976)

Fruits and leaves are edible.

(Jain, 1991)

Critical Notes

This species is widely used as a source of cattle fodder. The fruits are reported as edible in Jalgaon district. The leaves and fruits are reported already so. The species appears worthy of examining food and fodder value on modern scientific basis.

Family: Portulacaceae

Portulaca oleracea L.

Local Name

Ghol

Ethnobotanical Notes

Medicine

1. Extract of leaves, about one spoonful mixed in sugar, is given daily for a week to relief in strangury.
2. Paste of leaves is applied on burns and inflammation.
3. A spoonful of seed extract is administered orally twice or thrice a day to cure dysentery.
4. Juice of plant, about two or three drops, is dropped in ears for ear-ache.

Miscellaneous

1. Food: Leaves and tender shoots are used as vegetable.
2. Fodder: Plants constitute fodder for pet animals.

Notes from Classics

Leaves are applied externally in erysipelas. Infusion of leaf is diuretic in dysuria. Plant is cooling for burning fever, also has astringent properties. Plant and seeds are recommended for disease of kidney, gall bladder and lugs.

(Watt, 1889-1893)

Paste of leaves is applied to burns, scalds, swelling and erysipelas. Leaves and shoot are employed as antihaemorrhagic poultice. Plant is refrigerant, vulnerary, antiscorbutic, aperiant and diuretic. It is used in the treatment of scurvy, dysuria, haematuria, gonorrhoea, dysentery, sore nipple and ulcer of mouth. Plant is used to cure diseases of liver, spleen, kidney and gall bladder. It also cures cardiovascular diseases. Roasted seeds are eaten. They are diuretic and antidysenteric, also applied on burns and scalds.

(Anonymous, 1946-1976)

Plant is used as pot herb, used in the preparation of soups. Plant is good fodder for sheeps, cattle and pigs.

(Ambasta, 1986)

Leaves and stems are used as vegetable. Leaves are used in fever and burns. Whole plant is used as blood purifier, also for head-ache and prickly heat and burning sensation, applied on burns and taken orally for fever.

(Jain, 1991)

Critical Notes

Leaves in strangury and plant juice in ear-ache are not found reported in earlier literature. It is a common weed of waste lands and crop fields. It is exotic, native and Europe.

Portulaca quadrifida L.

Local Name

Barik-ghol, Bhaji, Ranghol

Ethnobotanical Notes

Miscellaneous

1. Food:Farming labours use the plant for vegetable.

Notes from Classics

Leaves are used for application on erysipelas. Infusion of leaves is diuretic in dysuria

(Watt, 1889-1899)

Herb is used as vegetable. Herb is used in asthma, cough, urinary discharge, inflammation and ulcer. Poultice of herb is applied in abdominal complaints, erysipelas and haemorrhoids

(Anonymous, 1948-1886)

Roasted seeds are eaten, considered diuretic, anti-dysenteric, also used as application for burns and scalds. Herbs are refrigerant, vulnerary, antiscobutic, aperient and diuretic, used iin scurvy, diseases of liver, spleen, kidney and bladder. Herb is also employed for cardio-vascular diseases, dysuria, haematuria, in dysentery, sore nipples and ulceration of mouth. Herb is used to stimulate gastric secretion. Herb is used as pot-herb. Fleshy stem is pickled and also used as fodder.

(Ambasta, 1986)

Leaves are antiscorbutic. They are applied on burns and ulcer as a cooling agent. Leaves are used as vegetable. Whole plant is used in skin diseases.

(Jain, 1991)

Critical Notes

The plant is common crop weed. It is commonly found in banana fields. It is removed manually by farm labours during weeding out. They collect and use the plant for preparation of vegetable, locally called 'bhaji'.

Family: Malvaceae

Abelmoschus esculentus (L.) Moench.

Local Name

Bhendi

Ethnobotanical Notes

Medicine

1. Raw fruits are consumed without using spices twice a day to control diarrhoea.
2. About 250 gm of raw fruits are sliced and boiled in about one litre water till the volume is reduced to half. The viscous water is taken about a glass orally in the morning for 21 consecutive days to treat urine stone.

3. A cup of fruit juice mixed with small quantity of sugar is administered in the morning for 10 to 12 days to cure dysuria.

Miscellaneous

1. Tribals extract the mucilage from the fruits and mixed in 'Mehandi" (Leaf powder of *Lawsonia inermis* L.) It is used to dye hand-palms for thin lining while designing on hand palms.

Notes from Classics

Bark of stem is made into ropes. Leaves are used as cattle fodder. Fruits are diuretic, used in cattarrhal affections. They are used in dysentery, also as a laxative in dropsical affection. Mucilage is used as aphrodisiac. Young pods are eaten for spermatorrhoea. Unripe fruits are used as vegetable. Flowers are used to prepare soup.

(Watt, 1889-1899)

Roots are used in syphilis. Seeds are stimulant, cordial and antispasmodic. Leaves are used externally as emollient poultice. Immature capsules are emollient, demulcent and diuretic, and are employed in the form of decoction in catarrhal affections, dysuria and gonorrhea.

(Anonymous, 1946-1976)

Stalks yield best fibre. Flowers are used to prepare soup. Tender pods are used as vegetable. Immature capsules are emollient, demulent and diuretic. Seeds are stimulant, cordial and antispasmodic. Seed yields fatty edible oil.

(Ambasta, 1986)

Fruits are used as abortifacient. Fruits and seeds are used medicinally as demulcent, in gonorrhoea and irritation of genito-urinary systems.

(Jain, 1991)

Critical Notes

Fruits are used to treat diarrhoea and urine stone. It is interesting to note use of mucilage of fruits while dying hand palms with 'Mehandi'. The plant species is cultivated. It is exotic, native of Africa.

Abelmoschus manihot (L.) Medik. subsp. *tetraphyllus* (Roxb. ex Hornem.) Borss.

Local Name

Jangli Bhendi, Ran Bhendi

Ethnobotanical Notes

Medicine

1. Extract of roots along with root bark is applied on abdomen to reduce abdominal pains.
2. Root powder is applied on burns. Leaves are spread beneath the sleeping bed and patients are advised to sleep on this bed.
3. Spoonful of root powder is taken with honey daily for three days. After 10 to 15 minutes, the patient is advised to drink water. It helps to control stomach complaints.

Miscellaneous

1. An instrument called 'Gophan' is made using fibres to drive off birds from crop fields.
2. Fibres are widely used for making straps called 'Jotye' for tying yoke of the cart to the neck of bullocks.

Notes from Classics

Plant yield tough fibre.

(Watt, 1889-1899)

Bark is emmanogague. Leaves and roots used as poultice for boils, sores, sprains and inflammation.

(Anonymous, 1946-1976)

Mucilage from roots is used in sizing paper. Bark is emmanogogue. Leaves are cooked and eaten. Seed yield fatty oil and fibre.

(Ambasta, 1986)

Roots are used against tuberculosis. Stem fibres are used for cordage. Fruits are used as vegetable.

(Jain, 1991)

Critical Notes

Roots are widely used to treat abdominal pains, stomach complaints and for burns. There medicinal uses are not hitherto reported. The present observations and earlier reports (*cf.* Watt, *loc.cit.*) warrant the species as good source of fibres from stem. It appears worth to bring it it under cultivation as an alternative source of fibres.

Abutilon indicum (L.) Sweet

Local Name

Mudra, Atti

Ethnobotanical Notes

Medicine

1. Infusion of roots, about two spoons, is drunk in fever till cure.
2. Mucilage extracted from 5 to 6 leaves is used as laxative.
3. Leaf paste is mixed with sesamum oil (*Sesamum orientale* L. and applied on joints against rheumatism.
4. Leaf extract, about half glass, is drunk twice a day for body-ache.
5. Extract of entire plant is administered orally and daily for tuberculosis till cure.
6. Extract of entire plants is applied and then massaged for facial and other form of paralysis.

Miscellaneous

1. Stem fibres are used for making ropes and cordage.

Notes from Classics

Infusion of roots is useful in strangury, haematuria and as a cooling medicine. It is also given in fever and leprosy. Stem bark is astringent and diuretic, and efficacious in haemorrhoids when given orally. Leaf extract is used as demulcent. Decoction of leaves is used for tooth-ache, gonorrhoea and inflammation of bladder. A lotion made from leaves is applied on joints in rheumatism. Seeds are laxative and demulcent, also given for cough. Fumes of seeds are allowed to treat kids affected with thread-worms. Seeds are employed in case of piles and are aphrodisiac.

(Watt, 1889-1899)

Infusion of roots is taken orally as cooling medicine. Mucilage of leaves is used as demulcent tonic. A lotion made from mucilage of leaves is applied on joints for rheumatism. Seeds are laxative and demulcent.

(Anonymous, 1948-1976)

Bark is astringent and diuretic. Root is nervine tonic and antipyretic, used in piles with chaulmugra oil, also used for leucoderma. Leaves are cooked and eaten. Their extract is diuretic and demulcent. Flowers are edible.

(Ambasta, 1986)

Decoction of roots is taken orally to facilitate child birth, also helps to cure cough. Root paste is applied on skin to cure leprosy. It is also used for massage in paralysis. Decoction of leaves in consumed for cough, fever, ulcer and gonorrhoea. Paste of leaves is applied on boils and wounds. Decoction of leaves is also taken orally for bronchitis, head-ache and urethritis. Plant extract is taken orally for abortion and also given for menorrhagia, spermatorrhoea and dysentery. Fruits are consumed for stomach complaints. Seeds are aphrodisiac.

(Jain, 1991)

Critical Notes

Mucilage and extract of leaves are employed as laxative and for body-ache respectively. Extract of the entire plant is useful for tuberculosis and paralysis. Also, the fibres obtained from the stems are useful for ropes and cordage. All these uses have remained unnoticed elsewhere in our country.

Gossypium herbaceum L.

Local Name

Kapus, Kapashi

Ethnobotanical Notes

Medicine

1. A spoonful powder of roots is mixed in a glass of water in which rice grains are washed. It is consumed at morning for two weeks to cure leucorrhoea..

2. Spoonful of root powder is drunk along with a half cup of juice of leaves of neem (*Azadirachta indica* A.Juss.) to prevent a woman from post-delivery complaints.
3. Cotton fibres mixed in few leaves of *Acacia nilotica* (L.) Del.subsp. *indica* (Benth.) Brenan and *Hibiscus rosa-sinensis* L.are chewed for mouth ulcer at morning till cure.

Notes from Classics

Root bark is emmenogogue and given for abortion. Decoction of roots is used for strangury and gravel. Roots and leaves are used in vapour bath in case of uterine colic. Juice of leaves is a remedy on dysentery. Leaves are applied on gout and joints. Flowers are prescribed for hypochondriasis. Poultice of fruits is applied on burns and scalds. Seeds in the form of emulsion are given in dysentery and supposed to be pectoral, also given as nervine tonic. They are prescribed for head-ache and brain affection. Seed oil is applied externally for clearing the spots and freckles on face. It is used as liniment for rheumatic affection.

(Watt, 1889-1893)

Root bark and seeds are abortifacient, also used as anticancer.

(Jain, 1981)

Critical Notes

Medicinal importance of roots are being reported for the first time to cure leucorrhoea and delivery complaints. It is a cultivated species.

Hibiscus cannabinus L.

Local Name

Ambadi

Ethnobotanical Notes

Miscellaneous

Food: Leaves are used as vegetable.

Fodder: Seeds are fed to milching cattle.

Fibres Fibres are obtained from stems.

Notes from Classics

Young leaves are employed as pot herb. Seeds yield edible oil. They are eaten when roasted. These are also used as cattle feed. Stems yield fibres of poor quality.

(Watt, 1989-1999)

Leaves are used as pot herb. Tender twigs are used as fodder for cattle.

(Anonymous, 1948-1976)

Ambasta (1986) reported similar uses as those of in Anonymous (*loc.cit.*).

From stems fibres are obtained. Oil is extracted from fruits. Leaves are used as vegetable.

(Jain, 1991)

Critical Notes

This species is also usually cultivated by the tribal farmers as a mixed crop.

Hibiscus caesius Garcke

Local Name

Kothum

Ethnobotanical Notes

Medicine

1. Leaf juice, about one spoonful, is given twice a day for head-ache till cure. Leaf paste is also applied on forehead for similar purpose.

Critical Notes

This species is being noted useful medicinally for the first time.

Hibiscus-rosa sinensis L.

Local Name

Jaswand

Ethnobotanical Notes

Medicine

1. Leaf extract, about half cup, is drunk at morning for about one month to cure leucorrhoea.

2. Paste of leaves and flowers mixed in equal proportion, is applied on scalp to promote hair growth.
3. Paste of petals exclusively is applied on skin of face to treat pimples.

Notes from Classics

Root is demulcent and used for cough, venereal disease and manorrhagia. Leaves are emollient, anodyne and laxative. Decoction of leaves is used in fever to facilitate expulsion of child-birth. Flowers are demulcent, emollient, refrigerant, aphrodisiac and emmengogue. Decoction of flowers is given in bronchial cattarh. Paste of flowers is applied to swellings and boils. Flowers are emollient and advised in the treatment of manorrhagia.

(Watt, 1889-1899)

Infusion of petals is given as demulcent in fever. It is given for strangury, irritable conditions of genito-urinary tract, and for hair growth. Seeds pounded in water are given for gonorrhoea.

(Anonymous, 1948-1976)

Flowers are eaten and pickled. Other medicinal properties are similarly described by Ambasta (1986) as in (Anonymous, *loc.cit.*).

Roots are prescribed for cough and gonorrhoea. Flowers are used against antifertility. It is reported as abortifacient, contraceptive and to facilitate child-birth. Flowers are used in the treatment of fever, inflammation of kidneys, menorrhagia, menstrual complaints and skin diseases.

(Jain, 1991)

Critical Notes

Leaves are useful in leucorrhoea. Leaf and flower paste together are helpful to promote hair growth. Flowers are beneficial while treating pimples. These uses are not reported in past literature. This species is usually planted in gardens as an ornamental. The garden plant is also tried by the local people for various purposes. It is exotic, native of China.

Hibiscus sabdariffa L.

Local Name

Ambadi

Ethnobotanical Notes

Miscellaneous

Food

1. Calyces are added in vegetable to make it sour.
2. Leaves are used as vegetable.
3. Stem yield fibres which are used for cordage.

Fodder

1. Seeds are ground and used as feed for cattle.

Notes from Classics

Fleshy calyces and capsules are made into jam and other preserved articles. Leaves are used in salad. Seeds are excellent feed for cattle.

(Watt, 1886-1896)

Tender leaves and stalks are eaten as salad. Fresh calyx is used in jellies and juices. Fruits are antiscorbutic. Residual oil cake is used for cattle feed.

(Anonymous, 1948-1976)

Ambasta (1986) gave similar uses as those in Anonymous (*loc.cit.*). Leaves and calyces are used as vegetable. Stem yield fibres which are used for cordage.

(Jain, 1991)

Critical Notes

The plant is usually cultivated as a mixed crop by tribal farmers. It is exotic, native of tropical Africa and Asia.

Kydia calycina Roxb.

Local Name

Sheenkar

Ethnobotanical Notes

Miscellaneous

Timber: Wood is used for hut construction.

Fodder: Leaves are used as fodder for pet animals.

Notes from Classics

Inner bark yields fibres used for making coarse ropes. Wood is used for buildings and making plough.

(Watt, 1889-1899)

Wood is useful for making planks, agricultural implements, carving, spoon, match-boxes, pencils, car. etc. Inner bark yield fibres.

(Anonymous, 1946-1976)

Leaves are lopped for fodder. Wood is used for making plough, agricultural implements, match-box, splints, packing cases, turnery, brush-backs, bobbins etc.

(Ambasta, 1986)

Plant is used for boils, diabetes and rheumatism. Bark fibre used for cordage.

(Jain, 1991)

Malvastrum coromandelianum (L.) Garcke

Local Name

Gul-tawdi

Ethnobotanical Notes

Medicine

1. Decoction of leaves is drunk, about 20 ml once only, to better urination.

Notes from Classics

Plant is bechic, emollient and has resolvent properties. Decoction of plant is given in dysentery. Leaves are applied to inflammed sore and wound as cooling and healing medicine. Flowers are pectoral and diaphoretic. Stem yields fibres which are used in making brooms.

(Anonymous, 1948-1976)

Ambasta (1986) mentioned similar uses as those in Anonymous (*loc.cit.*).

Roots are aphrodisiac. Leaves are used for jaundice, rheumatism and sprain, sores and wounds. Twigs are used for making brooms.

(Jain, 1991)

Critical Notes

Use of herbs to increase urination is not noted in past literature.

Sida acuta Burm.f.

Local Name

Chikana

Ethnobotanical Notes

Medicine

1. Spoonful of root powder is mixed in milk once a day for two weeks to cure bleeding leucorrhoea.

Notes from Classics

Decoction of roots is diaphoretic, antipyretic, stomachic and tonic. Roots are used in febrile affection, dyspepsia and debility. Juice of roots is employed in intestinal worms. Infusion of roots promote perspiration and increase appetite. Infusion of roots is cooling, tonic and astringent, useful in nervous and urinary diseases and fever. Leaves with gingelly oil hasten suppuration. Leaves are useful in case of opthalmia. They are aphrodisiac. Leaves are diuretic and useful in rheumatic affection and demulcent in gonorrhoea.

(Watt, 1889-1899)

Roots are cooling, tonic, astringent, stomachic, diaphoretic and have antipyretic properties. Roots are useful in nervous and urinary diseases. Roots are useful in treating disorders of blood and bile, and in chronic bowel complaints. Fresh juice of roots is applied to wounds and ulcer, also used as vermifuge. Strong decoction of roots is given in debility. Leaves are abortifacient, anthelmintic applied externally on cuts. Leaves are demulcent and diuretic, used in rheumatic problems. Leaves with gingelly oil are applied on ulcer. Leaves are boiled in oil and applied to testicular swellings, in elephantiasis, and also on sores. Leaves are abortifacient. Decoction of leaves and roots is used in haemorrhoids and importancy. Leaves are vermicide. Leaf juice is given in chest pain. Fruits are applied on boils and burns.

(Anonymous, 1946-1976)

Roots are also used for expelling worms.

(Ambasta, 1986)

Roots are cooling, tonic, useful for sotomach-ache, back-pains and as an antidote for snake-bite. Leaves are used as abortifacient, anthelmintic applied on boils, burns, cuts, as haemostat. Leaves are used to treat gastric diseases. They are vermifuge. Stem fibres are used to make brooms.

(Jain, 1991)

Critical Notes

Roots for curing leucorrhoea have not been documented in past literature.

Sida alba L.

Local Name

Khareti

Ethnobotanical Notes

Miscellaneous

1. *Fodder*: Leaves are used as fodder for goats.

Notes from Classics

Roots are diaphoretic, used in debility, fever and also cattle diseases. Leaves are demulcent, refrigerant and used in gonorrhoea.

(Watt, 1889-1899)

Leaves are cooked and eaten.

(Anonymous, 1946-1976)

Roots are tonic, diaphoretic, used in debility and fever. Decoction of roots is given as demulcent in bladder and gonorrhoea.

(Ambasta, 1986)

Roots are used for antifertility.

(Jain, 1991)

Critical Notes

Leaves constitute a good source of fodder for goats. This is not reported in classic literature.

Salmalia malabarica (DC.) Schott and Endl.

Local Name

Lal sawar, Kate-sawar, Hawar

Ethnobotanical Notes

Medicine

1. Root extract about halt cup is given against snake-bite immediately after bite.
2. Decoction of stem bark, about half cup prepared in goat milk, is drunk for general weakness.
3. Decoction of stem bark with some sugar is taken orally daily in the morning to reduce body heat till cure. Tubercles are rubbed in water, this paste is applied for early cure of pimples by tribals.

Veterinary Medicine

1. Decoction of stem bark, about one litre, is administered to domestic animals twice a day for dysentery.

Miscellaneous

1. Tubercles are chewed as mouth turns red, an amusment by cow-boy.
2. ***Food***: Flowers are used for vegetable..

Notes from Classics

Roots act as stimulant, tonic and emetic. Root powder is an aphrodisiac. It is also given in impotancy. Tap root of young plant is used for gonorrhoea. Gum is also given as laxative to children. Leaves are used as external application for glandular swellings. Dried flowers are used as demulcent and on haemorrhoids.

(Watt, 1889-1899)

Gum is astringent, stimulant and tonic. It is alterative styptic and demulcent. Seeds are used for dysentery, haemoptysis of pulmonary tuberculosis, influenza and menorrhagia. Floss is suitable for stuffing life belts, cushions and mattresses. It is used as insulating material and surgical dressing.

(Anonymous, 1946-1976)

Bark is demulcent, tonic and styptic. Tender leaves are eaten, also lopped for fodder. Tree yield gum. It is demulcent, tonic and styptic, also used for caulking boat. Wood is used in match industry, suitable for toys, well-curbs, tea cheast, plywood, pencils, pen holders and frames. Floss is used for stuffing lifebelts, mattresses, cushions and pillows, also used for insulating material. Fruits are stimulant, expectorant and diuretic. They are used in calculus affections and ulceration of bladder and Kidneys. Seeds yield edible fatty oil, also used for soap making and as illuminant. Flower buds and fleshy calyces are eaten as vegetable.

(Ambasta, 1986)

Roots are used to cure abdominal pain, cough, dysentery and impotence. Stem bark is applied to cure pimples, boils, bone fracture, skin diseases. It is useful in weakness of sex, blood dysentery and syphilis. It is good tonic. Leaves are used in antifertility. Flowers are edible, used as antidote to snake-bite. Seeds are advised in the treatment of chicken-pox, small-pox. Plant cure antrophy, cholera, fever, gum, trouble, hematuria, inflammation, leprosy, menorrhoea, pain in legs, rheumatism, scabies, sore in mouth, spermatorrhoea, tooth-ache, ulcer and fistula. It is also cures pneumonia. Plant is used as antidote against scorpion and snake-bite.

(Jain, 1991)

Critical Notes

Roots are used as antidote for snake-bite. Stem bark controls dysentery. These uses are untapped from other regions of our country.

Family: Sterculiaceae

Helicteres isora L.

Local Name

Murud sheng

Ethnobotanical Notes

Medicine

1. Stem is pounded in water. A spoonful of it is administered to children for flatulence only once a day till cure.

2. Few drops of stem bark extract are dropped in ears for ear-ache.
3. Seeds extracted in water are applied on body for body paint.

Miscellaneous

1. Stem bark is used as fibre to bind partition walls of huts, bathrooms and stacks of fodder grasses etc.
2. Wood is used as fuel.

Notes from Classics

Fruit is employed for colic, flatulence, diarrhoea and intestinal strangulation. It is a remedy for snake-bite and dog-bite. Fruit powder is mixed with castor oil for discharge from ears. It is demulcent, astringent, tonic and stomachic. A piece of fruit is tied to neck of a newly born infant to protect from bowl complaints. Fruits are tied upon the wrist of bride and bridegroom with *Randia dumotorum* is Vaisy caste.

(Watt, 1889-1893)

Root juice is beneficial in empyema, stomach affections and diabetes. Bark is prescribed for diarrhoea, dysentery and biliousness. Dried fruits are useful in intestinal complaints and as antigalctogogue. Rope and cordage is made from bark. Wood is used as fuel. Bark yields fibres, which are used for canvas and cordage. Fruits are used in diarrhoea, chronic dysentery, flatulence and improves appetite.

(Ambasta, 1986)

Root is abortifacient, astringent, antigalactogogue demulcent and expectorant. It is administered for asthma, cold, cuts, diabetes and sciatica. Root is antidote against snake-bite and also applied on wounds. Stem bark is used for cholera, diarrhoea, ear complaints and anemia. From bark fibre is obtained. Twigs are placed in hut of women for minimize the labour pains. Leaves, stem and flowers cure anemia. Fruit is anthelmintic and used for cold, colic, cough, dysentery, flatulence, fever, ear and gastro-intestinal complaints. Fruits also help to expel gastro-intestinal worms. It is useful for rickets of babies, sores, stomach-ache and vomiting. Seeds are purgative and give relief in flatulence. Seed oil is applied on body for body pain.

(Jain, 1991)

Sterculia urens Roxb.

Local Name

Kadai

Ethnobotanical Notes

1. Paste of roots is applied on the fractures of bone for quick mending.
2. Decoction of bark, about half cup, is given to a pregnant woman to facilitate delivery.
3. Twenty gm of gum mixed in a cup of water and drunk at morning only for flatulence for three days.

Miscellaneous

Food

1. Gum is edible and tonic.
2. Seeds are eaten raw or after roasting or cooking.
3. Pulp of newly growing roots is edible. It is eaten raw by tribals.

Fodder

1. Leaves are good fodder for cattle.

Notes from Classics

Leaves and tender branches steeped in water yield mucilaginous extract which is employed for pleuropneumonia of cattle. Plant is useful medicine in throat affection.

(Watt, 1989-99)

Tender roots are reported to be eaten after cooking. Pounded bark is given to a pregnant woman to facilitate delivery. Bark is used for cordage. Leaves and tender shoots are steeped in water yield mucilaginous extract which is employed for pleuropheumonia of cattle. Leaves are nutritive fodder. A glue made from seeds can be used as adhesive in making plywood. Seeds are eaten after cooking or roasting.

(Anonymous, 1946-1976)

Bark yield fibres for cordage. Seeds are eaten after roasting and cooking. They yield edible oil, it is used for making soap. Wood is used for toys, paking cases, slate frame, furniture, match boxes etc.

(Ambasta, 1986)

Roots are administered for bone dislocations and bone fractures. They facilitate child-birth. Seeds and gum are edible. Seeds are used as vegetable.

Critical Notes

Roots are useful for bone fracture. This is not noted earlier.

Family: Tiliaceae

Corchorus aestuans Linn.

Local Name

Chuncha

Ethnobotanical Notes

Miscellaneous

Young leaves are used as vegetable.

Notes from Classics

It is wild species rarely used for extracting fibres.

(Anonymous, 1948- 1976)

Seeds are stomachic.

(Ambasta, 1986)

Leaves are used for stomach-ache and fever.

(Jain, 1991)

Critical Notes

Use of leaves as vegetable is not reported earlier.

Corchorus depressus (L.) Stocks.

Local Name

Haran khadi, Haran-khuri

Ethnobotanical Notes

Medicine

1. Paste of roots and leaves is applied on joints for rheumatism.
2. Root extract, about half cup, is advised for two to four months for spermatorrhoea till cure.
3. Root and leaf paste together is applied locally on eczema.
4. Equal amount of roots and leaves and pounded in honey and a spoonful of it is taken orally for twenty-one consecutive days for facial paralysis.
5. Entire plants and seeds in equal proportion are powdered. One spoonful powder homogenized in honey is consumed as a general tonic.

Notes from Classics

The plant is cooling medicine. It is used as demulcent in the treatment of gonorrhoea.

(Watt, 1889-1899)

Leaves are emollient. Seeds are tonic.

(Ambasta, 1986)

Critical Notes

Paste of leaves and roots together for rheumatism and eczema, root decoction against spermatorrhoea are newly added medicinal uses. Likewise, use of entire plant as a general body tonic is being reported for the first time. The utility of seeds for this purpose is already documented which is also observed in this district.

Grewia flavescens A. Juss.

Local Name

Tel-chopada, Dhaman

Ethnobotanical Notes

Medicine

1. Spoonful of root powder is drunk with water twice per day for about one month for lochia.
2. Ripe fruits are sweet and edible.

Notes from Classics

Roots are used in althea. Plant is a remedy on leprosy. It yields fibres from which ropes are made. Fruits are eaten by natives.

(Watt, 1889-1899)

Drupes are edible.

(Anonymous, 1948-1976)

Fruits are edible. Branches are used for making baskets. Leaves are branches are used as fodder.

(Ambasta, 1986)

Critical Notes

In the treatment of lochia, the roots are being reported useful for the first time.

Corchorus trilocularis L.

Local Name

Chukanda

Ethnobotanical Notes

Medicine

1. Extract of two or three seeds is prepared in mother's milk and given orally to children for stomach-ache.

Notes from Classics

Bitter seeds are administered in fever, rheumatism and pains in abdominal vicera. Mucilage is demulcent.

(Watt, 1889-1899)

Leaves are used as vegetable.

(Jain, 1991)

Critical Notes

Medicinal utility for curing stomach-ache is not found in earlier literature. It is a crop weed.

Grewia tiliaefolia Vahl

Local Name

Dhaman

Ethnobotanical Notes

Miscellaneous

1. Stem bark yield fibres which is used in the construction of huts.
2. Wood is useful for making poles of huts and as fuel.
3. ***Food***: Fruits when ripe are edible.
4. ***Fodder***: Leaves are good fodder for cattle.

Notes from Classics

Small drupes are of agreeable acid flavour and eaten by natives. Leaves and twigs are lopped for cattle fodder.

(Watt, 1886-1896)

Timber is used for shafts, poles, frames, panels, mats, oars, tool handles, agricultural implements and spokes, also employed for cooperage, golf shafts and boils. Bark yield fibres which are used for cordage. Fruit is edible. Leaves are twigs lopped for fibre.

(Anonymous, 1948-1976)

Leaves are used as vegetable.

(Ambasta, 1986)

Fruits and leaves are edible.

(Jain, 1991)

Critical Notes

In afforestation programmes, especially in tribal areas, this species of *Grewia* may be given priority. This will suffice the needs of tribals.

Grewia villosa Willd.

Local Name

Dal-lodga

Ethnobotanical Notes

Medicine

1. Ripe fruits are consumed by ladies for conception.

Notes from Classics

Roots are employed for diarrhoea. Fruits are eaten in Punjab.

(Watt, 1889-1899)

Roots are employed for diarrhoea. Drupes and seeds are edible.

(Anonymous, 1946-1976)

Roots are used in diarrhoea. Stems are used for spear shaft walking sticks and bows. Bark yields cordage fibre. Fruits and seeds are edible. Seed contain fatty oil. Plant is useful for carbuncle, cholera, dysentery, sores and wounds.

(Jain, 1991)

Critical Notes

Edibility of fruit is widely documented but for the sake of conception is a new report from Jalgaon district.

Triumfetta rotundifolia Lamk.

Local Name

Zhila

Ethnobotanical Notes

Medicine

1. Leaves are chewed for stomatitis by cowherds.

Notes from Classics

Fruits are edible.

(Watt, 1889-1899)

The plant is used as demulcent.

(Anonymous, 1948-1976)

Seeds yield fatty oil.

(Ambasta, 1986)

Critical Notes

Use of leaves for stomatitis is unreported earlier.

Family: Oxalidaceae

Biophytum sensitivum (L.) DC.

Local Name

Sharmi, Lajualu, Lajro

Ethnobotanical Notes

Medicine

1. Leaf paste, about one spoonful, is consumed daily once for urinary disorder till cure.
2. Flowers of this species, bark of *Butea monosperma* (Lamk.) Taub. and *Syzygium cumini* (L.) Skeels are ground together. Two spoons of this powder is consumed daily once for urinary disorder till cure.

Notes from Classics

Decoction of root is considerd useful in gravellish complaints. Roots in powder form are given for piles and fistula. Roots are resolvent, alternative and useful in bilious affection, fever, jaundice, leprosy, scabies and small pox. Leaf paste is applied on hydrocele. Leaves and equal quantity of horse urine is made into 'anjan' which is used to remove film of conjuctiva.

(Watt, 1889-1893)

Roots are styptic. Decoction of leaves is given for diabetes, asthma and phthisis. Powdered seeds are applied to abscesses to promote suppuration.

(Anonymous, 1948-1976)

Ambasta (1986) records uses similar to that of Anonymous (*loc.cit.*).

Roots are used to decrease sexual vigour. Leaves are useful in bilious fever and removes giddiness. They are useful for head-ache, muscular and rhemumatic pains. Leaves are good tonic, used against anti-fertility.

(Jain, 1991)

Critical Notes

Medicinal utility of flowers for curing leucorrhoea and that of leaves for urinary disorders are not documented in past.

Family: Rutaceae

Aegle marmelos (L.) Corr.

Local Name

Bel

Ethnobotanical Notes

Medicine

1. Root bark (5 gm), dried leaves (5 gm) of *Ocimum tenuiflorum* L. and four to five cloves [*Syzygium aromaticum* L.] are boiled in about one litre water to reduce the volume 1/8th of the total volume, and given one spoonful twice a day for a week to cure typhoid fever.
2. Root paste is applied to mother's chest to purify milk.
3. Extract of stem bark, about two to three spoons, with honey once a day prevent vomiting.
4. Leaf extract, about two or three spoons, is taken orally for few months and it is also applied around eyes as a remedy against night blindness.
5. Regular consumption of juice of ten leaves mixed with black pepper (*Piper nigrum* L.) helps to cure diabetes.
6. Leaf juice, about half cup, is drunk with honey once a day for six months against male sterility.
7. Decoction of leaves, about half cup with a pinch of black pepper powder, is drunk once a day till cure jaundice.
8. A spoonful of leaf extract is given orally in case of intestinal ulcer till cure.
9. Fruit paste of ripe onse is applied regularly on bleeding pile.
10. Decoction of unripe fruits about half cup, once a day is administered to cure chronic diarrhoea and dysentery.
11. Ripe fruit pulp is thought cooling and digestive. Hence 'sharbat' prepared from the pulp is drunk at night.
12. Pulp of fruits about spoonful is given to children to check dysentery.
13. Pulp of ripe fruit and dried fruit pericarp of this plant is consumed for a week to cure constipation.

14. Pulp of unripe fruit is mixed with pickled mango (*Mangifera indica* L.) and lemon [*Citrus aurantifolia* Christm (SW.)] and eaten for jaundice.
15. Few drops of seed oil are dropped in nose to stop bleeding.

Veterinary Medicine

1. Pulp of five to six ripe fruit is given against dysentery till cure.

Notes from Classics

Unripe fruit is cut, sun-dried and sold as astringent. It is digestive and stomachic.

(Watt, 1886-1876)

Root bark and stem bark is used in intermittent fever. It is also used as fish poison. Unripe fruits are astringent, stomachic and digestive. They are used against diarrhoea. It is also stomachic. Ripe fruits are sweet, aromatic and cooling. The young fruits are used for necklace.

(Anonymous, 1948-1976)

Roots are used for bite of mad dogs and for gastric trouble. Stem is used for stomach disorder and dysentery. Leaves are used for abscesses, back-ache diabetes, eye complaints, fever, heat in abdomen, jaundice and vomiting. They are used for healing wounds in cats and cattle. Fruits are used as astringent, laxative and tonic. They are used for remedy on bite of mad dog, constipation, diarrhoea, dysentery and gastric trouble. They cure stomach disorder.

(Jain, 199)

Gummy substance around seed serves as an adhesive. This gummy substance also used as varnish for pictures.

(Ambasta, 1991)

Plant is useful for abdominal and breast pain, cholera, constipation, cramps, dysentery, nausea, night fever, post natal care, puerperas fever and snake-bite.

(Jain, 1991)

Critical Notes

The notable uses such as root bark in typhoid fever, leaves against, night-blindness, stem bark for vomiting, leaf extract in intestinal ulcer, ripe fruit as coolant and digestive, and seed oil for bleeding nose are not documented else where.

Citrus aurantifolia (Christm.) SW.

Local Name

Nimboo, Limboo

Ethnobotanical Notes

Medicine

1. Fruit juice is mixed with rose-water and dropped in eyes three to four times a day to cure conjunctivitis.
2. Fruit juice mixed with rose water and glycerine is applied on face to remove wrinkles.
3. Fruit juice mixed with small quantity of curd is applied on face to remove darkness and pimples.
4. Excessive salvation in kids can be prevented by taking one spoonful of juice and honey at morning for a month or so.
5. Seeds of cardmom (*Elettaria cardamomum* Maton.) about two to three with one spoonful of juice of fruit after meal improve digestion.

Notes from Classics

Fruit is refrigerant and astringent. Rind of fruit is hot, dry and tonic. Fruit pulp is cooling in nature.

(Watt, 1889-1899)

Fruit is appetizer and stomachic. It is also antiscrobutic and anthelmintic. Fruit checks biliousness and helps in digestion.

(Anonymous, 1948-1976)

Fruit is a good source of vitamin-C and extensively used for culinary purposes. Fruits are also used for flavouring jam and jellies..

(Ambasta, 1986)

Fruits are used to cure diarrhoea. They are edible.

(Jain, 1991)

Critical Notes

Use of seeds to improve digestion is hitherto unrecorded. Likewise, although the fruit or their juice are commonly used for purposes domestically mentioned above, do not appear documented. It is cultivated for its fruits.

Limonia acidissima L.

Local Name

Kavath

Ethnobotanical Notes

Medicine

1. Spoonful of powder of leaves drunk with one glass of butter milk once at night to treat habitual constipation.
2. Fresh leaf extract about two spoonful with a pinch of table sugar and few drops of lime juice [*Citrus aurantifolia* (Christm.) SW.] is given to stop vomiting.
3. Equal amount of pulp of ripe fruit of this plant and *Tamarindus indica* L. are mixed homogeneously and a 'sharbat' is drunk once a day to stop hiccup.
4. Spoonful of powder of rind is mixed in milk and cow-ghee. This paste is applied on swelling and burned body part.
5. Spoonful of fruit pulp mixed in honey and given to kids once a day to stop dysentery till cure.

Notes from Classics

Roots are purgative, sydorific, cure colic disorder and cardiologic tonic. Dried fruits are tonic, diminishes intestinal fermentation..

(Watt, 1889-1899)

Plant is resistant to small pox, useful malignant fever and antidote to various poison.

(Anonymous, 1948-1976)

Fruit is tonic, stomachic, poor purgative, sudorific and useful in malignant fever. Wood is useful for making axles of cart, oil pressers, shafts, rice pounders, scales, hammers, walking sticks and tool handles.

(Ambasta, 1986)

Root powder is used to cure body pain. Bark is used on itch. Stem is used for fermentation of *Coryota liquer*. Fruits control diarrhoea, dysentery and digestive complaints. Plant is used to treat asthma and bronchitis. Fruits are used in preparing jellies.

(Jain, 1991)

Critical Notes

Leaves are being reported for the first time useful for constipation and to stop vomiting. Likewise, pulp of fruits for hiccup, and rind of fruit for body swelling and burning are also being reported for the first time. It is planted at some places.

Murraya koenigii (L.) Spreng.

Local Name

Kadhi-patta

Ethnobotanical Notes

Medicine

1. Fresh young green leaves are consumed to cure dysentery.

Notes from Classics

Root, bark and leaves are tonic and stomachic, also used to cure eruption and bites of poisonous animals. Leaves are entiperoidic, eaten raw to cure dysentery, also on bruises and externally applied on eruption. Infusion of leaves stops vomiting and it is febrifuge. Leaves either fresh or dry are used to flavour curry. Plant is antibilious. Wood is used for agricultural implements.

(Watt, 1889-1899)

Root juice is taken to relieve pain in kidneys. Leaves, roots, and bark are tonic, stomachic and carminative. Leaves are used in dysentery, diarrhoea, also for checking vomiting. They are applied externally on bruises and eruptions. Wood is used for making agricultural implements.

(Anonymous, 1946-1976)

Ambasta (1986) wrote similar uses and properties as those in Anonymous (*loc. cit.*).

Root is laxative and used for eruptions. Root bark is also used for eruptions. Leaves are anthelmintic, used for diarrhoea dysentery and malarial fever. It gives flavour to curries and it used as a spice. Whole plant is used for stomach complaints.

(Jain, 1991)

Critical Notes

Leaves are already reported useful to control dysentery. So it is planted in household gardens especially for foliage which is generally useful for the flavouring curries, vegetable and other food preparation. It is a good home remedy.

Family: Simaroubaceae

Ailanthus excelsa Roxb.

Local Name

Maruk, Maharukh

Ethnobotanical Notes

Medicine

1. Decoction of stem bark is mixed in bath water by woman to cure post-delivery pains.
2. Leaf paste is applied on scalp to kill lice by the aborigines.

Veterinary Medicine

1. Leaf extract is applied on the body of domestic animals to kill lices.

Miscellaneous

1. *Fodder*: Leaves and young shoots are used as fodder.

Notes from Classics

Stem bark is aromatic and used as a tonic. Stem bark is febrifuge in cases of debility. It is antispasmodic and expectorant, cures chronic bronchitis and useful in dyspeptic complaints.

(Watt, 1886-1896)

Leaves are also used a tonic. Bark is anthelmintic, febrifuge, expectorant and antispasmodic, used for asthma, bronchitis and dysentery.

(Anonymous, 1946-1976)

Wood is used formaking small boats, packing cases, sword handles, oys and matches. Leaves and shoots are used as fodder.

(Ambasta, 1991)

Bark is aborficacient, also used for bronchitis, cough and cold. It is febrifuge, also cause infertility in women. Stem bark helps to cure post natal problems and also gives relief in skin diseases. Leaves are used against rheumatism, skin diseases and for healing wound. It is a good tonic.

(Jain, 1991)

Critical Notes

The medicinal uses noted in Jalgaon district are not prevalent in other regions of India.

Family: Balanitaceae

Balanites aegyptiaca (L.) Del.

Local Name

Hiingan bet, Hingornya

Ethnobotanical Notes

Medicine

1. A pinch of stem bark powder in a cup of milk is taken to kill intestinal worms.
2. Paste of unripe fruit is applied on itch.
3. Seed powder five to six grams is taken with water to cure constipation.

Veterinary Medicine

Paste of unripe fruit is applied on wounds of domestic animals.

Food

(i) Seeds are fried and eaten, (ii) seed oil is used in vegetable preparation.

Miscellaneous

1. Fruits are used to kill lice in hair.
2. Fruit is also employed as detergent to wash clothes.
3. Stem bark is used as fish poison.
4. Wood is used for fuel.

Notes from Classics

Bark is given to cattle as an anthehelmintic. Bark, unripe fruits and leaves are anthehelmintic and purgative. Unripe drupes have strong Cathartic and anthelmintic properties. Seeds are used as an expectorant and employed in colic. Leaves and twigs are used as fodder for cattle. Ripe fruits are eaten and fermented in liquor. Wood is used for making walking sticks and for fuel.

(Watt, 1889-1899)

Bark is used as an anthehelmintic. Unripe fruit is cathartic. Ripe ones are used for whooping cough and skin diseases. Seeds are expectorant, used in whooping cough and in colic. Seed oil is used for burns, freckles and for soap making. Wood is used for walking sticks and shoemaker board.

(Anonymous, 1948-1976)

Similar uses as those in Anonymous (*loc.cit.*) are noted by Ambasta (1986).

Roots are used against insect bite. Stem bark is anthelmintic, used for cough, cold, swelling face and urine complaints. It is vermifuge. Fruits are anthelmintic, used for cough, cold, pneumonia and as purgative. They are applied as paste on skin diseases and for snake-bite. Fruits are vermifuge, used as a detergent to wash stains on clothes. Fruit shell is used for fire works. Seed oil is applied on burns. Roots and stems are used as fish poison.

(Jain, 1991)

Critical Notes

Stem bark against intestinal worms, fruits on itch and seeds for constipation are being noticed for the first time. Veterinary use of fruits is also new report. Consumption of seed oil by the aborigines of Satpura mountain is likewise noteworthy and hitherto unnoticed elsewhere.

Family: Burseraceae

Boswellia serrata Roxb. ex Coleb.

Local Name

Salie guggul

Ethnobotanical Notes

Medicine

1. Paste of stem bark is applied for rheumatism and mascular pains. Gum is also applied for similar purpose.
2. Leaf decoction, about half cup is taken orally for seven days early morning to cure conjunctivitis.
3. Leaf paste is applied on strips of cotton cloth. These strips are placed on eyes for burning sensation.

Notes from Classics

Flowers and seeds are eaten by Bhils. Gum is used against amenorrhoea, dysmenorrhoea and gonorrhoea.

(Watt, 1889-1899)

Gum contains digestive enzymes.

(Anonymous, 1948-1976)

Bark is used against diarrhoea and skin diseases. Gum is considered expectorant, diuretic and stomachic and used in diarrhoea, dysentery, pulmonary affections and cutaneous trouble. Flowers and seeds are eaten. Wood is used for making news prints and paper. It is also used for plywood, hard-boards furniture, packing, cases, toys etc. Tree exudes oleo gum-resin called 'salia guggul' varnish can be made from it. Gum is used in calico-printing and textile industry. Gum is source of fur-fural used in manufacture of nylon, plastic etc.

(Ambasta, 1986)

Stem bark is antiseptic used for cough, cold, diarrhoea and dysentery. It is antidote for snake bite and scorpion sting. Stem bark is also used for eye diseases, somach complaints and tooth-ache. Leaves give relief in eye disease. Resin is useful on cough, cold and skin eruptions, sore, stomach complaints and pulmonary diseases.

Seeds are applied for growth of hair, on snake-bite and scorpion sting. Plant is useful for epilepsy, snake-bite and stomach complaints.

(Jain, 1991)

Critical Notes

Curing of burning sensation of eyes by leaves and utility of stem bark against rheumatism and muscular pains have not been reported earlier.

Garuga pinnata Roxb.

Local Name

Kakad, Kakod

Ethnobotanical Notes

Medicine

1. After cutting young branch three to four drops of juice and dropped in eyes. This is practiced for three days at morning to cure cataract.
2. Half cup of decoction of leaves is drunk twice a day to control body temperature in fever.

Miscellaneous

1. Bark powder is used as fish poison.
2. *Food*: Drupes are pickled.
3. *Fodder*: Leaves and tender shoots are used as fodder.

Notes from Classics

Juice of stem is dropped in eyes to cure opacities of the conjunctiva. Juice of leaves with leaves of *Adhathoda zeylanica* mixed with honey and given in asthma. Fruits are pickled and eaten as a cooling and stomachic remedy. Shoots and leaves are used as fodder.

(Watt, 1889-1899)

Decoction of roots is used for pulmonary affection. Stem is remedy for opacity and conjunctivitis. Drupes are eaten row, cooled and pickled, posses cooling and digestive properties.

(Anonymous, 1948-1976)

Heart wood is suitable for furniture, sapwood for boarding. Timber is used for planking, drums and cabinet work. Leaves and shoots are used as fodder.

(Ambasta, 1986)

Roots are applied for skin and veneral diseases. Bark is prescribed for diabetes, eye diseases and to cure wounds. It is fish poison. Leaves are astringent and administered for asthma.

(Jain, 1991)

Critical Notes

Plant juice is beneficial in cataract. Decoction of leaves is useful to reduce body temperature in fever. Either of uses are not reported earlier.

Family: Meliaceae

Azadirachta indica A.Juss.

Local Name

Neem, Limb Kadu-neem

Ethnobotanical Notes

Medicine

1. A cup of leaf extract is taken orally for fifteen to twenty-one days for body cooling.
2. About two spoons of leaf juice is taken orally to avoid vomiting.
3. A paste of leaves and turmeric powder (*Cucuma longa* L.) is applied on swollen chick in mumps.
4. Leaf extract about 50 ml is drunk to four or five days to control malaria.
5. Regular application of extract of leaves is on face to remove scars of measles.
6. Hair wash with leaf extract is preservative for premature hair fall.
7. Few leaves are boiled in water, the vapours are allowed to pass into ears for ear-ache.

8. A cup of leaf extract mixed with some sugar is taken orally for two or three weeks in syphilis.
9. Consumption of five to seven leaves regularly in the morning helps to cure acidity.
10. Leaf extract, about 10 ml, is given to a delivered woman to reduce the increased size of uterus.
11. Extract of tender leaves is mixed with rock salt. A glass of it is consumed to kill worms in digestive system.
12. Root pieces are tied around waist of a pregnant women to reduce labour pain.
13. Stem bark powder is mixed in milk and applied on the pimples.
14. Ripe fruits mixed in jaggery are consumed for twice a day for 14 days to cure piles.
15. Ripe fruits are mixed with common salt or rock salt to improve digestion. About 10 gm of this mixture is advised daily in the morning.
16. Decoction of inner bark of stem, about 20 ml, is drunk as a remedy against stomach-ache.
17. Seed powder is consumed about 5 gm daily for the three months, it improves the function of uterus.
18. Few drops of seed oil are administered orally to cure piles. It is advised till cure.

Miscellaneous

1. Leaves are kept in grain storage for preservation.
2. Young stem is used as tooth brush.
3. Timber (wood) is employed for house building and ploughs.
4. *Fodder*: Leaves are used as cattle fodder, especially for goats.
5. Dried leaves are burnt in houses to repel mosquitoes.

Notes from Classics

Stem bark, root bark and young fruits are useful in general debility. Bark is used as an astringent, tonic, antiperiodic and febrifuge. Ash of bark is useful in eczema. Leaf pulp is applied externally on the mammae as a lactifuge. Decoction of leaves is used as a wash for ulcer. Leaves are also used for skin diseases, sprains and glandular swellings. Fresh juice of leaves is given with salt for intestinal worms and with honey for skin diseases and jaundice. Hot infusion of leaves is used for formenting swollen glands, sprains and appears to be anodyne. Powdered and fried leaves with ghee are made into ointment, which is very efficacious in sinuses. Tender leaves are taken internally to cure leprosy. Flowers are used in atonic dyspepsia. Seed oil is proved as a local stimulating and antiseptic, used in skin diseases, ulcer, urticaria, eczerna, leprosy, scabies and rheumatism. Oil is useful in cholera.

(Watt, 1886-1896)

Bark is tonic and astringent. It is antiperiodic, used in malarial fever, also applied on skin diseases. Leaves are used as poultice on boils, whereas there decoction is used on ulcer and eczema. Berries are used as purgative and anthelmintic. They are emolliant. Seed oil is used for skin diseases, ulcer, sores and ring worm. It is anthelmintic. Oil is used for message or rheumatic joints.

(Anonymous, 1948-1976)

Flowers are tonic and stomatich. Smell of burning leaves kill insects. Seed oil is used for making soaps. Timber is employed for house building, for boats, panels, toys and ploughs. Fresh tender twigs are used to clean teeth.

(Ambasta, 1986)

Bark is remedy on anti-fertility of males. It is useful against boils, jaundice, liver complaints, skin diseases and wounds. Leaves are used as antipyretic and blood purifier, used against diarrhoea, dysentery, malaria, fever, liver and heart complaints, skin diseases and scabies. Flowers are used for dyspepsia, wounds and debility. Fruits are used as febrifuge, leprosy, piles, skn diseases and tuberculosis. Seeds are antiseptic. They are efficacious in debility. Seed oil used on piles, scabies and post natal care. Seed destroy lice. The plant is good remedy on measles, rheumatism, small pox, spleen complaints syphilis, tumour and ulcer. Flowers, fruits and leaves are used as vegetable. Fibre is obtained from bark. Seed cake is used as pesticide. Exudate from tree is considered sacred.

(Jain, 1991)

Critical Notes

Roots are effective in reducing labour pains. Stem bark is effective on pimples. It is useful for stomach-ache. However, its use in liver complaint is already documented. Leaves are important medicine locally for body cooling, to avoid vomiting, to treat swollen cheeks in mumps, to prevent hair fall, to remove scars of measles, for ear-ache, syphilis, to reduce acidity, to treat increased size of uterus and to kill intestinal worms. Fruits are useful to treat piles and digestive complaints. Seeds are medicinal for improving function of uterus and curing piles. The miscellaneous uses, for example, leaves for preservation of food grains, as mosquito repellant, as fodder for goats are not on records.

Melia azedarch L.

Local Name

Bakam, Bakam-neem

Ethnobotanical Notes

Medicine

1. Equal amount of leaves and rock-salt are mixed to make pellets. Single pellet is taken daily, for about a month to suppress body-tumours.

Notes from Classics

Root bark, stem bark and fruits are tonic and antipyriodic. Bark is bitter, tonic, astringent, useful in fever, nausea, vomiting and skin diseases. Root bark and young fruits are used in intermittent fever and debility. Fruit juice of leaves with salt is used against intestinal worms with honey in skin diseases and jaundice. Leaf pulp is applied on mammae as lactifuge. Paste of leaves prepared in cow-ghee is applied on psoriasis. Poultice of leaves is used to suppress inflammation. They are anthelmintic. Seed-oil is used for skin diseases, ulcer, rheumatism and sprains. Seed oil is applied on eczema, leprotic wounds and excoriation of scalp. A milky fluid is exuded from old trees is considered alterative tonic. Seed oil and leaves are stimulant, insecticidal and antipyretic. Gum is demulcent and tonic.

(Watt, 1889-1899)

Fruits are tonic. Seed-oil is suitable for soap making and hair oil. Wood is used for toys, packing, museum cases and agricultural implements.

(Ambasta, 1986)

Infusion of bark is effective against ascariasia. Leaf juice is anthelmintic, diuretic and emmenogogue. Decoction of leaves is regarded astringent and stomachic. A poultice of flower is applied for eruptive skin diseases. Fruits are febrifuge and disinfectant. Seeds are prescribed for rheumatism. Gum is used in spleen enlargement.

(Anonymous, 1968-76)

Leaves are antiseptic, insecticidal, preven hair fall and applied on wound. Leaves are useful in head-ache, promote conception and used to treat rheumatism and gout. Shoots are edible. Fruits are anthelmintic and used for skin diseases. Seed are used as vermifuge.

(Jain, 1991)

Critical Notes

Leaves are found useful to treat tumours, a use not noticed in past. Usually it is planted in the district. It is exotic native of Iran and Turkey. But the local people have tried this taxon for medicinal use.

Family: Celastraceae

Cassine albens (Retz.) Kosterm.

Local Name

Bhutaypalas

Ethnobotanical Notes

Medicine

1. Infusion of leaves, about half cup, is given twice a day to lower down body temperature till cure.
2. Gum is applied for eczema till cure.

Miscellaneous

1. Stem bark is used for fish killing.
2. Wood is used as timber.

Notes from Classics

Roots are used as antidote for snake-bite. Decoction of fresh bark of root is applied over swellings. Dried leaves act as sternutatory and also used as fumigatory to women for hysterical fits. Snuff of leaves is used for head-ache.

(Watt, 1889-1899)

Roots are used for snake-bite, used in the treatment of inflammation of belly. Root bark is emetic and applied for swelling. It also controls urinogenital diseases. It is also employed to remove pimples. Stem bark is useful in diarrhoea, epilepsy, gastric complaints and hysteria. Decoction of leaves is used in bath water after child-birth to prevent rheumatic pains. Plant is considered useful for cholera, diarrhoea and epilepsy.

(Jain, 1991)

Critical Notes

Leaves are useful to lower down body temperature. Gum is applied for eczema. Stem bark is employed to stupify fishes. These uses have not been documented earlier.

Celastrus paniculatus Willd.

Local Name

Malkangni

Ethnobotanical Notes

Medicine

1. Juice of leaf is applied all over the body for body swellings.
2. Leaf juice, about half cup, is taken orally at every morning for a week for body-ache.
3. Seed oil is slightly warmed and applied with massage on paralysis of legs. It is followed till cure.

Notes from Classics

Seeds are given in rheumatism and paralysis in cattle. Seeds are aphrodisiac and stimulant. They are also used in leprosy.

(Watt, 1889-1899)

Bark is abortifacient. Leaves are emmanogogue. Leaf sap is used as antidote on opium poisoning. Seeds are bitter, laxative, emetic and tonic. They are used as stimulant and diaphoretic in rheumatism, gout and various fevers. Seed oil is powerful stimulant.

(Anonymous, 1948-1976)

Fruits, seeds and arils yield fatty oil.

(Ambasta, 1991)

Stem bark is abortifacient. It is applied locally on bone fractures and used in bronchitis. Seeds are reputed for its use in back-ache, gout, cold, cough, leprosy, paralysis and rheumatism. It is tonic and improves memory. Seeds are applied locally on scabies, wound and skin diseases. Seed oil is stimulant, used for stomach-ache and swollen vein can cured if applied locally. Plant is useful for anemia, colic, diarrhoea and sores. Fruits are beneficial for skin diseases. It is edible. Flowers and fruits are used as vegetable.

Critical Notes

Utility of leaves for body-ache and swelling have not gone down into literature yet. These are first reports.

Maytenus senegalensis (Lam.) Excell.

Local Name

Henkal, Kakedya

Ethnobotanical Notes

Medicine

1. Leaf extract, about half cup, is taken orally for twice a day for three to four days to cure cough.
2. Leves are chewed for stomatitis.

Miscellaneous

Thorny dried branches are used for fencing to prevent entry in cropped fields.

Notes from Classics

Leaves are used as fodder. Branches are applied for roof of houses. Wood is used for making boxes.

(Watt, 1989-79)

Paste of bark with mustard oil is used to kill lice in hair. Bark is used in the gastro-intestinal trouble. Leaves with milk are given to children as vermifuge. Leafy twig is used to relieves tooth-ache. Leaf ash with ghee is used as ointment.

(Anonymous, 1946-1976)

Bark is ground to paste and applied with mustard oil to destroy pediculi. Decoction of shoot is used for colic, dysentery and diarrhoea.

(Ambasta, 1986)

Root is used in veterinary medicine as regenerator. Leaves are used to cure cough, and eye complaints. Fruit are edible and blood purifier.

(Jain, 1991)

Critical Notes

Leaves are found beneficial in the treatment of cough and stomatitis. These uses did not find place in past literature.

Family: Rhamnaceae

Ventilago denticulata Willd.

Local Name

Arshisli.

Ethnobotanical Notes

Miscellaneous

1. Leaves and stem bark are crushed and this paste is used for fish killing.

Notes from Classics

Bark is used for body pain due to malarial fever. Ring made up of tendril is worn against tooth-ache.

(Watt, 1889-1899)

Bark yields cordge fibre. Wood is used for fuel. Seed oil is used for cooking.

(Anonymous, 1946-1976)

Pulverized bark is applied on sprains. Seeds yield fatty oil. Stem yield cordage fibre.

(Ambasta, 1986)

Seed oil is edible. Stem yields cordage fibre.

(Jain, 1991)

Critical Notes

Leaves and stem bark are employed for stupifying fishes by tribals. These uses have not been found published earlier.

Ziziphus caracutta Roxb.

Local Name

Chotibor, Jali

Ethnobotanical Notes

Miscellaneous

1. ***Fodder***: Leaves are used as fodder for goats.
2. ***Food***: Ripe fruits are eaten.
3. ***Wood***: Wood is used for making agricultural implements.

Notes from Classics

Young shoots, leaves and fruits are eaten by cattle and goats.

(Watt, 1889-1899)

Leaves are used as fodder. Wood is useful for agricultural implements, cart wheels and tool handles. Wood is used as fuel.

(Anonymous, 1946-1976)

Leaves are good fodder for cattle. Kernels are eaten. Wood is used for making agricultural implements, cart wheels, tool handles in-lay-work, frames shoe heels and pencils.

(Ambasta, 1986)

Seeds are edible. From fruits dye is obtained.

(Jain, 1991)

Ziziphus glaberrima (Sedgw.) Santapau

Local Name

Bor

Ethnobotanical Notes

Miscellaneous

1. *Fodder*: Leaves are used as fodder.
2. *Wood*: Wood is used for making agricultural implements.
3. Wood is also used as fuel.
4. Branches are used for fencing around farm yards and crop fields.

Notes from Classics

Bark powder mixed with ghee is applied to the swelling in the cheek caused by tooth-ache and mouth ulcer. Bark is used as astringent in diarrhoea. Flower is given for menorrhagia. Fruits are eaten by natives. Branches are used for fencing.

(Watt, 1889-1899)

Fruits are eaten. Branches are used for fencing. Wood is used as fuel.

(Anonymous, 1946-1976)

Leaves are used as fodder. Wood is used for making agricultural implements, all types of handles, frame and in-lay work. Wood is also used as fuel.

(Ambasta, 1986)

Roots are used for dysentery. Bark is used for diarrhoea. Flowers are used for syphilis and ulcer of tongue. Plants are useful for carbuncle and gum trouble. Fruits are edible.

(Jain, 1991)

Ziziphus nummularia (Burm.*f.*) Ait. and Arn.

Local Name

Bor

Ethnobotanical Notes

Medicine

1. Fresh tender leaves are boiled in water alongwith a pinch of rock salt. Gargling with this water 2 to 3 times a day cure sore throat.
2. Powder of leaves is applied locally on scabies.

3. Half cup of infusion of crushed green fruits is advised twice a day in the treatment of dysentery till cure.
4. Decoction of 50 gm of ripe fruits and 10 gms of black pepper (*Piper nigrum* L.) is given. Half cup per day is drunk for cough and cold will cure.

Miscellaneous

1. Leaves are used as fodder for goats.
2. Wood is used as fuel.

Notes from Classics

Bark is employed to make poultice for sores. Decoction of it is used for gargling in sore throat and ulcerated gum. Fruits are cool, astringent and used in bilious affections. Leaves are most valuable fodder for camels, goats, buffaloes and cows.

(Watt, 1889-1899)

Wood is used as fuel. Leaves are used in scabies and skin diseases. Fumes of dried leaves are inhaled for cough and cold. Fruits are cooling, astringent and used in bilious affection.

(Anonymous, 1946-1976)

Ambasta (1986) reports similar properties as in Anonymous (*loc.cit.*).

Roots cure diarrhoea and indigestion. Roots are tonic. In veterinary medicine, roots are useful in calf. Bark is useful for dysentery and inflammation of gums. Leaves are applied locally on boils.

(Jain, 1991)

Critical Notes

Leaves are useful on sore throat and scabies as prescribed above. Fruits are beneficial in dysentery, cough and cold. These have not gone down in past ethnobotanical literature.

Ziziphus mauritiana Lamk.

Local Name

Bor

Ethnobotanical Notes

Medicine

1. A spoonful of powder of fruit alongwith jaggery is advised to consume twice a day to cure bleeding leucorrhoea.
2. About half cup of decoction of fruit is taken twice a day for fever in tonsillitis.
3. Equal amount of leaves and rock salt is fried, mixed homogenously and applied over wound for quick healing.
4. Paste of seed is applied on pimples.
5. Ash of seeds is applied around eyes at night to cure conjunctivitis.

Miscellaneous

1. Wood is used for handle of axe etc.
2. Wood is used as fuel.
3. *Fodder*: Leaves are served as fodder for goats.
4. Dried branches are employed for hedging around farmyards and fields.

Notes from Classics

Roots and bark are astringent, used with babul bark for gargles. Roots are also used in fever, delirium and strangury. Powder of bark is used for dressing of wound of ulcer. Leaves are a remedy for diarrhoea. Young leaves are pounded with bark of those of *Ficus glomerata* Roxb. and are applied on scorpion sting. Tender leaves and twigs are useful on boils, abscesses and carbuncles. Fruits are cooling, aperient, aphrodisiac and nourishing. They are employed in bilious affection and constipation. Seeds are also used for dressing wounds. Kernels are antibilious, useful in nausia, thirst, fever and to purify blood.

(Watt, 1889-1899)

Decoction of bark is used in diarrhoea and dysentery. Bark is astringent. Leaves are astringent and aphrodisiac, used in typhoid fever in children. Fruits are cooling, anodyne and tonic. Fruits are useful in nausia, vomiting and give relief in abdominal pains in pregnancy. Fruits are used as poultice for dressing of wounds.

(Anonymous, 1946-1976)

Ambasta (1986) reported similar properties as those in Anonymous (*loc.cit*).

Roots cure ulcer wounds. Stem bark is used to cure diarrhoea. Leaves are useful in conjunctivitis and eye diseases. They also help to prevent hair fall. Leaves are used as antidote to scorpion sting. They cure fever, rheumatism and whooping cough. Plants are used in the treatment of diarrhoea, dysentery, head-ache, oedema, mouth sores and spleen complaints.

(Jain, 1991)

Critical Notes

Fruits are noted useful for leucorrhoea and tonsillitis. Seeds are used for pimples and conjunctivitis. These local medical uses have not bee noted elsewhere. It is cultivated for its edible fruits.

Ziziphus rugosa Lam.

Local Name

Toran

Ethnobotanical Notes

Food: Fruits are eaten.

Notes from Classics

Fruits are eaten by natives in all localities in which plant grows. Branches are used to make heaped-up fences.

(Watt, 1889-1899)

Bark is used as astringent. Flowers are used against menorrhagia. Fruits are eaten. Wood is used as fuel.

(Anonymous, 1946-1976)

Fruits are eaten. Crushed, defatted seeds yield starch. Bark is given in diarrhoea. Flowers are used for menorrhagia. Branches are used for fencing. Wood is used for fuel.

(Ambasta, 1980)

Roots are used for dysentery. Bark is used for diarrhoea and dysentery. Plant is used for carbuncle, gum trouble, syphilis and ulcer of tongue. Fruits are edible.

(Jain, 1991)

Family: Vitaceae

Ampelocissus latifolia (Roxb.) Planch.

Local Name

Panivel, Jangli-angur

Ethnobotanical Notes

Medicine

Root powder is mixed in water. After about one hour, this paste is applied on body for inflammation.

Notes from Classics

The plant is important as a new source of grapes.

(Watt, 1889-1899)

Decoction of roots is given in dysentery. Juice of tender leaves is used against dental troubles and as a detergent for indolent ulcers.

(Ambasta, 1986)

Root powder is used for boils, bone fractures and dysentery. It is also useful on fever, inflammation and menstrual complaints. Roots are also used against snake bite, wounds and stomach pains. Plant is useful in pneumonia, snake bite and sores.

(Jain, 1991)

Cissus quadrangula L.

Local Name

Kand-wel

Ethnobotanical Notes

Medicine

1. Few drops of stem extract are dropped in nose to stop bleeding nose.
2. Few drops of stem extract are also dropped in ears to cure ear-ache.
3. Infusion of stem is topically applied on bone fractures.
4. Plant extract, about two spoonful, is taken orally thrice a day for the treatment of spondylitis.

Notes from Classics

Young shoots and leaves are powdered and given in bowel complaints. Juice of stem is dropped in ears for otorrhoea and nose in epistaxis. Juice of stem is remedy in scurvy and irregular menstruation. Stem also cures stomach-ache. It is good appetizer. Shoot is used in dyspepsia and cattle diseases.

(Watt, 1889-1899)

Powdered roots are used to apply on fractured bones. Stem paste is used to cure asthma. Fresh shoot is pounded and applied to burns and wounds. Dried shoot is powdered and consumed for digestive troubles. Infusion of plant is purgative. It is beneficial to scurvy.

(Anonymous, 1946-1976)

Young stem and root yield fibre. Young shoot is used in curries and in preparation of papad.

(Ambasta, 1986)

Roots are used for bone fractures, cuts and wounds. Paste of stem is applied on bone fracture. Stem controls, stomach-ache and swellings. Leaves reduce pain in child-birth and stomach ache. Plant is useful for bone fractures and wound. Stems are pickled and used as vegetable also.

(Jain, 1986)

Critical Notes

To treat spondylitis, this species is beneficial, a use hither to not reported. It is sometimes cultivated in gardens.

Cayratia trifolia (L.) Domin

Local Name

Bail-mal

Ethnobotanical Notes

Medicine

Root bark is applied externally to reduce muscle pains.

Miscellaneous

Food

1. Fruits are edible.
2. Leaves are used as vegetable.

Veterinary Medicine

This climber is wrapped around the neck of a frantic bullock.

Notes from Classics

Roots with black pepper are applied to boils. Leaf poultice is used generally for yoke sores on the neck of bullocks.

(Watt, 1889-1999)

Roots are astringent, ground with black pepper and applied as poultice on boils. A poultice of leaves is rubefacient, applied to ulcer and yoke sores of bullocks.

(Anonymous, 1946-1976)

Ambasta (1986) records similar properties as in Anonymous (*loc.cit.*).

Roots help to reduce anemic condition and used as antidote on snake bite. Leaf paste is applied externally on boils and also to stop bleeding. Leaves are applied on sores of cattle. Fruits are edible. Leaves are used as vegetable.

(Jain, 1991)

Critical Notes

Root bark is found to reduce muscle pains. Entire climber is used for frantic bullock. These uses are additional to the earlier ones.

Leea asiatica (L.) Ridsdale

Local Name

Lahan Deni

Ethnobotanical Notes

Miscellaneous

1. Fruits are edible

Notes from Classics

Small black succulant fruits are eaten by natives.

(Watt, 1889-1899)

Berries are eaten.

(Anonymous, 1946-1976)

Fruits are eaten.

(Ambasta, 1986)

Fruits are edible.

(Jain, 1991)

Family: Sapindaceae

Cardiospermum halicacabum L.

Local Name

Kapal-phodi, Kan-phuti

Ethnobotanical Notes

Medicine

1. Few drops of leaf juice are dropped in ears to cure ear-ache.
2. Root extract, about two or three drops, also cure ear-ache.
3. Few drops of leaf extract are poured on wound to prevent pus formation.

Notes from Classics

Roots are used as emetic, laxative and stomachic. They are diaphoretic, diuretic and tonic. Roots are prescribed for rheumatism, piles and nervous disorders. Leaves and castor-oil are applied locally on rheumatic swellings. Seeds are also used as tonic in fever. Seeds and leaves are cooked as vegetable. Whole plant is steeped in milk and used externally to reduce swellings and tumours.

(Watt, 1889-1899)

Roots are diuretic, diaphoretic and emetic. They are laxative rubefacient and powerful emmenagogue. Leaves are rubefacient and cure ear-ache.

(Anonymous, 1946-1976)

Roots are used for rheumatism, lumbago and nervous diseases. Leaves are useful as a poultice in rheumatism.

(Ambasta, 1986)

Roots reduce dizziness. They are used in rheumatism and lung disease. Leaves are used as anasaraca. They reduce bulkness and stiffness of limbs and have a property of healing wounds. Leaves are effective in the treatment of lung diseases, ear-ache and eye sores. Seed paste is used to kill lice in cattle. Leaves are used as vegetable.

(Jain, 1991)

Critical Notes

Roots are also used for ear-ache. This is newly recorded observation. It is exotic, native of South America.

Sapindus emarginatus Vahl

Local Name

Ritha

Ethnobotanical Notes

Medicine

1. A pinch of fruit powder is mixed in spoonful of a lemon juice [*Citrus aurantifolia* (Christm.) SW.], few drops are poured in nose to cure epileptic fit.
2. Fruit paste is applied thyroid gland tuber, it suppress the tuber.
3. Fruits are pounded in water and few drops of filtrate are dropped in nose to treat migraine.
4. Fruits are used to wash hair.

Notes from Classics

Root is expectorant. Pulp of fruit is sweetish, hot, dry, tonic, alexipharmic and purgative. Fruit is rubbed in water and given as antidote for venomous reptile's bite, also used for diarrhoea and cholera. Pessaries made of the kernels are used to stimulate the uterus in child-birth and amenorrhoea. Pericarp or pulp of kernel of the fruit is emetic, nauseant and expectorant. Thick watery solution of fruit pulp is dropped in nostril for hemicrania, hysteria, epilepsy, coma and delirium. Kernels are useful to bring out abortion.

(Watt, 1889-1899)

Roots and bark are mild expectorant and demulcent. Decoction of bark is given to cattle for ulcer, used as fish poison. Fruits are emetic, tonic, astringent, anthelmintic, used in the treatment of asthma, colic due to indigestion and diarrhoea, also for cholera, tubercular gland, paralysis of limbs and lumbago. Few drops of solution of pulp of mesocarp dropped in nose give relief in hemicrania, epileptic fits. Wood is used for making carts, axe, boxes, comb, tool handles.

(Anonymous, 1946-1976)

Roots and barks are expectorant and demulcent. Root bark is used as detergent and also as a fish poison. Kernels yield fixed oil which is used for soap manufacture. Fruits are emetic, tonic, astringent, anthelmintic used in asthma, colic due to indigestion, diarrhoea, cholera, paralysis of limb. Wood is used for making carts cabinet work, turnery and also employed for construction work.

(Ambasta, 1986)

Bark is used for body-ache and head-ache. It also act as antidote for snake bite. Garland of bark pieces are placed in neck of cattle to cure wound of calf laying. Fruit pulp is used as detergent. Seeds are given to pregnant women for labour pains. Plant is used for epileptic fits and expectorant.

(Jain, 1991)

Critical Notes

Fruits are used to treat migraine. This is not noted earlier.

Family: Anacardiaceae

Buchanania lanzan Spreng.

Local Name

Charoli

Ethnobotanical Notes

Miscellaneous

1. *Food*: Seeds are edible and nutritive, used in local sweet-meats.
2. *Fodder*: Leaves are used as fodder during famine period only.
3. *Timber*: Wood is used for window frames.

Notes from Classics

Bark and fruits are used to furnish a natural varnish. Leaves are used as fodder. Fruit is eaten by hill tribe.

(Watt, 1889-1899)

Leaves are used as fodder. Seeds are eaten. Wood is used for small beams, furniture, boxes and match-boxes. Tree yields gum suitable for dressing textile.

(Anonymous, 1948-76)

Uses similar to that of Anonymous (*loc.cit.*) are noted by Ambasta (1986).

Musical instrument is prepared from fresh wood.

(Jain, 1991)

Critical Notes

This tree species is planted by tribals. They use its wood very judiciously. This tree species is worth suggesting for captive plantation because of its valuable seeds.

Mangifera indica L.

Local Name

Amba

Ethnobotanical Notes

Medicine

1. Half cup of decoction of bark is given daily for a week to cure diptheria.
2. Few drops of juice of young leaves are dropped in nose for bleeding.
3. Two or three leaves are pounded in sugar. It is given to a patient suffering from loose motion till cure.
4. Spoonful of powder of inflorescence is consumed at night daily to maintain the sugar level in case of diabetic patient.
5. Spoonful of seed powder is taken at morning once a day for a week to increase urination.
6. Seed coat powder about spoonful is consumed to treat dysentery till cure.

Miscellaneous

1. Cotyledons are steamed, salted and dried. They are eaten after dinner as a substitute for betel nut.

Notes from Classics

Bark is used in diarrhoea and hemorrhage from uterus, lungs and intestine. Leaves are used in disease of mouth. The midrib of the leaves calcined is used to remove wart of eyelids. Fruits are restorative, tonic, laxative, diuretic and fattening. Unripe fruits are astringent, acidic, stomachic and appetizing. Seeds are anthelmintic, used in bleeding piles, menorrhagia, asthma, worms and loosness belly. Gum resin mixed with leaf juice is used in scabies and cutaneous affections.

(Watt, 1889-1899)

Bark is astringent, used in diptheria, rheumatism. It is believed to possess a tonic action on mucus membrane. Ash of leaves is applied on burns and scalds. Leaves are masticated to give tone for gum. Fumes from burning leaves are inhaled for relief from hiccup and affection of throat. Dried flowers are astringent, given for diarrhoea, chronic dysentery and action on mucous membrane. Dried Kerneis are anthelmintic and astringent in bleeding pile. Ripe fruit considered as invigorating refreshing and fattening. Fruit juice is tonic containing vitamin-A.

(Anonymous, 1946-1976)

Bark is used for uterine hemorrhage. Fruit is laxative, diuretic. Unripe fruit is pickled and chutney is prepared from it. Ripe fruits are preserved by canning, juice or squash, jam or gellis preparation. Seeds are used to treat asthma. Wood is used to making inferior quality furniture, floor boards, window frames etc.

(Ambasta, 1986)

Bark is abortifacient, act against anti-fertility. It cures jaundice and stomach-ache. Leaves and flowers are used for cancer. Fruits are used for diarrhoea. They are digestive, laxative tonic. Fruits and kernels are pickled. Fruits are used in porridge. Latex is a remedy for scabies. Gum is used for tooth-ache. Plant is used for ansarca, cholera, dropsy, dysentery, eye complaints, fever, rheumatism and for ulcer of tongue. Wood is used for musical instrument.

(Jain, 1991)

Critical Notes

Leaves are medicinal for nose bleeding and loose motion, whereas flowers are useful to treat diabetes. Seeds are beneficial in increasing urination and to control dysentery. Cotyledons are used as masticatory after meals or tea. All these uses are not on record yet.

Lannea coromandelica (Houtt.) Merr.

Local Name

Mohin

Ethnobotanical Notes

Medicine

1. Paste of bark powder is applied daily on sprain till cure.
2. Half cup of diluted gum is drunk twice a day for one month in the treatment of leucorrhoea.

Miscellaneous

1. Fruits are crushed in water and used as fish poison.

Notes from Classics

Bark is astringent. Decoction of bark is used locally for cutaneous eruption and obstinate ulceration. Leaves are useful for asthma. These are cordial to women. Gum beaten up with coconut milk is applied on sprains and bruises. Bark is acrid, astringent, used as lotion on wounds, sore ulcer and sore eyes. It is stimulant, used in gout for apthae of mouth and tooth-ache. Boiled leaves are applied on sprain, swelling and in elephantiasis. Juice of green branches is mixed in tamarind, as an emetic, in coma and insensibility produced by opium. Leaves are good fodder.

(Anonymous, 1948-1976)

Bark extract is used for dying and textile printing. Wood is used for house building, packing cases, wheel spokes, ploughs, combs and wooden jars.

(Ambasta, 1986)

Bark is applied on eruptions, wound and also as antidote on snake-bite. From bark resin and dye are obtained. Leaves are prescribed for elephantiasis, sprain and swellings. Paste of fruit is applied on bone fractures. Plant is used to cure carbuncle, cholera, diarrhoea and dysentery. Wood is used for making tobacco pouch.

(Jain, 1991)

Critical Notes

Bark for curing fissure of pile and gum for leucorrhoea are being reported for the first time.

Rhus sinuata Thunb.

Local Name

Amoni

Ethnobotanical Notes

Miscellaneous

1. ***Food***:Fruits are sour and edible. They are sold sometimes near schools and common places by the natives. School boys enjoy them.
2. ***Fodder***: Young shoots are used as fodder, especially for goats.

Notes from Classics

Bark is largely used for tanning. Fresh fruits are eaten. Plants offer poor quality fodder.

(Watt, 1889-1899)

Fruits are sourish, bitter in taste and eaten either cooked or roasted. Leaves are eaten, also used as fodder. Wood is used for fence post, fuel and tool handles.

(Anonymous, 1946-1976)

Fruits are eaten, cooked or roasted. Leaves are eaten and also used as fodder. Wood is used for utility articles like small tool handles and fence posts.

(Ambasta, 1986)

Fruits are edible.

(Jain, 1991)

Critical Notes

Fruits are collected by rural people. They are sour and enjoyed by the boys and girls. They are sold especially near schools.

Semecarpus anacardium L.f.

Local Name

Bhilawa, Biba

Ethnobotanical Notes

Medicine

1. A needle is pierced through a seed and burnt. The oil is thereby expressed. Three to four drops of oil are added in a cup of curd. It is consumed for three days at morning to cure piles.
2. A spoonful of extract of seeds is given with honey once daily for 5-6 days in the treatment of asthma.
3. To get relief from rheumatic pains, seed oil is applied on knee joints.
4. Half seed and half spoonful of turmeric powder (*Cucuma longa* L.) is consumed for a week to cure cholera.
5. Warm seed oil applied over injury caused by thorn. To avoid infection, immediately hot jaggery is placed on it.
6. Paste of seeds is applied over the piles till cure.

Notes from Classics

Juice of pericarp is resicant. Ripe fruits are acrid, stimulant, digestive, nervine and used in dyspepsias, skin diseases, piles and nervous debility. They also used in scrofulous, veneral and leprous affections. Nut oil is used in rheumatism and sprains. Juice of pericarp is hot and dry, useful in all skin diseases, palsy, epilepsy and diseases of nervous system. Juice of pericarp is applied for piles. Fruit is smoked for impotency. Fruit juice is used for abortion. Juice of pericarp is used for rheumatism and tooth-ache.

(Watt, 1889-1899)

Juice of pericarp is counter irritant and vasicant. It cures painful blisters. It is used for tattooing and by mahuts for chobing elephant feet. It also has antibacterial properties. Fruit is acrid, hot, anthelmintic, beneficial in ascities, tumours, warts, acute rheumatism, asthma, neuralgia, epilepsy and psoriasis.

(Anonymous, 1946-1976)

Fleshy orange cup of fruit is eaten when ripe. Kernels are edible, yield semidry oil, used in preservative and lubricant. Trees exude gum resin used in leprosy and nervous debility. Juice of pericarp is used in preparation of ink.

(Ambasta, 1986)

Roots are used in case of antisterility, in impotency and wounds. Bark cures gonorrhoea. Receptacles are applied on burns and scalds. Kernels are used for tongue diseases. Pericarp is applied on scores in feet of cows. Latex is applied on head-ache and skin diseases. Leaves are used as vegetable. Fruits are abortifacient, cure asthma, cough, cuts, dyspepsia, epilepsy, nervous debility, neuralgia, piles and psoriasis. They are also employed for rheumatism, skin and throat diseases. They also help to expel worms.

(Jain, 1991)

Critical Notes

Seeds are used in the treatment of cholera and seed oil as disinfectant if injured. It is to be noted that live trees are not found in this district. The seeds are available in local shops.

Family: Moringaceae

Moringa concanensis Nimmo ex Dalz. & Gibs.

Local Name

Ran-shewga, Kadu-Shewga

Ethnobotanical Notes

Medicine

1. Extract of bark is applied onto wounds caused by honey-bee sting to stop burning pains.
2. Powder of flowers, about one spoonful, is consumed at night to decrease the sugar level in case of diabetic patients.

Notes from Classics

Unripe fruits and flowers are eaten as pot herb.

(Watt, 1889-1899)

Various parts of the tree is used indigeneous medicine, in the some ways as *Moringa olerifera*.

(Anonymous, 1946-1976)

All parts of tree are used in the treatment of ascites, venomous bite, rheumatism and as cardiac and circulatory stimulants. Roots are rubefacient and vesicant. Leaves are useful in scurvy and catarrhal affection. They are emetic. Flowers are tonic, diuretic and cholagogue. Seeds are antipyretic, seed oil is useful in rheumtism. Oil-cake is used as fodder. Gum is useful for calico-printing.

(Ambasta, 1986)

Roots are used to remove giddiness and as tonic. Seed oil is used for sore throat.

(Jain, 1981)

Critical Notes

Bark is a remedy in case of sting of honey bee and flowers are used to decrease sugar level. These uses have not noted yet.

Moringa olerifera Lam.

Local Name

God Shewga

Ethnobotanical Notes

Medicine

1. Three to four seeds are crushed in boiled rice water. Half cup of it is taken orally for chest pains caused due to excessive cough.
2. Five to six flowers are crushed to paste and applied on sores for quick healing.

Notes from Classics

Bark is used as carminative. It is used against flatulence of colic. Leaves are used for dog-bite. Flowers are boiled in milk, are said to be aphrodisiac. Leaves, flowers and pods are eaten. Trees are largely lopped for fodder.

(Watt, 1889-1899)

Roots are root-bark are rubefacient. Leaves are rich in vitamin-A and vitamin-C, so useful in scurvy and catarrhal affection. Leaves are emetic and applied externally on wounds. Flowers are tonic, diuretic and cholagogue. Seeds are antipyretic. Seed oil is applied on rheumatic joints and gout.

(Anonymous, 1946-1976)

Root is acrid, pungent, stimulant and diuretic. Decoction of root bark is used in the treatment of ascites, enlarged spleen and liver inflammation. Fresh juice of the root bark relieves otalgia in ear. Fresh root is used in epilepsy and hysteria. Root bark is employed to relieve spasm. Fresh root is stimulant and diuretic, also used in dropsical affection, in gargle for horse-ness. It gives relief in sore throat. Juice of root is digestive and useful in asthma. Poultice of root reduces swellings. Flowers dispers swelling. They are tonic, diuretic and increases flow of bile and used in catarrhal affections. Fruit are administered in a affection of liver, spleen, tetanus, delibility, paralysis and leprosy. Gum is used for abortion.

(Ambasta, 1986)

Roots give relief in asthma, delivery, epilepsy, liver complaints, paralysis, rheumatism and spleen enlargement. They are used as an antidote on scorpion-bite. Bark is useful in baldness, caries dental problem, fever and menstrual complaints. It is also prescribed for stomach-ache and typhoid. Leaves are administered for catarrh, menstrual complaint, post-natal care, scurvy and wounds. Fruits are appetizer and help to expel intestinal worms. They are stimulant and tonic. Gum is used for ear complaints. They are abortifacient. Plant is used baldness, cholera, madness, pneumonia, scabies, scrofula, sores, spleen enlargement, syphilis and urine complaints. Flower and fruits are used as vegetable. Whole plant is used as fodder.

Critical Notes

Seeds are good remedy against cough. Flowers are helpful to heal injuries. Both medicinal uses are addition to the science of ethnobotany. It is cultivated near human habitation or bandhs of fields.

Family: Papilionaceae

Abrus precatorius L.

Local Name

Gunj, Charmtadi

Ethnobotanical Notes

Medicine

1. Root powder of *Abrus* and *Asparagus racemosus* Willd. are mixed in equal amount in warm water. Two spoonful of it is drunk daily for a month as an aphrodisiac by men or women.
2. Few leaves are eaten for the treatment of bronchitis.
3. Leaves of white seeded variety are crushed and juice is extracted with milk or water. Two spoonful of juice is given thrice a day for hoarseness of voice.
4. Fruit rind of *Punica granatum* L. are dried and powdered. It is mixed in pomegranate fruit juice, seed coat powder of red seeded variety is also mixed in it. The whole paste is burned and ash obtained. It is applied on eyes for cataract.
5. Paste of *Abrus* seeds and flowers of *Hibiscus rosa-sinensis* L. in equal amount are homogenized. It is applied to scalp at night regularly to promote hair growth on bald.
6. Two seeds of white variety are powdered and mixed in a cup of water. It is given orally to a pregnant woman to increase labour pains.
7. An half seed pounded in spoonful of honey is given to infant for stomach disorders.

Miscellaneous

1. Seeds were used for weighing gold by Gold-smiths. The practice has now ceased.
2. Red seeds are strung in 'tongal', an ornament of ears of Banjara woman.
3. Leaves are used as masticatory.

Notes from Classics

Roots are useful in whooping cough and leucorrhoea. Leaves are consumed for local body pains. Seeds are employed in tonic for nervous system and also used for cornea of eyes. Seed powder is used as snuff for head-ache. Seed powder is consumed as aphrodisiac and emetic. Seeds are boiled and then used for consumption. Seeds were used for weighing gold by Goldsmiths.

(Watt, 1889-1899)

Decoction of roots is used for cough, cold and pain. Leaves are used for hoarseness of voice. Seed poultice is tied on abdomen, it helps for abortion.

(Anonymous, 1946-1976)

Roots are diuretic, tonic and emetic. Seeds are administered in affections of nervous system and seed paste is applied on sciatica, stiffness of shoulders or joints and paralysis.

(Ambasta, 1986)

Roots are used as antiseptic and blood purifier. They are also prescribed for bronchitis and as a febrifuge. Root powder is applied in leucodermal diseases, nervous disorders and to promote conception. They are also used as antidote on scorpion sting. Leaves are used for body pain, skin diseases, rheumatism, scabies, sore throat, swelling and urine complaints. They are useful in veterinary medicine on wounds and also as masticatory. The plant is used for convulsions, diarrhoea, eye complaints, leucorrhoea, menstrual problems, muscular pains and to facilitate child-birth.

(Jain, 1991)

Critical Notes

Roots are aphrodisiac, leaves in the treatment of bronchitis and seeds for cataract, to promote hair growth on bald and stomach disorders are being noted for the first time. Seeds strung in an ornament also did not find place in the past literature.

Alysicarpus longifolius Wt. and Arn.

Local Name

Shewara

Ethnobotanical Notes

Miscellaneous

1. *Fodder*: Leaves and young shoots are used as fodder for cattle.

Notes from Classics

A good fodder plant, are used in both green and dried state.

(Anonymous, 1946-1976)

A good fodder plant, used in both green and dried state. Roots are sweet, substitute for liquorice.

(Ambasta, 1986)

Seeds are used as vegetable.

(Jain, 1991)

Critical Notes

It is found on waste places as well as a weed in cropped fields. It is removed from fields manually and fed especially to milching animals. It is said to be of good fodder value. It may be worth trying to cultivate as fodder crop.

Arachis hypogaea L.

Local Name

Bhuimoog

Ethnobotanical Notes

Medicine

1. Fresh unripe fruits are considered as good lactogogue and given to women to increase lactation.
2. Seed oil is applied for massage which gives good complexion.

Notes from Classics

Seed oil is efficient substitute for olive oil. It is useful in catarrh of bladder. Ground nut is bilious and edible.

(Watt, 1889-1899)

In the form of emulsion, it is used for feeding children. It is considered aperient and an emollient. In veterinary medicine, it is used as nutritive, laxative and emollient.

(Anonymous, 1946-1976)

Seeds are used in various food and confectionary, used as 'Vanaspati'. Cake is used as feed for cattle and also as manure. Commercial products like groundnut milk, peanut ice-cream, massage oil are prepared from it.

(Ambasta, 1986)

Critical Notes

Fresh unripe fruits are consumed by ladies specially to increase lactation. This is not recorded earlier. It is a crop plant. It is exotic, native of Brazil.

Butea monosperma (Lam.) Taub.

Local Name

Palas, Khakra

Ethnobotanical Notes

Medicine

1. Root powder is applied on injury of snake-bite. Spoonful of root powder mixed with water is drunk as an antidote on snake-bite.
2. Paste of stem bark is applied in case of swellings of body.
3. Petiole is chewed constantly and juice is sucked. It cures cough, cold and stomatitis.
4. Leaf juice is dropped in eyes to cure conjunctivitis.
5. Leaf powder, about two spoonful per day for a month, is drunk with a cup of water to cure diabetes.
6. Extract of fresh leaves are employed to kill the intestinal worms.
7. Leaf extract is used to gargle in case of sore throat.
8. Leaf extract, about a cup is taken regularly at night for a month. It decreases sugar level in diabetes.
9. Leaf extract, about three to four spoons if drunk, at night for two to three months, checks irregular bleeding during menstruation.

10. Flowers are crushed in milk and sugar is added. Three to four spoon⁻ of drunk per day for a month helps to reduce body heat and chronic fever.
11. Flowers are soaked in water overnight and a cup of this infusion is drunk every morning for leucorrhoea till cure. It is also useful for strangury.
12. One gm of seed powder is mixed in half cup of milk and given thrice for a week to kill intestinal worms.
13. Seeds are crushed in milk and this mixture about two spoons is taken orally to treat urinal complaints and also for urine stone.
14. Gum is applied for cracks on foot sole while sleeping at night.
15. Two spoons of diluted gum is adviced for dysentery till cure.

Miscellaneous

1. ***Food***: Flowers and young fruits are used as vegetable.
2. ***Fodder***: Green leaves are good fodder for domestic animals.
3. Fresh green leaves are used to make dinner plates and small bowls.
4. From stem bark fibres are obtained which is used for making cordage.
5. Stem bark powder is used to stupefy fishes.

Notes from Classics

Leaves are astringent and tonic. They cure boils and pimples. Leaf extract is taken to cure pain in colic, piles and expels worms. Leaf poultice is tied to dispel tumours and haemorrhoids. Flowers are depurative, diuretic and aphrodisiac. They are used as poultice. They promote menstrual flow and cure urine complaints. Seeds are anthelmintic and rubefacient. Gum is astringent. It is used in haemorrhagic affections. Gum acts against ulcer and sore throat, also used for diarrhoea and dyspepsia.

(Watt, 1889-1899)

Seeds are anthelmintic and used for round worms and tape-worms. Seeds are rubefacient. They cure herpes called 'dhobies itch'.

(Anonymous, 1948-1976)

Root bark yields fibres. Stem bark is astringent, used for piles, tumours and menstrual disorders. Flowers yield yellow colouring matter. Seeds yield fatty oil. Wood is mainly used for well-curbs. Leaves are used mainly for dinner plates and fodder. Wood pulp is suitable for news print paper manufacture.

(Ambasta, 1986)

Roots are used for snake-bite and tuberculosis. Leaves are used on boils, inflammation, skin diseases and delirium. Flowers are abortifacient, anthelmintic and contraceptive. They are useful against antifertility, diabetes and dysentery. Flowers cure eye, stomach and urine complaints. Sunstroke effect is minimized after drinking flower extract. Flowers are useful for piles and dog-bite. Gum is used against antifertility, diarrhoea and dysentery. It is tonic as well. The plant is febrifuge and useful for menorrhoea. Flowers and fruits are used as vegetable. Leaves are used for making dinner plates and bowls. Stem bark yields fibre, also used as fish-poison. Dye is obtained from flowers.

(Jain, 1991)

Critical Notes

Utility of petiole for cough, cold and stomatitis; root for checking biliousness, bark for body swellings; leaves for conjunctivitis, diabetes, sore throat, bleeding menstruation; flowers for reducing body heat, for chronic fever, leucorrhoea; seeds for urinal complaints and urine stone; gum for foot cracks etc. are all not noticed elsewhere. Leaves are collected on large scale for making dining plates and dishes. They are sold in local markets. It is good species for cottage industry. Due to illicit cuttings their number has reduced considerably. This should be checked especially to maintain the supply of leaves.

Cajanus cajan (L.) Millsp.

Local Name

Tur, Tor

Ethnobotanical Notes

Medicine

1. Some black bluish grains are used to prepare paste in water. It is applied on eyes for conjunctivitis.

2. About 25 gm of pulse is boiled in about one litre of water. The extract about a cup is taken in the morning to cure fistula till cure.

Miscellaneous

1. Stalks are used for making partition walls of hut or roof.
2. Stalks are employed for making baskets to be used for agricultural work.
3. Stalks are tied together and used for sweeping ground and farm-yard.
4. Green and dried leaves are good fodder, especially for milching animals.
5. Stalks are good fuel.

Notes from Classics

Leaves are used in disease of mouth. Pulse and leaves are made into paste, warmed and applied to mother's chest of secretion of milk. Tender leaves are chewed in case of apthae and spongy gums. A poultice of seeds is employed to check swellings.

(Watt, 1889-1899)

Green leaves are used as fodder. Dried stalks are used as fuel.

(Anonymous, 1946-1976)

Pulses are eaten as dal. Green pods are used as vegetable. Husks and green leaves are used as fodder.

(Ambasta, 1986)

Leaves are used for gargle. Leaf paste is applied on skin for burns, sore mouth and tongue. Seeds are useful to revive taste.

(Jain, 1991)

Critical Notes

Use of seeds to treat conjunctiva and the pulse for fistula are unreported. All parts are very useful for various purposes, apart from food. It is exotic and native of Africa. It is a crop plant.

Cicer arientinum L.

Local Name

Chana, Harbara

Ethnobotanical Notes

Medicine

1. About twenty five gm of seeds or 'dal' is obtained in about half litre of water. This mixture is called 'kat' or 'ush'. It is taken, half cup twice a day for three days, against cough and cold.
2. Flour mixed with milk is applied on face for half an hour. This helps to remove scars on face.

Notes from Classics

Boiled leaves are applied as poultice to sprin and dislocated limb. Leaves are also used for dismennorrhoea. Roasted seeds are aphrodisiac. They are used for retention of urine and enlarged spleen.

(Watt, 1889-1899)

Seeds are consumed as dal. Germinated gram is used as a prophylactic against scurvy.

(Ambasta, 1986)

Seeds are edible and tonic. Stem is used for jaundice.

(Jain, 1991)

Critical Notes

Use of seed flour for cough and cold and also for fair complexion are not documented earlier. It is a crop species. It is exotic, native and Europe.

Clitoria ternatea L.

Local Name

Gokarni

Ethnobotanical Notes

Medicine

1. Decoction of roots, about half cup, is given twice a day as febrifuge.
2. Roots are used as brush for tooth-ache.
3. Powder of roots mixed in sugarcane juice (*Saccharum officinarum* L.) and kept in iron vessel for one month. This paste is applied on scalp to turn hair black.

4. Paste of leaves is applied over the wound caused by the bite of rabid dog.
5. Seed powder, about two spoonful, is taken with warm water for seven days to cure abdominal dropsy.

Miscellaneous

1. Plants are used as fodder of goats.

Notes from Classics

Roots are cathartic, used in dropsy and enlargement of abdominal viscera. Seed powder is used as purgative and anthelmintic. Seeds are also useful in weakness of sight and sore throat.

(Watt, 1889-1899)

Roots are acrid, cathartic and diuretic. Root bark is used as diuretic and laxative. Powdered seeds mixed in ginger are used as purgative.

(Anonymous, 1946-1976)

Leaves are used as fodder. Fruits yield blue dye.

(Ambasta, 1986)

Roots are considered as abortifacient. They are used for the treatment of goiter, leprosy and stomach ache. Roots are purgative. Seeds are used as laxative. Leaves are applied on swellings. Plant is used as an antidote on snake bite. Roots are also used on abdominal swellings of animals.

(Jain, 1991)

Critical Notes

Roots are effective as febrifuge and for tooth-ache. Interestingly, they help to blacken hair. Leaves and seeds are good remedy for rabid dog and abdominal dropsy respectively. The present investigator's medicinal reports are noteworthy and new. It is exotic, native of Central and South America. It is planted in household gardens and also naturalized on hedges.

Dalbergia sissoo Roxb. ex DC.

Local Name

Shisam

Ethnobotanical Notes

Medicine

1. Gum and roots are crushed together in water. One spoonful of filtrate is taken at night to treat asthma.

Veterinary Medicine

Saw dust of this plant is burnt and the fumes are allowed to pass over the bodies of domestic animals for unwillingness to eat fodder during fever.

Miscellaneous

1. Wood is used to make furniture especially to make 'Devhara', a small temple in home.
2. *Fodder*: Leaves are used as fodder for cattle and goats.

Notes from Classics

Root is astringent. Bark powder is mixed in ginger and is taken to check cholera. Decoction of leaves is given in acute stage of gonorrhoea. Leaves with any oil is applied for eruption. Wood oil is applied externally in cutaneous affection. Raspings of woods are officinal and regarded as alterative. They are considered hot, used in boil, leprosy, eruptions and to alloy vomiting.

(Watt, 1889-1899)

Roots are astringent. Leaves are bitter and stimulant. Decoction of leaves is useful in gonorrhoea. Leaf mucilage with groundnut oil is applied in excoriations. Wood is useful in cutaneous affections.

(Anonymous, 1946-1976)

Roots are astringent. Leaves are bitter and stimulant. Decoction of leaves is useful in gonorrhoea. Leaf mucilage with ground nut oil is applied in excoriations. Wood is useful in cutaneous affections.

(Anonymous, 1948-1976)

Leaves are used as fodder. Wood is used for furniture and cabinet. It is also used for railway slippers, musical instruments, charpai legs and carvings. Wood yield fixed oil. Wood is suitable for charcoal making.

(Ambasta, 1986)

Stem is prescribed for head-ache, leprosy and used as ornament for ears. Leaves are employed for dysentery, gonorrhoea and skin diseses. Leaves are anthelmintic and applied on eruptions.

(Jain, 1991)

Critical Notes

Medicinal use of gum and roots to treat asthma appear be the first report. Passing of fumes of saw dust of wood of this taxon when the pet animals are reluctant to eat fodder is new and interesting.

Dalbergia paniculata Roxb.

Local Name

Shisum, Satpudya

Ethnobotanical Notes

Medicine

1. Leaf paste is applied on boils. It checks burning and painful sensation.

Miscellaneous

1. ***Fodder***: Leaves are used as fodder for cattle.

Notes from Classics

Leaves and twigs are used as field manure in Chennai.

(Watt, 1889-1899)

Wood is used for building purposes and employed for musical instrument.

(Anonymous, 1946-1976)

Ambasta (1986) noted similar uses as those in Anonymous (*loc.cit.*)

Bark cures body pain and skin diseases. Leaves are remedy on colic pains. Tree is a host for lac.

(Jain, 1991)

Critical Notes

Leaves are useful medicinally for boils and burning sensation. These uses have not been yet noted elsewhere. Utility of leaves for fodder for cattle is likewise new observation.

Desmodium gangeticum (L.) DC.

Local Name

Salparni

Ethnobotanical Notes

Medicine

1. Roots are boiled in milk. Half cup of it is administered for seven days or even more for flatulence.
2. Roots are boiled in sugar and milk. This is taken orally twice a day for 7-15 days for leucorrhoea.

Notes from Classics

The plant is febrifuge and anticattarrhal. It is one of the ingredient of 'dasmula'.

(Watt, 1889-1899)

Plant is used as green manure. Fiberous stem is suitable for making paper.

(Anonymous, 1946-1976)

Roots are administered for asthma, cough, dysentery, fever and to stop vomiting. Roots act as an antidote on snake poison. Stem fibres are used for making paper. Whole plant is prescribed for bronchitis, fever and tonic. Roots are used as febrifuge, expectorant and diuretic.

(Ambasta, 1986)

Critical Notes

Medicinal utility of roots for flatulence and leucorrhoea are being noted for the first time. It is to be noted that the roots of this species are one of the constituent of an Ayurvedic drug 'Dasmula'.

Erythrina variegata L.

Local Name

Pangara, Palas

Ethnobotanical Notes

Medicine

1. Decoction of stem bark, about half cup, is drunk at morning and evening for six days to cure fever due to cough and cold.

2. Paste of leaf is applied locally on joints for rheumatic swellings.
3. Powder of flowers is applied locally on face to remove pimples.

Miscellaneous

1. ***Food***: Seeds are eaten after boiling or roasting.
2. ***Fodder***: Leaves are used as fodder.
3. ***Wood***: Wood is used as fuel.

Notes from Classics

Bark is antibilious, febrifuge and anthelmintic. It is also useful as collyrium in opthalmia. Leaves are vermifuge and cathartic. They are applied externally to disperse venereal buboes and relieves pain in joints. Leaf juice is dropped in ears for ear-ache and anodyne in tooth-ache.

(Watt, 1889-1899)

Bark is astringent, used as febrifuge, antibilious and anthelmintic. Leaves are laxative, diuretic, anthelmintic, galctogogue and emmenogogue, applied externally for dispersing venereal buboes and for relieving joint pains. Fresh juice of leaves is used for the relief of ear-ache and as an anodyne in tooth-ache. Juice is also used for killing worms in sores. Leaves and tender shoots are eaten as pot herb. Leaves are good fodder. Wood is used for rafts, floats and paper pulp.

Ambasta (1986) recorded uses similar to those in Anonymous (*loc.cit.*). Bark is anthelmintic, used to cure cold and cough, eye complaints, fever, rheumatism and skin disease. Leaves are given for the treatment of ulcer and cough. Plants are prescribed for convulsion, menorrhagia, paralysis, pimples and snake bite. Fruits are used as vegetable. Seeds are used as beads.

(Jain, 1995)

Critical Notes

Leaves are medicinally important for rheumatic swellings. Flowers are employed to treat pimples. These uses are have remained untapped from other parts of our country. It is usually planted.

Lablab purpurea (L.) SW.

Local Name

Wal

Ethnobotanical Notes

Medicine

1. Two spoonful of leaf juice is taken in milk or water regularly twice a day to cure arthritis.

Notes from Classics

Roots are used for poisoning the wild animals.

(Watt, 1889-1899)

Seeds are febriguge, stomachic, antispasmodic and aphrodisiac. Bean is a popular vegetable. Plant is used as fodder for cattle.

(Anonymous, 1946-1976)

Leaves are used as fodder and popular vegetable. Seeds are febrifuge, stomachic, antispasmodic, and aphrodisiac.

(Ambasta, 1986)

Roots are used against antifertility and also vermicide. Leaves are remedy on ring-work, also used for cough. Fruits are used as vegetable. Pulses are edible.

(Jain, 1991)

Critical Notes

Leaves are being reported for the first time as a remedy against arthritis.

Medicago sativa L.

Local Name

Lasoon ghas, Ghodaghas

Ethnobotanical Notes

Medicine

1. Extract of leaves about spoonful with honey is given for a week to treat polyurea.

2. Decoction of leaves, about half cup, is given daily once for a week to women after delivery for back-ache.
3. Paste of seeds is applied to scalp at night about a week to promote hair growth.

Notes from Classics

The plant is extensively used as legume fodder.

(Anonymous, 1948-1976)

Tender leaves are consumed as vegetable. Seeds yield oil is suitable for paints and varnishes and contain yellow dye.

(Ambasta, 1986)

Leaves are used for kidney troubles.

(Jain, 1991)

Critical Notes

Usefulness of leaves to treat polyurea, back-ache especially after delivery and of seeds for hair growth are not reported in past. It is exotic, native of west Temperate Asia.

Melilotus alba Medik.

Local Name

Safed-senji

Ethnobotanical Notes

Miscellaneous

1. *Food*: Leaves and young shoots are used as vegetable.
2. *Fodder*: Plant is used as fodder for domestic animals.

Notes from Classics

Seeds are used for bowel complaints, infantile diarrhoea and given as gruel.

(Watt, 1889-1899)

The plant is cultivated for fodder in Madhya Pradesh and Uttar Pradesh.

(Anonymous, 1946-1976)

The plant is cultivated as fodder. Seeds yield oil which is used in paints and varnish. Plant is used as cattle feed.

(Ambasta, 1986)

Leaves and young shoots are used as vegetable.

(Jain, 1991)

Critical Notes

The species is generally found as a weed in crop fields and waste places. This weedy species is found useful as fodder and vegetable. It is crop weed. It is exotic, native of Europe and Western Temperate Asia.

Milletia extensa Benth.

Local Name

Agari

Ethnobotanical Notes

Miscellaneous

1. Roots are crushed and used as fish-poison.

Notes from Classics

Roots are applied to sore in cattle to kill vermin, also used as poison of fish.

(Watt, 1889-1899)

Leaves and twigs are lopped for fodder. Roots possests insect repellant properties. It is used as fish poison and applied to sores of cattle to kill vermin. An infusion of powdered root is rubbed on cattle and horses to keep off flies. Bark and stem yield for rough cordage. They are beaten and used as brushes.

(Anonymous, 1946-1976)

Ambasta (1986) mentioned similar properties as in Anonymous (*loc.cit.*).

Root juice is used to kill lice, ticks, bed bugs and as fish-poison. Roots are used as veterinary medicine. Roots are applied to sores. Bark yield fibres. Plant is used for tooth-ache, spleen disease and fodder.

(Jain, 1991)

Critical Notes

Very few lanes of this species are observed in forest areas, entire plants are rooted out and thus destroyed. This is to be checked stringently.

Mucuna pruriens (L.) DC.

Local Name

Khaj-kuiri

Ethnobotanical Notes

Miscellaneous

1. *Food*: Fresh seeds are used as vegetable.
2. Tender leaves are also used as vegetable.
3. *Fodder*: Leaves are used as fodder for goats.

Notes from Classics

Pods and young leaves are used as vegetable and as fodder respectively.

(Anonymous, 1998-1976)

Pods are eaten after boiling during scarcity.

(Ambasta, 1986)

Fruits and seeds are edible.

(Jain, 1991)

Critical Notes

Use of seeds and leaves as vegetable are also noted widely from other regions of our country. It appears that they are substantially valuable. Food value should be tested on modern scientific line. The stinging hairs of pods are sometimes spread in and around sitting places as a mischief.

Ougeinia oojeinensis (Roxb.) Hochr.

Local Name

Tiwas, Kala-palas

Ethnobotanical Notes

Miscellaneous

1. Leaves are good fodder for domestic animals.

2. Flowers are cooked and used as vegetable. Stem bark is used to stupify fishes.
3. Wood is used especially for making bullock carts.

Notes from Classics

Plants yield valuable timber for making carts, shaft, axles, agricultural implements, tool handles, legs of bed, tent poles, windows and door frames.

(Anonymous, 1946-1976)

Bark is used as fish poison. Plants are lopped for fodder. Stem yields fibre for cordage.

(Ambasta, 1986)

Flowers are edible. Stem bark is used for fish poison.

(Jain, 1991)

Critical Notes

The plant is used for making bullock-carts and as fodder. This has led the taxon in a precarious state. Very few good trees are observed in Satpura mountain. If control measures to protect it are not taken, it would be threatened in near future. Moreover, eating the flowers as vegetable reduces the possibility of formation of fruits. This should be checked urgently. Only captive plantation can meet the local timber demand.

Pongamia pinnaa (L.) Pierre

Local Name

Karanj

Ethnobotanical Notes

Medicine

1. Bark powder, about five gm mixed in spoonful of honey, is taken twice a day about fifteen days to cure typhoid.
2. Half spoon of paste of leaves is given with warm water to purify mothers milk.
3. Leaf paste is applied on scabies.
4. Leaf powder mixed in wax is applied on cracks of feet.

5. Seeds are crushed in cow-urine and applied over the skin for ventilago.
6. Seed oil is locally applied on rheumatic joints.
7. Two or three drops of seed oil are dropped in nose at night to cure cough and cold till cure.
8. Two or three drops of seed oil are dropped in ears for ear-ache.

Miscellaneous

1. *Fodder*: Leaves are good fodder for pet animals.
2. Stem pieces are used as tooth brush.
3. Seed oil is used to burn lamps.

Notes from Classics

Juice of root is applied for cleaning foul ulcer and closing fistulous ulcer. Juice of root with coconut oil and lime water is used for gonorrhoea. Also, juice of root with gingelly oil and sulphate of copper is given to cure sinuses. Juice of stem and root possess antiseptic properties. Leaves are used against leprosy, epilepsy, abdominal enlargement and bleeding piles. Powder of rind is useful in whooping cough. Seed oil is used for rheumatism. Lininent of oil is employed to cure prurigo, psoriasis and pithyriasis.

(Watt, 1889-1899)

Root extract mixed with coconut milk is a remedy for gonorrhoea. Poultice of leaves is used for ulcer, intestinal worms. Juice of leaves is applied on fistulous sore. Leaves are used in flatulence, dyspepsia and diarrhoea. Young leaves applied on bleeding piles. Leaves are fodder and galactogogue. Flowers are remedy on diabetes. Pods are worn around neck to protect against whooping cough. Seeds are used in skin diseases, leprosy, epilepsy and abdominal enlargement. Seed oil is used on skin diseases and rheumatic pains. Oil with lemon juice is employed on pithyriasis, psoriasis prurigo and versicolor capitis. Fruits are eaten.

(Anonymous, 1946-1976)

Ambasta (1986) gave similar properties as those in Anonymous (*loc.cit.*).

Roots are used in the treatment of fistula, sore and ulcer. Bark is administered for malaria, piles, wound and sore. Stem is used as tooth brush. Flowers are employed for diabetes. Fruits are prescribed for cough. Seeds are employed for bronchitis, chest pain, cold, rheumatism and whooping cough. Seed oil is used as bird lime and fish poison.

(Jain, 1991)

Critical Notes

Use of bark to treat typhoid; of leaves to purify mother's milk and to cure cracks of feet; of seeds or seed-oil against ventilago, cough, cold and ear-ache are not documented earlier. This plant species is a reputed medicinal plant. It found suitable on modern scientific ground, it may be used for human disease or ailments in future.

Psoralia corylifolia L.

Local Name

Bawchi

Ethnobotanical Notes

Medicine

1. Half spoon of seed powder of this plant mixed with half spoon of fruit powder of *Emblica officinalis* Gaertn. and half spoon bark powder of *Acacia catechu* (Roxb. ex. Rottl.) Willd. mixed thoroughly and this mixture is taken with water orally twice a day for about six months to cure ventilago.
2. About twenty five gm of seed powder is mixed with twenty five gm of turmeric powder (*Curcuma longa* L.) with twenty five gm of rock salt and are homogenized in honey. A spoonful of it is consumed per day to cure scabies.
3. Seed powder is applied locally on leprotic wound.

Notes from Classics

Seeds are cold dry laxative, stimulant and spasmodic, recommended in leprosy and skin disease by vitiated state of blood. They are used in febrile and bilious afection. They are diuretic, anthelmintic and astomachic and deobstruent. Seeds are useful in leucoderma, leprosy and cutaneous affection.

(Watt, 1889-1899)

Seeds are laxative aphrodisiac, anthelmintic, diuretic, diaphoretic in febrile conditions. They are specially recommended in leucoderma, leprosy, psoriasis and inflammatory diseases. Roots are used in carries of teeth. Leaves are used in diarrhoea.

(Anonymous, 1946-1976)

Seeds are used in leucoderma and syphilis. Seed cake is suitable for manure and feed.

(Ambasta, 1986)

Seeds are reputed for skin diseases such as itch, leprosy psoriasis, also employed for skin disease in veterinary medicine.

(Jain, 1991)

Critical Notes

Seeds of this species along with the parts mentioned above are useful to cure ventilago, a use hitherto unnoticed. Other uses important from the view- point of cross-country observation.

Pterocarpus marsupium Roxb.

Local Name

Biva

Ethnobotanical Notes

Medicine

1. *Fodder*: Infusion of bark, about half cup, is advised once a day for three day to cure dysentery.
2. Paste of leaves is applied on sores.
3. Spoonful of wood powder is taken twice a day with honey to cure asthma till cure.

Miscellaneous

1. Leaves are used as fodder.
2. Wood is used for making doors and windows.

Notes from Classics

Bark is useful for tooth-ache and it is valuable astringent. Leaves are useful as external application in boils, sores and skin diseases. Leaves are used as fodder. Gum cures diarrhoea. Wood is used for door and window frames, furniture, boat building, carts, agricultural implements and employed for making sleepers. Bark is astringent and it is used in tooth-ache. Leaves are useful in external application, for sores, boils and skin diseases. Leaves are excellent fodder. Flowers are used in fever. Tree yield gum-kino which is a powerful astringent, used in the treatment of diarrhoea, dysentery and tooth-ache. Gum kino is used for leucorrhoea and haemorrhage. Gum kino finds application in dying and tanning. An aqueous infusion of wood is used in diabetes. Water stored in vessel made up of wood is reputed to have antidiabetic qualities.

(Anonymous, 1948-1976)

Wood is used for building construction and making doors, windows, wagons, boats, tool handles, picture frames, cabins, furniture and agricultural implements.

(Ambasta, 1986)

Bark is employed for dysentery, tongue diseases, tooth-ache and urine complaints. Bark cures dysentery in animals. Leaves are applied on sores and skin diseases. Flowers are used in fever. Wood is used in the treatment of asthma, body and chest pains, and digestion. Resin has cooling effect and cures diarrhoea.

(Jain, 1991)

Pueraria tuberosa Roxb. ex (Willd.) DC.

Local Name

Bhul Kohla

Ethnobotanical Notes

Medicine

1. Powder of tuber about two spoonful mixed in half cup of wine increase the lactation of mother.
2. Powder of tuber about spoonful is drunk with milk of cow on first month of pregnancy for settlement of foetus.
3. Spoonful of fruit powder mixed in cup of condensed milk and drunk regularly for sexual vigour.

4. Spoonful of juice of fresh tuber and paste of one bulblet of garlic (*Allium sativum* L.) is consumed once a day for migraine.
5. Oil is extracted from a mixture of fruits of *Pueraria tuberosa* (Roxb. ex Willd.) DC., *Abutilon indicum* (L.) Sweet and gum of *Boswellia serrata* Roxb. ex.Coleb. A spoonful is given daily for hysteria.
6. A spoonful of powder of tuber is consumed daily to treat eruption of skin caused due to impure blood.

Miscellaneous

1. ***Food***: Tubers are ground to flour to make bread as a famine food.
2. ***Fodder***: The green foliage and tender twigs are used as fodder for cattle.

Notes from Classics

Pounded tuber is made into poultice and applied to reduced inflammatory swelling of joints. Tubers are cooling, emetic, lactogogue and tonic.

(Watt, 1889-1899)

Roots are demulcent and refrigerant in fever. They act as cataplasm for swellings of gout. The dried roots are sold as drug and used as lactogogue. Green foliage and tender twigs are nutrient and palatable fodder.

(Anonymous, 1946-1976)

Root are demulcent and eaten raw or boiled. Leaves are used as fodder for horses and cattle.

(Ambasta, 1986)

Roots are used for asthma and body-ache. Tubers are used for abdominal and chest pain, cholera, diarrhoea, fever, rheumatism, swelling as tonic for lactation. Seed paste is applying for skin diseases. Plant is useful in the treatment of ulcer.

(Jain, 1991)

Critical Notes

Tubers are useful to treat settlement of foetus and migraine. Fruits are beneficial in case of hysteria. These uses are still untapped.

Removal of tubers exploits the total plant. This has negative impact on its populations. People should be provided from its collection.

Sesbania sesban (L.) Merr.

Local Name

Shewari

Ethnobotanical Notes

Medicine

1. Leaves are used as fodder for domestic animals.
2. It is planted in rows some times around crops fields as a wind breaker.

Notes from Classics

Bark yields fibres which are used for ropes. Leaves and young branches are lopped for fodder. Wood is grown to furnish poles and to support the piper vine, and various cucurbitaceous plants. Wood is used as fuel to boil Jaggery. It is planted in Bengal as hedge plant.

(Watt, 1889-1899)

Tree is grown for shade and support crop of sweet orange, tea, turmeric and coffee. Leaves and fruits are eaten. Young leaves are lopped for fodder. Bark is used for making ropes. Pith is dried and used for making mats. Wood is used for toys.

(Anonymous, 1946-1976)

Crown of tree is used to shade and support paper and bettle. Leaves and fruits are eaten. Young leaves and branches are lopped for fodder. Bark is a source of fibre which is used for making ropes. Pith is dried and used in making mats. Wood is used for dye.

(Ambasta, 1986)

Flowers and leaves are used for chronic eruptions and ulcer. Plant is used for colic pains, pleurisy pneumonia and urine complaints.

(Jain, 1986)

Critical Notes

It is planted as wind breaker around crop fields. This is not mentioned in the past literature.

Stylosanthes fruticosa (Retz.) Alst.

Local Name

Stylograss

Ethnobotanical Notes

Medicine

1. Decoction of leaves, about a cup, is taken twice a day for malarial fever till cure.
2. Infusion of leaves, about a cup, is advised twice a day for pregnant women for seven days for smooth delivery.
3. Juice of leaf, about half cup, is given immediately after snake-bite.

Miscellaneous

1. ***Fodder***: The plans are used as a fodder for domestic animals.

Notes from Classics

Leaves are used as medicine for diarrhoea in children. They are smoked as tobacco. Infusion of plant is taken for cold. The plant is aphrodisiac. The plant is used as fodder for animals.

(Anonymous, 1948-1976)

Ambasta (1986) reported the uses similar as in Anonymous (*loc.cit.*).

Critical Notes

It is good fodder plant, it may be advised for cultivation exclusively for fodder. Its use against snake-bite is additional.

Trigonella foenum-graecum L.

Local Name

Methi

Ethnobotanical Notes

Medicine

1. Seeds are kept in the leaf pulp of *Aloe vera* (L.) Burm. *f.* over night. These are used to wash and improve hair growth.

2. Five or six seeds are taken with jaggary every morning for rheumatism.

Notes from Classics

Leaves are fried in ghee and used in dysentery. Leaves are applied as poultice in case of contusion. Seeds are boiled in milk and given internally in bleeding piles. Seeds are tonic, carminative, aphrodisiac, recommended for dyspepsia, diarrhoea and rheumatism. Flour of seeds is used as poultice and applied on skin as cosmetic. Seeds are tonic, stimulant and used in the treatment of colic, flatulance and dysentery. They are also suppurative, aperient, diuretic, emmenogogue, useful in dropsy, chronic cough, enlargement of spleen and liver, external and internal swellings, burns and to prevent hair fall. Plant is used as fodder, believed to be heating and lactifuge.

(Watt, 1889-1899)

Seeds are aromatic, carminative, tonic, galactogogue, used externally for pile, as poultice for abscesses and ulcers and internally as emollient for inflammation of intestinal tract. Seeds are applied for veterinary medicine. Seeds are the constituent of condation powder, for cattle, horses and ship.

(Anonymous, 1946-1976)

Seeds are aromatic, carminative, tonic and galactogogue Seeds are useful internally to cure boils, ulcer and as emollient. They are used to cure intestinal inflammation.

(Ambasta, 1986)

Seeds are administered for gas trouble, piles and rheumatism. They are used as tonic and spice. Plants are used for baldness, gonorrhoea, oedema, sores and syphilis. Shoots are used as vegetable.

(Jain, 1991)

Critical Notes

It is cultivated for leafy vegetable and for seeds. It is exotic, native of south Europe.

Vigna radiata (L.) Wilezek. var.sublobata (Roxb.) Verdc.

Local Name

Moog

Ethnobotanical Notes

Medicine

1. Decoction of 50 gm seeds with turmeric powder is given once a day to treat acidity and cough till cure.
2. Poultice of seed paste is applied on eyes to cure swellings of eyes.

Notes from Classics

Decoction of seeds is given in beri-beri, vartigo. They are diuretic. Mung extract is said to have curative properties in *Polyneuritis gallinarum*. Tender pods are consumed as vegetable.

(Watt, 1889-1899)

It is believed that consumption of seeds prevent flow of milk.

(Anonymous, 1946-1976)

Decoction of seeds given in beri-beri as diuretic. Seed used for vertigo. Extract of seed given in Polyneuritis gallinarum.

(Ambasta, 1986)

Seeds are edible.

(Jain, 1991)

Critical Notes

Usefulness of seeds as remedy against acidity, cough and swellings of eyes is not documented earlier. It is cultivated for grains in this district.

Vigna unguiculata (L.) Walp.

Local Name

Chawli

Ethnobotanical Notes

Medicine

1. Fifty gm of seeds alongwith seed-coats are boiled and salted. They are consumed once a day for diarrhoea for three days.
2. Leaf juice is applied on skin diseases, locally called 'anjuri'.

Notes from Classics

Seeds are hot and dry, diuretic, difficult to digestion. Seeds strengthen the stomach.

(Watt, 1889-1899)

Pods and seeds used as food. Seed flour cooked for sweetmeats, cakes and porridge. Leaves and seeds are eaten as pot-herb. Plant serves as valuable feed for live-stock.

(Anonymous, 1946-1976)

Seeds are antibillious, useful in hepatic troubles and jaundice. Plant is useful for small-pox.

(Jain, 1991)

Critical Notes

Medicinal utility of seeds to treat diarrhoea and of leaves in case of 'anjuri' are interesting and new. The species is cultivated.

Family: Caesalpiniaceae

Bauhinia racemosa Lamk.

Local Name

Apta, Sone

Ethnobotanical Notes

Medicine

1. Extract of leaves about five ml mixed with sugar and milk is given four times a day for normal urination.
2. Leaves are soaked in water over night. On the next day morning two spoonful extract of leaves is mixed in sugar. It is taken orally twice a day for gonorrhoea till cure.

Miscellaneous

1. Leaves are used by the tribals for bidi – wrapping.
2. Green fruits and foliage constitute good fodder for goats.
3. The dried branches are used as fuel.

Notes from Classics

Decoction of leaves is used to relieve head-ache in malarial fever.

(Watt, 1889-1899)

Bark is astringent, used in dysentery. Bark yields strong cordage fibre. Leaves are given with onions for diarrhoea. These are anthelmintic. Decoction of leaves is used in malaria.

(Anonymous, 1946-1976)

Ambasta (1886) notes similar uses as those of Anonymous (*loc.cit.*). Stem bark is used in diarrhoea and dysentery. Fruits and seeds are used as vegetable. Stem fibres are employed for making basketary. Leaves are used for bidi-making.

(Jain, 1991)

Critical Notes

Medicinal utility of the leaves for urinal complaints and gonorrhoea are additional and noteworthy reports from this district. Green fruits and leaves are used to feed goats. This is also not recorded earlier.

Bauhinia purpurea L.

Local Name

Kanchan

Ethnobotanical Notes

Medicine

1. Stem bark, about 40 gm, is boiled in about one litre water. It is boiled till the volume reduces to $1/8^{th}$. About two spoons of this decoction alongwith some honey is taken for a week to cure lymph adenities.
2. Dried leaves and bark are powdered. This powder is taken orally as one spoonful for a month twice a day to reduce tumours in body.

Miscellaneous

1. *Food*: Fresh, young leaves are used as vegetable.
2. *Fodder*: Leaves are used as fodder.

Notes from Classics

Roots are tonic and carminative in nature. Bark is astringent. Decoction of bark is useful to cure ulcer wounds. Bark is also employed in diarrhoea. Leaves are useful as fodder. Flowers are laxative and are used as pot herb.

(Watt, 1889-1899)

Tannin is obtained from bark. Leaves are used as fodder. Flowers buds are used as pot herb. Wood is useful for agricultural implements. Tree yield gum.

(Anonymous, 1946-1976)

Roots are carminative. Bark is used in diarrhoea, also yields fibres. Leaves are used as fodder. Flower buds are eaten as a pot herb. They are also pickled. Flower buds are laxative and anthelmintic.

(Ambasta, 1986)

Bark helps to cure small-pox and tumour in stomach. It also heals wounds quickly if applied locally. Stem is used for bone-fracture. Decoction of leaves is drunk for jaundice. Flowers are used to cure indigestion. Whole plant is used as antidote on bite of snake, scorpion and tiger. It is also used as an antidote in poisoning of 'Datura'. Whole plant is used to reduce body swellings. Leaves, fruits, buds are used as vegetable. Fibres obtained from stem is used for cordage. Leaves are used for roof thatching.

(Jain, 1991)

Critical Notes

Fruits and bark for lymph gland and lymph adenities are used respectively. Leaf and bark powder together are employed to reduce tumours in body. All these new observations are yet to be noted elsewhere. It is planted by forest department.

Caesalpinia bonduc (L.) Roxb.

Local Name

Sagargota

Ethnobotanical Notes

Medicine

1. Paste of leaves is applied onto boils.

2. About 50 gm of pulp of young green fruits with 25 gm of jaggary and a pinch of lime powder mixed and small tablets are prepared. A single tablet twice a day is advised for stomach-ach for three days.

Miscellaneous

1. The plant species is planted around fields as live hedge.
2. Boys and girls use the seeds in playing local games.

Notes from Classics

Seeds are tonic and febrifuge. They are useful for asthma and colic, seeds are used for dispersing swelling and applied on ulcer. Seeds restrain hemorrhage and helps in keeping off infections diseases. Burnt seeds are used in spongy gums and boils also.

(Watt, 1889-1899)

Leaves and bark are emmenogogue, ferifuge and anthelmintic. Leaves and seeds are used for external application for inflammatory swellings from face. Seed oil is emollient. It is used as embrocation to remove freckles.

(Anonymous, 1946-1976)

Seed and fatty oil is used for discharge from ears and also useful in colic.

(Ambasta, 1986)

Roots are used on pulmonary tubercular problems. Infusion of seeds are used in colic, fever and intermittent fever. It gives relief in rheumatism and skin disease. Seeds control the menstrual complaints and also helps in healing wounds.

(Jain, 1991)

Critical Notes

Leaves cure boils, whereas young fruits are helpful to stop stomach-ache. These uses along with miscellaneous ones yet need documentation.

Caesalpinia decapetala (Roth) Alst.

Local Name

Chillar

Ethnobotanical Notes

Miscellaneous

1. This species is planted as live hedge around the crop fields.

Notes from Classics

This plant makes impenetrable fence.

(Watt, 1889-1899)

Thorny climbing shrub is well known as a hedge plant.

(Anonymous, 1946-1976)

Bark is used for tanning and roots are purgative.

(Ambasta, 1986)

Leaves are used to cure mouth sores of domestic animals.

(Jain, 1991)

Cassia auriculata L.

Local Name

Awali, Walanya

Ethnobotanical Notes

Medicine

1. Bark powder is applied on gum for tooth-ache. Stems are used as tooth brushes.
2. Leaves and black soil are boiled in water. The mixture is applied on swellings as poultice.
3. Decoction of flowers, about half cup, is given once a day to control sugar level in diabetes.
4. Paste made from equal amount of flower buds and parched gram (*Cicer arientinum* L.) in given orally to cure jaundice. About 10 gm of paste is consumed per day till cure.
5. Ash of flowers mixed in citrus juice [*Citrus aurantifolia* (Christm.) L.] is applied locally for eczema.
6. A pinch of gum powder with half spoon latex of *Ficus benghalensis* L. is taken twice a day for a week to remove urine stone.

Miscellaneous

1. Plant is used as fuel.

Notes from Classics

Bark is reported to be astringent. Infusion of leaves is used as cooling medicine. Leaves are eaten as vegetable. Whole plant is used in diuresis and diabetes. Seeds are applied on eye for opthelmia.

(Watt, 1889-1899)

Roots are used locally in the treatment of skin diseases. Bark is astringent. Leaves and fruits are anthelmintic.

(Anonymous, 1946-1976)

Bark is astringent. Leaves and fruits are anthelmintic. Roots are employed in skin troubles. Seeds are used in eye troubles, diabetes and chylous urine.

(Ambasta, 1986)

Roots are used for rheumatism. Stem twigs are used for tooth-ache and stem-bark is used as a fish-poison. Stem bark is also used for tanning. Whole plant is used for dysentery. Seeds are prescribed for asthma and cough.

(Jain, 1991)

Critical Notes

Leaves are useful to reduce swellings. Flower buds are used to cure jaundice and eczema. Gum along with latex of *Ficus benghalensis* L. is medicinal for urine stone. These uses are not reported in past.

Cassia fistula L.

Local Name

Bawa, Bahava, Ramdanda

Ethnobotanical Notes

Medicine

1. Paste of leaves prepared in cow-urine is applied locally on scabies regularly till cure.
2. Leaves of *Cassia fistula* L., *Celasrus paniculata* Willd. and *Nerium indicum* Mill. are pounded homogenously in butter milk and applied on skin for ventilago.

3. Leaves are pounded in river water and paste is applied on skin for eczema.
4. Decoction of fruit pulp, about two spoonful, is given twice a day to cure paralysis of legs.
5. Paste of fruit pulp is applied on bladder for a week for manorrhagia.
6. Fruit is rubbed in mother's milk. Half spoon of it is administered to a kid orally for a day for flatulence in colic.
7. Infusion of seeds, about spoonful, is given at morning for three days to treat stomach-ache.

Miscellaneous

1. ***Fodder***: Leaves are lopped for fodder.
2. ***Wood***: Wood is used for ploughs, wheel and tool handles.

Notes from Classics

Bark and leaves are applied externally on pustules and skin diseases. Tender leaves are laxative and applied around the navel of kids to produce motion. Flowers are febrifuge. Pulp of fruit is purgative and cathartic. Externally the pulp is used for gout and rheumatism.

(Watt, 1889-1899)

Root bark, seeds and leaves are laxative.

(Anonymous, 1946-1976)

Root bark is given in the treatment of black water fever. Dried fruits are used as purgative, laxative and for habitual constipation. Wood is used for house posts, ploughs, tool handles and wheels.

(Ambasta, 1986)

Leaves are used as fodder. Seeds are dispersed near holes of rats in crop fields to keep them off. Fumes of pods cure indigestion. Wood is used as limber.

(Jain, 1991)

Critical Notes

Leaves are medicinally important for scabies and eczema. Fruits are served to cure paralysis, manorrhagia and flatulence in colic.

Seeds are useful to treat stomach-ache. All of these uses are not hitherto noted elsewhere.

Cassia tora L.

Local Name

Tarota

Ethnobotanical Notes

Medicine

1. Paste of roots homogenized with lemon juice [*Citrus aurantifolia* (Christm.) Sw.] is applied to cure a wound caused byhorns.
2. Juice of leaves mixed in coconut oil (*Cocus nucifera* L.) is applied locally on scabies.
3. Paste of leaves is applied locally on joints to minimize the rheumatic pains.
4. Powder of seed of this species and fruits of *Triblus terrestris* L. are mixed in equal amount. A spoonful per day of it is consumed with honey for a week till cure asthma.

Miscellaneous

1. *Food*: Young leaves are used as vegetable.

Notes from Classics

Root is rubbed with lime juice and applied on ring-worm. Leaves and seeds are good remedy on skin diseases such as leprosy and psoriasis. Tender leaves are boiled and eaten as pot herb.

(Watt, 1889-1899)

Leaves are purgative, used to expel ring-worm. They are also used for many skin diseases.

(Anonymous, 1946-1976)

Seeds are used as substitute for coffee and as a mordant in dyeing.

(Ambasta, 1986)

Roots are used to expel ring-worm. They are good remedy on fever and scorpion bite. Leaves are anthelmintic, antiseptic and vermicide. Leaf paste is applied on boils, bone fracture, cuts, wounds and skin diseases. Leaves are used in the treatment of indigestion, intestinal disorder, jaundice, night blindness and rheumatism. Seeds are used against scabies, eczema, itch, fever, stomach ache and antidote for scorpion bite. Seeds are also useful as tonic. Plant is used for epilepsy, scabies and sores. Leaves and fruits are edible.

(Jain, 1991)

Critical Notes

Medicinal use of seeds to treat asthma has remained unnoticed earlier.

Delonix regia (Boj.ex Hook.) Raf.

Local Name

Gulmohar

Ethnobotanical Notes

Medicine

1. Bark infusion in whey is given for burning sensation of entire body. About half cup is taken orally till cure.

Notes from Classics

Seeds contain gum which may find use in food and textile industry.

(Watt, 1889-1899)

Ambasta (1986) reported similar properties as given in Anonymous (*loc.cit.*).

Critical Notes

Employment of bark against burning sensation of entire body is not reported in past. The species is planted as road side tree and exotic, native of Madagascar, it finds place in local uses. It appears that the local rural people have tried it so in modern times.

Hardwickia binata Roxb.

Local Name

Anjan

Ethnobotanical Notes

Medicine

1. Leaves are added in bath water to reduce body heat of kids.

Miscellaneous

1. Leaves are used as green fodder especially for milching cows and buffaloes.
2. Ploughs and tool handles are made from wood.
3. Leafy twigs are used for thatching huts.
4. It is used also as fuel wood.

Notes from Classics

Leaves are used as fodder. Wood is used for making bridge, sleepers and house posts.

(Watt, 1889-1899)

Leaves are used as cattle fodder. Wood is utilized for making ploughs, carvings, screw benches, tool handles, pegs, bricks blocks.

(Anonymous, 1946-1976)

Bark yields fibres which are used for ropes.

(Ambasta, 1986)

Critical Notes

Plant is lopped for fodder. Wood is also valuable as timber and fuel wood. Leaves are sometimes sold in local market. The illicit cutting has reduced this species to considerable extent. It should be given priority in afforestration programmes. It will certainly help the ambient society as large as far as fodder, fuel and small timber is concerned. Moreover, it thrives best in any soil, being hardly species.

Parkinsonia aculeata Linn.

Local Name

Vilayati babul

Ethnobotanical Notes

Miscellaneous

1. ***Food***: Seeds are edible.
2. ***Fodder***: Leaves are used for fodder of goats.
3. Wood is used as fuel.

Notes from Classics

Bark yields fibres. In Punjab, tree is lopped for its small branches given to goats as fodder. Wood is used as fuel. Plant is adopted for fuel.

(Watt, 1889-1899)

Young branches are lopped for feeding goats and sheep. Bark yield fibres, suitable for mixing with other pulp in making paper. Seeds are edible.

(Anonymous, 1946-1976)

Ambasta (1986) reported similar uses as those in Anonymous (*loc.cit.*).

Tamarindus indica L.

Local Name

Chinch

Ethnobotanical Notes

Medicine

1. Spoonful of stem bark powder with honey is taken thrice after two hours intervals, as antidote on snake-bite.
2. Poultice of boiled leaves is tied on eyes to reduce swellings of eyes.
3. Half cup of flower extract is prepared in water. It is taken at every night for one month to reduce pain of piles.
4. A cupful of ripe fruit extract is taken daily for four days to treat cholera.
5. Decoction of fruit pulp, about half cup, is administered twice a day for eight days to a patient suffering from gastritis.
6. Decoction of ripe fruit pulp, about a cup, is given after meal to improve indigestion.
7. Young pods are burnt and spoonful ash mixed in water is drunk orally for a week to cure acidity.
8. Rind of pods is burnt. Three gm of ash mixed in spoonful of cow-urine is taken orally for three to four times a day to cure jaundice.

9. Thick paste of pod mixed with salt common is applied locally onto swelling of foot, caused due to long walk.

Veterinary Medicine

1. Leaves are mixed with black soil and boiled. This soil in plastered on bone fracture of domestic animals.

Miscellaneous

1. Dried branches are used as fuel.
2. ***Food***: Ripe and unripe fruits are edible. Also used as condiments.
3. ***Wood***: Wood is widely used for agricultural implements and house hold articles.
4. Seeds are roasted and consumed as a masticatory by tribals.

Notes from Classics

Roots with black pepper are used in dysentery. Bark is astringent and tonic. Ash of bark is given for digestion. It is cooling, antiscorbutic, laxative, useful for febrile and bilious affection. Leaves are used to wash indolent ulcer. Heated leaves are used for dysentery, they are vermifuge. Flowers are used in inflammation in conjunctiva. Juice of it is given for bleeding piles. Fruits are refrigerant, digestive, carminative, laxative, used for burning of body, constivness, intoxication from spirituous liquors of *Datura*. Fruits are refrigerant used for febrile and inflammatory affection, also for diarrhoea and dysentery. Shell of fruit are burnt and used as medicine. Fruit pulp is cardiac, astringent, aperient and used for bilious affection. A gargle of tamarind water is given for sore throat. Fruit is useful in preventive scurvy. It is antiascorbutic. Seeds used to cure boils.

(Watt, 1989-1996)

Wood is used for agricultural implements, tool handles, tent pegs, rice pounder, in oil-mills etc. Fruit pulp is refrigerant, carminative and laxative, used for febrile diseases and bilious disorder. Pulp is cathartic and recommended for sore throat. Pulp possess antiseptic properties and applied as a poultice to inflammatory swellings. Kernels are fried or boiled and eaten. Seeds are astringent, used in diarrhoea and dysentery. Kernels pounded in water are given in cough, also useful for colitis, diarrhoea and intestinal disorders.

(Anonymous, 1948-1976)

From young stem bark fibre is obtained. Kernels are used as food in times of scarcity. Wood is used in agricultural implements, tool handles, tent poles and for construction. Tender leaves, flowers and seeds are eaten as vegetable. Fruit pulp is acidic, used in sauces, chutneys and beverages. Pulp is refrigerant, carminative, laxative, given as infusion in biliousness and febrile conditions. Fruit pulp is used in jam and jellies. Seed oil is used in paints and varnishes.

(Ambasta, 1986)

Bark is used for stomach-ache. Leaves are anthelmintic, used for boils, eye complaints, fever, swellings, tooth-ache and urine complaints. Fruits are used for sunstroke and digestion. Seeds are used in soft drink and for boils. Plant is used in anasarca, cold, cough, dropsy, eye complaints, muscular pains, scorpion bite, snake bite and syphilis. Seeds, fruits and leaves are edible.

(Jain, 1994)

Critical Notes

Stem bark is useful as an antidote for snake-bite. Fruits are beneficial in case of cholera, gastritis, jaundice and to cure acidity. These are not reported in past literature. Use of leaves for fractured bones is also not reported to date.

Family: Mimosaceae

Acacia catechu (L.*f.*) Willd.

Local Name

Khair

Ethnobotanical Notes

Medicine

1. Stem bark is boiled and this concentrate is used for gargling to cure bleeding gums.
2. Gargling with bark extract cures stomatitis.
3. Stem bark of this plant and that of *Bridelia airy-shawii* P.T.Li are boiled in water. This concentrate is mixed with the

fruit powder of (a) *Emblica officinalis* Gaertn. (b) *Terminalia bellirica* (Gaertn.) Roxb. (c) *Terminalia chebula* Retz. This mixture is also used in fistula.

4. Extract of stem-bark, about two spoon, is drunk twice a day for cough and cold. This is followed for three consecutive days.
5. Root, stem bark, leaves, flowers and fruits are boiled together. Half cup of this concentrate is taken orally once a day for leprosy till cure.
6. Catechu powder is applied on wounds of leprosy.

Notes from Classics

Stem bark powder is astringent. Infusion of bark is used for gargling in case of sore throat and stomatitis, also cures chronic dysentery. Decoction of stem bark is used for gonorrhoea. Strong infusion of bark is used for bleeding haemorrhoids. Leaf juice extracted from young leaves is dropped in eyes for conjunctivitis. Gum is given to delivered women as a tonic. Gum is used against diarrhoea. Pods are used for in preparation and catechu is made from read heart wood.

(Watt, 1889-1899)

Old tree produce a substance called 'kher-sal' which is white powder or crystalline deposit found in cavity of wood. It is used for tooth-ache, cough and sore throat.

(Anonymous, 1946-1976)

Catechu is used for printing and dying purposes. Wood is used for oil, sugarcane crusher, bows and spear handles, plough, carts, bed post legs, tool handles, tent pegs spokes and hubs of wheels.

(Ambasta, 1986)

Roots are used for rheumatism and tooth-ache. Stem bark is good astringent and used against haemoptysis. It cures dysentery and leucorrhoea, it also facilitate child-birth and stops menstrual complaints. Leaves are used in dysentery and gonorrhoea.

(Jain, 1991)

Critical Notes

Uses of stem bark for bleeding gums, fistula, cough and cold are new reports. Likewise, concentrate obtained from all parts of this tree is found useful to leprosy patients. It should be assessed on scientific grounds.

Acacia leucophloea (Roxb.) Willd.

Local Name

Hiwar

Ethnobotanical Notes

Medicine

1. Root is chewed for hiccups.

Miscellaneous

1. ***Food***: Young pods and seeds are eaten by tribals.
2. ***Fodder***: Green pods are used as fodder for domestic animals.
3. Wood is used for agricultural implements.

Notes from Classics

Bark is astringent. Bark powder is mixed in flour of bajra (*Pennisetum americanum* Schum.) or Maize (*Zea mays* L.) during famine. Gum is medicinally used. It is nutritive. Young pods and seeds are used for consumption, also as fodder.

(Watt, 1889-1899)

Gum is used in indigenous medicine. Bark is ground and mixed with flour. Pods and seeds are eaten, pods also used as fodder.

(Anonymous, 1946-1976)

Bark is used in bronchitis and biliousness. Bark fibre is used for cordage and fishing net, also employed for making spirits. Wood is used for agricultural implements, carts and turnery. Tree yields gum which is used as emulcifing agent.

(Ambasta, 1986)

Roots are used as abortifacient. Bark is used against ulcer. It is astringent. Pods are edible.

(Jain, 1991)

Critical Notes

Chewing of roots for hiccup is being noted for the first time.

Acacia nilotica Willd. ex Del. *subsp. indica* (Bth.) Brenan

Local Name

Bhabul, Sadhi babhul.

Ethnobotanical Notes

Medicine

1. Stem bark extract is employed to clean mouth and strengthen gums.
2. Extract of stem bark taken orally about fifteen to twenty ml once a day for a week helps to reduce acidity and also cure cough.
3. Stem bark powder, about five gm, is consumed once a day as remedy against stomach-ache.
4. Stem bark extract, about a cup, is mixed with some sugar. It is drunk once a day to cure dysurea.
5. Decoction of thorns, about ten ml per day, is used to stop hiccups.
6. Leaves of this plant are mixed with that of a *Clematis heynei* M.A. Rao and juice is extracted. Two or three drops of juice is dropped in eyes to cure conjunctivitis.
7. Equal quantity of leaf juice and milk is mixed. A glass of it is drunk for two weeks to cure diuresis.
8. Dried leaves are powdered. This powder is applied on wounds for quick healing.
9. If tender leaves are consumed every day upto nine month by a pregnant woman, the child born is of fair colour, though the parents are black in colour.
10. Tender pods are dried in shade and powdered. About 5 gm powder is homogenized with cow-ghee and is taken orally twice a day to cure diuresis.
11. Seed powder is taken with honey. This increases strength of the fractured bones.
12. Wound on account of syphilis is filled with catechu of this tree. It cures early.

Miscellaneous

1. ***Food***: Pods when green are used as fodder for goats.

2. ***Wood***: Wood is used for making axles, shafts, yolks and agricultural implements.

Notes from Classics

Stem bark and seeds are burnt and the ash used as tooth powder. Infusion of stem bark is used to gargle for sore throat and stomatitis. Decoction of stem bark is drunk in dysentery and diarrhoea. It is used by women in menorrhagia and caries of teeth. Stem bark is a valuable remedy on prolapsus anus. Stem bark is used as an applicant in leucorrhoea and as poultice for ulcer. Gum is administered to a woman as a tonic after delivery, also employed for dysentery and diabetes. Mucilage is used to allay the irritation of gonorrhoea. Extract of pods is astringent, styptic and tonic, used internally and externally in relaxed condition of mucus. Gum fried in ghee is used as poultice over eyelid for conjunctivitis.

(Watt, 1889-1899)

Pods are used for urinogenital complaints.

(Anonymous, 1946-1976)

Wood is used for wheel work, spokes, axles, shafts and yoke. Also, it is used for agricultural implements, well curbs, tent pegs etc. Gum is used for calico-printing, sizing material and dying, and in paper manufacture. Pods are used as fodder. Decoction of bark is used as gargle.

(Ambasta, 1986)

Stem bark is used for asthma, bronchitis and diabetes. Dysentery is controlled by bark powder. This is also used against skin disease. Bark is edible. Leaves are prescribed for eye complaints, head-ache and throat infection. Urine complaints are cured by employing leaves. Flowers are used for dysentery, ear complaints and tonic. Gum is used as masticatory. Plant is useful for ear complaints, leprosy, hair-fall and syphilis.

(Jain, 1991)

Critical Notes

Nearly all the medicinal uses recorded from Jalgaon district are different from those documented earlier in the literature. It is also interesting that more or less all parts of this plant species are useful.

Acacia pennata (L.) Willd.

Local Name

Chilari

Ethnobotanical Notes

Miscellaneous

1. Bark powder is used for fish killing.

Notes from Classics

Bark is used for tanning; also for making fishing nets.

(Anonymous, 1946-1976)

Bark contains tannin which is used for tanning fishing nets. Decoction of leaves is febrifuge. Fruit pulp is pesticidal.

(Ambasta, 1986)

Roots are used for digestive complaints. Bark is used as detergent, also as antidote for snake-bite, and fish-poison. Bark is also used in tanning. Leaves are febrifuge, used for body-ache and head-ache. Latex is also used as fish poison.

(Jain, 1991)

Acacia polycantha Willd.

Local Name

Teliya babul, Son-khair

Ethnobotanical Notes

Medicine

1. Young branches are rubbed in water. Three glasses of this extract is given to bullocks, cows and buffaloes on abdominal swelling.

Miscellaneous

1. Wood is valuable as small timber.
2. It is used as fuel wood.
3. It is useful for making agricultural instruments, handles of axe etc.

Critical Notes

Medicinal utility of young branches to treat abdominal swellings of pet animals is not recorded earlier.

Albizia lebbek (L.) Willd.

Local Name

Shirish

Ethnobotanical Notes

Medicine

1. Leaves and flowers are crushed together and the paste is applied on chest for burning pains and on neck for neck stiffing.
2. Paste of flowers and wheat grains is applied on mammary glands of a mother to purify milk.
3. Flower paste is applied on face for some time to improve fairness and complexion.
4. Leaves constitute a good fodder for domestic animals but used only in times of scarcity.

Notes from Classics

Root bark powder is used to cure spongy and ulcered gums. Stem bark gives relief in eye injuries and is useful for sore eyes. It is used for diarrhoea, dysentery and gonorrhoea. Flowers are externally applied on boils and swellings. Seeds are used as 'anjan' for ophthalmic diseases. They are astringent, given in piles, diarrhoea, and gonorrhoea. Oil is applied on leprosy.

(Watt, 1889-1899)

Wood is used for furniture, construction purpose, agricultural implements, oil pressures, cane crushers, carts, carriages etc. It is good fuel wood.

(Anonymous, 1946-1976)

Leaves and seeds are used for eye-troubles. Bark is used for boils

(Ambasta, 1986)

Root bark is used for sore eyes. Stem bark is employed as detergent, useful for diarrhoea, dysentery and gonorrhoea. Fresh stem twigs are used as tooth brush. Young leaves are used for night blindness and used as fodder for cattle. Flowers are used for boils and swellings. Seeds are useful in piles, diarrhoea and dysentery.

(Jain, 1991)

Critical Notes

Leaves and flowers together used for burning pains in chest, flowers for purification of milk of a mother and for better facial complexion are new and noteworthy uses. It is planted and also grows wild. It is exotic, native of North Australia and tropical Asia.

Albizia procera Benth.

Local Name

Pandhara shirish

Ethnobotanical Notes

Miscellaneous

1. *Fodder*: Leaves are lopped for fodder.
2. *Wood*: Wood is used for making agricultural implements.

Notes from Classics

Leaves are used as fodder. Trees yield large quantity of gum. Wood is used for rice pounder, sugar cane crusher, wheels, agricultural implements, bridges, house post etc.

(Watt, 1889-1899)

Wood is used for construction purposes, furniture, for carts and carriages. It is also used for dug-outs, oars cane-crushers, rice pounders.

(Anonymous, 1946-1976)

Ambasta (1986) wrote similar uses as those in Anonymous (*loc.cit.*)

Stem bark is used for back ache. It is a fish poison. Seeds are used for body pains.

(Jain, 1991)

Dichrostachys cinerea (L.) Wt. and Arn.

Local Name

Yellatur

Ethnobotanical Notes

Medicine

1. Root infusion two to three spoonful is given for daily for three days against dysentery.
2. Root extract about, half cup is given twice a day for seven days to cure anuria.
3. Extract of stem bark about two spoonfull is used as purgative.
4. Leaf extract about 25 ml is mixed in a cup of whey and consumed for seven days in case of lochia.
5. Stem bark powder is applied on gums. It cures tooth-ache, also used as vermifuge if taken internally in water.

Miscellaneous

1. Leaves are used as fodder for goats.

Notes from Classics

Young shoots are bruised and applied to eye in case of opthalmia. Leaves mixed in corn are given to horses to free from worms. Leaves are also used as fodder.

(Watt, 1889-1899)

Tender shoots are bruised and applied to eyes in case of opthalmia. Roots are astringent. They are used in rheumatism urinary calculi and renal problems.

(Anonymous, 1946-1976)

Wood is used for cog-wheels, walking sticks and tent poles. Bark yields fibres. Leaves are used as fodder.

(Ambasta, 1886)

Tender shoots are bruised and applied to eyes in case of opthalmia. Roots are used in veterinary medicine and also employed for colic disorders. Leaves are good remedy for leucorrhoea and haenorrhea. Fruits are used as galactogogue.

(Jain, 1991)

Critical Notes

Stem bark as a vermifuge and purgative, roots against dysentery and anuria, and leaves for lochia are new medicinal utilities hitherto unrecorded.

Laeucaena latisiliqua (L.) Gillis.

Local Name

Subabhul

Ethnobotanical Notes

Medicine

1. Gum is applied to gums to treat tooth-ache.

Notes from Classics

Seeds are employed in Madras for making bags, baskets and ornaments.

(Watt, 1889-1899)

Plant is used as hedge in villages. Tops, leaves, pods, seed or plants are relished by cattle, sheep and goats. Wood is used for fuel, and paper manufacture.

(Anonymous, 1946-1976)

Plant affords fodder for cattle. Wood is a source of paper pulp. Young shoots and pods are used as vegetable. Seed yields fatty oil, they are parched and eaten.

(Ambasta, 1986)

Fruits and tender twigs used as vegetable.

(Jain, 1991)

Critical Notes

Use of gum to treat tooth-ache is unreported in classic literature. It is exotic, planted by farmers and also found place in afforestation programmes of social forestry.

Mimosa hamata Willd.

Ethnobotanical Notes

Medicine

1. A glass of infusion of fresh-stem is administered against snake-bite.

Miscellaneous

1. *Fodder*: Plant is used as fodder for goats.
2. Stem is used as tooth brush.

3. Spiny, dried branches are used for fencing around cultivated fields to prevent entry of cattle.

Notes from Classics

Browsed by cattle, may be tried as a manure.

(Ambasta, 1986)

Seeds are blood purifier. Leaves are used for chronic cough and pharyngitis.

(Jain, 1991)

Critical Notes

Stem is useful against snake-bite and as tooth brush. These uses are not published in past literature. Spiny branches, however, are obviously used by farmers to prohibit cattle from entering in crop fields.

Pithecellobium dulce Roxb.

Local Name

Vilayati chinch

Ethnobotanical Notes

Miscellaneous

1. ***Food***: Fruits are edible.
2. ***Fodder***: Leaves are lopped for fodder especially for goats.

Notes from Classics

Pulpy aril is eaten by poorer classes. Plant is cultivated by poorer class.

(Watt, 1889-1899)

Pods are used as fodder for cattle, sheep and goats. Seeds are eaten raw or in curries. Plant is mostly grown in India as hedge and for fuel.

(Anonymous, 1946-1976)

Wood is used for construction and packing cases, fence posts, carts, building, fuel and agricultural implements.

(Ambasta, 1986)

Roots are used as antirabic in dog-bite. Leaves are used against the leprosy and promote hair growth. Fruits are edible.

(Jain, 1991)

Critical Notes

It is planted. It is exotic and native of Mexico.

Prosopis cineraria (L.) Druce

Local Name

Shami, Saundad

Ethnobotanical Notes

Medicine

1. Pieces of roots are mixed in fodder to increase lactation in case of domestic animals.

Miscellaneous

1. Dry bark is sweet in taste, used in the form of flour during famine.
2. Pods are eaten when green.
3. Wood is used for making agricultural implements and foot handles.

Notes from Classics

Bark is only article for food. Pods are largely consumed by natives of Deccan Gujarat and Punjab during famine. Pods are good fodder for goats and sheeps.

(Watt, 1889-1899)

Pods are good fodder, eaten when green. Wood is suitable for internal construction of columns, roots, doors, windows, tool handles, well-curbs etc.

(Anonymous, 1946-1976)

Fruits are edible, astringent, demulcent and pectoral. Pods are used as fodder. Wood is suitable for interior construction, wheels of cars, agricultural implements, tool handles and well curb. It is also used for fuel. Bark is used for famine food.

(Ambasta, 1986)

Roots and flowers are used to check miscarriage. Bark, shoots and gum are edible. Fruits and leaves are used as vegetable.

(Jain, 1991)

Critical Notes

Roots are not reported in past as useful for increasing lactation in case of pet animals.

Prosopis juliflora DC.

Local Name

Vilayati Babul

Ethnobotanical Notes

Miscellaneous

1. ***Fodder***: Leaves and fruits are fed to goats.
2. ***Wood***: Wood is used as excellent fuel.

Notes from Classics

Tree exude gum, forms adhesive mucilage and can be used as emulsifying agent, used in confectionary and employed for making pottery.

(Anonymous, 1946-1976)

Spongy walls of ripe pods are nutritive, important as stock feed. Seeds ground into flour, is used for preparing bread. Foliage can be fed to live-stock. Wood is used for house building, railway, furniture, turnery and fence post.

(Ambasta, 1986)

Fruits are edible in thirst.

(Jain, 1991)

Critical Notes

It is exotic, native of Mexico and central America. It has largely naturalized on waste lands and has practically occupied each such places. Poor people cut and collect for fuel. Actually, it is a good source of fuel. Moreover, it's leaves and fruits are good fodder for goats. Although, the taxon is weedy in nature, is found very useful locally.

Family: Crassulaceae

Kalanchoe pinnata (Lamk.) Pers.

Local Name

Ponfuti

Ethnobotanical Notes

Medicine

1. Two leaves mixed with jaggary are consumed. It is followed for seven to eight days to cure pile.

Notes from Classics

Leaves are warmed on fire and applied to wounds, boils bites of insects, also used as poultice for ulcer.

(Watt, 1889-1899)

Toasted leaves are applied for wounds, bruises, boils and bites of venomous insects. In the form of poultice and powder they are used for sloughing ulcer. Leaf paste is applied to wound, it encourages papillation, also beneficial in veterinary cases.

(Anonymous, 1946-1976)

Toasted leaves are applied to wounds, bruises, boils and bite of venomous insects, swelling and discolouration of affected part are prevented. The healing of injured portions is accelerated in the form of poultice. Powder they is applied to sloughing ulcer.

(Ambasta, 1986)

Leaves are used to cure dysentery, body heat, boils, sores, wounds, gastric complaints, head-ache, influenza, insect bite, kidney stone and other diseases.

(Jain, 1991)

Critical Notes

Leaves are reported for the first time useful for piles. It is planted as an ornamental but the local people found it useful. It is exotic, native of Africa.

Family: Combretaceae

Anogeissus latifolia Bedd.

Local Name

Dhawda, Dhamda

Ethnobotanical Notes

Medicine

1. Decoction of bark, two spoon daily, is drunk as a remedy against cough till cure.
2. Decoction of leaf, about half cup, is taken orally for 8-10 days to treat epileptic fits, called 'Sad'.
3. Gum is used as tonic. It is mixed in sweet balls and generally consumed after delivery.

Miscellaneous

1. ***Timber***: Wood is used for making agricultural implements and for construction of huts.

Notes from Classics

Gum is sold for calico printing. Leaves yield black dye useful in tanning.

(Watt, 1889-1899)

Gum is used for sizing paper and calico printing. Wood is used for making boats, furniture, fuel-wood, agricultural implements, shafts, axles, poles, tool handles etc.

(Anonymous, 1946-1976)

Tasar silk worm is fed on its leaves.

(Ambasta, 1986)

Stem bark is useful in diarrhoea, dysentery and dysuria. It is also used in cough, colic and liver complaints. Stem bark is important against snake bite, scorpion sting and skin diseases. Seeds are used as antidote on snake bite. The plant is remedy on cancer of face, carbuncle, cholera, cold, excess perspiration and spleen complaints.

(Jain, 1991)

Critical Notes

Gum used as tonic and leaves as a remedy against epilethic. These fits are being recorded for the first time.

Combretum albidum G.Don

Local Name

Walzira, Welia, Dewala

Ethnobotanical Notes

Miscellaneous

1. *Fodder*: Leaves are used as fodder for goats.
2. Branches are used to tie partition walls of huts and for fishing nets.

Notes from Classics

Plants are eaten by buffalos.

(Watt, 1889-1899)

Terminalia bellirica (Gaertn.) Roxb.

Local Name

Bhehada

Ethnobotanical Notes

Medicine

1. Equal amount of fruit powder of this species and of *Emblica officinalis* Gaertn. bark of *Azadirachta indica* A. Juss. and *Acacia catechu* (L.*f.*) Willd. are mixed in lukewarm water. Spoonful decoction of it twice a day for two days is administered to cure vitiligo.
2. Fruits of this species mixed with fruits of *Terminallia chebula* Retz. and *Emblica officinalis* Gaertn. powdered and burnt. Ash is applied on wounds of syphilis to cure them.
3. Tender twigs are used as tooth brush during malarial fever.
4. *Food*: Kernels are edible.

Miscellaneous

1. Wood is used for making ploughs and bullock carts. Leaves are used for making dinning plates and dishes.

Notes from Classics

Kernels is said to be narcotic and astringent. Kernel with betel nut and leaf cures dyspepsia. Kernels are applied locally on inflammation. Fruit is astringent, laxative, digestive, attenuant and aperiant. They are employed to cure dropsy, piles, diarrhoea, leprosy, fever eye-inflammations and hoarseness.

(Watt, 1889-1899)

Kernels possess narcotic properties, eaten with betel nut and leaf for dyspepsia. Ripe fruit is astringent, employed for dropsy, piles, diarrhoea, leprosy and in fever. Half ripe fruit is used as purgative. Fruit pulp mixed in honey is used as application in case of opthalmia. Oil is applied to hair and rheumatic joints. Gum is demulcent and purgative. Fruits are used in ink making.

(Anonymous, 1946-1976)

Bark is diuretic. Ripe fruits are astringent and half ripe fruits are purgative. Pulp mixed in honey is applied opthalmia. Wood is used for shafts, rafler, boats and ploughs. Gum is demulcent and purgative.

(Ambasta, 1986)

Bark is employed for fever and cold. Exudates of stems stop itching by ticks. Fruits are laxative, purgative, cooling, employed for asthma, bronchitis, constipation, cough, diabetes, diarrhoea, dropsy, fever, piles, intestinal complaints, leprosy, liver complaints, stomach-ache and respiratory disorder. Plant is used for cholera, cramps, dysuria, muscular pain and antidote to snake-bite. Seeds are employed for gastric complaints. Fruits seeds or kernels are edible. Fruit is used as a vegetable. Tannin is obtained from fruits.

(Jain, 1991)

Critical Notes

Fruits are used to better vitiligo and syphillis. Tender twigs are used as tooth brushes during malarial fever. These local medicinal uses await for documentation. Leaves are used for preparing dinning dishes and bowls. This tree species should be given priority in the afforestation programme. This will help the poor tribals being highly demanded medicinal fruit as well as for cottage industry.

Terminalia chebula Retz.

Local Name

Hirda

Ethnobotanical Notes

Medicine

1. Consumption of two fruits boiled in cow urine at morning for three days cure jaundice.
2. Paste of fruits boiled in cow urine is applied on scabies and swelling of body till cure.
3. Spoonful of powder of fruit mixed in milk is taken thrice a day upto five days to control cough.
4. Ash of fruit about spoonful is consumed for 3 to 4 days with warm water to cure lymph adenitis.
5. Equal amount of fruit powder and rhizome of ginger (*Zingiber offincinalis* Rosc.) mixed homogeneously and taken single spoonful only once to control vomiting.
6. Spoonful of fruit-ash is taken in honey for 3-4 days to cure mouth ulcer.

Notes from Classics

Fruit is laxative, stomachic, tonic and alterative. It gives strength and prevents ageing of life. Fruits cure cough, asthma, urinary diseases, piles, diarrhoea, intestinal worms, flatulence, vomiting, hiccup, heart diseases, enlarged spleen, liver ascites and skin diseases.

(Watt, 1889-1899)

Fruits have laxative, stomachic and alterative properties. Fruit pulp is used to cure bleeding and ulceration of gums. Fruits are soaked for overnight and it is a good cooling wash for eyes. Fruit gives good relief in conjunctivitis. Fruit powder is smoked in asthma.

(Anonymous, 1946-1976)

Ambasta (1986) reported properties as those in Anonymous (*loc.cit.*).

Bark is diuretic and cardiac tonic and also controls blood pressure. Bark powder is applied on eczema. Fruits are used in cough, cold, colic, diabetes, dysentery, eye diseases, indigestion, menstrual complaints, pneumonia, sore throat, bronchitis, constipation, diarrhoea, and spleen complaints.

(Jain, 1991)

Critical Notes

Fruits are being reported for the first time useful for jaundice, body swelling and lymph adenitis. This tree species should be given priority during afforestation especially in tribal areas to boost their economy. Fruits are in great demand in market being of significant medicinal value.

Terminalia crenulata Roth

Local Name

Ragat roda

Ethnobotanical Notes

Medicine

1. Decoction of bark, about half cup, is drunk at morning for three days to cure diarrhoea.
2. Infusion of bark, about half cup, is given twice a day for five days to cure dysentery.
3. Paste of stem bark is applied on bone fracture of night till cure.
4. A spoonful of powder of stem bark is consumed at morning per day to cure anemia.

Miscellaneous

1. *Fodder*: Leaves relished by goats, cows and buffaloes.
2. Bark is used to stupefy fishes.
3. Wood is used to make ploughs and cultivators

Notes from Classics

Bark is a remedy for atonic diarrhoea. Bark is applied on callous ulcer. Ash of bark is used as a substitute for lime in betal leaf or pan. Leaves are lopped for fodder. Bark is poisonous to fish.

(Watt, 1889-1899)

Wood is used for construction work and agricultural implements.

(Anonymous, 1946-1976)

Ambasta (1986) reported similar properties as those in Anonymous (*loc.cit.*).

Bark is used for diarrhoea and vomiting. Gum is edible.

(Jain, 1991)

Critical Notes

Bark is used to control dysentery. It is also useful for bone fracture and to cure anaemia. These uses have not been published in past.

Family: Myrtaceae

Eucalyptus globulus Labill.

Local Name

Nilgiri

Ethnobotanical Notes

Miscellaneous

1. Wood is used for making plough bars and handles of agricultural implements.
2. Wood is also used for fuel.

Notes from Classics

Oil of leaves and young shoot is stimulating, narcotic and antispasmodic. It is useful in migraine, neuralgia, asthma and chronic bronchitis. Oil is antiseptic ointment is used for leprosy. Infusion of leaves is useful for throat ulcer. Roots are purgative.

(Watt, 1889-1899)

Oil of leaves is antiseptic, rubefacient and expectorant. Ointment is used for burns and skin disease. Dried leaves inform of tincture is used for asthma, phthisis and in chronic bronchitis. Leaf decoction is used as insect repellent.

(Anonymous, 1946-1976)

Wood is suitable for boat building, agricultural implements, spokes of wheel, plough bars, axe handles and for wood pulp. Plant is mostly used as fuel.

(Ambasta, 1986)

Leaf decoction is prescribed for body ache head-ache, malaria and swellings.

(Jain, 1991)

Critical Notes

This plant is introduced in this region by the department of forests. They are usually planted on bandhs of fields, near wells and fallow lands. The wood is sold in markets. But it is used locally only for miscellaneous purposes. The people of this district are well known for medicinal utility of oil obtained from leaves but do not know its method of extraction. Planting just few trees does not help the local people. Therefore, if it is to be planted, only captive plantation should be planned.

Psidium guajava L.

Local Name

Peru, Jam

Ethnobotanical Notes

Medicine

1. Bark of stem is fried and about spoonful of it is mixed in a cup of butter milk. It is drunk to cure dysentery. This is followed till cure.
2. Poultice of leaves is placed on eyes at night to cure swelling and pains of eyes.

Notes from Classics

Root bark is employed for chronic diarrhoea and applied on prolapsus anni of children. In scurvy, decoction of bark is employed as a mouth wash for swollen gums. Bark is useful in cholera, vomiting and diarrhoea. For ulcer bark is employed as poultice. Ripe fruits are edible and aperient.

(Watt, 1889-1899)

Roots are astringent. Bark is astringent, tonic, used in diarrhoea and ash is caustic. Leaves are used on wounds and ulcer. They are also tonic and astringent. Decoction of leaves is used in cholera, arresting vomiting and diarrhoea. Infusion of leaves and shoot is febrifuge and antispasmodic, used in cerebral affection, nephratis and cochexia. Pounded leaves are applied in rheumatism, also used in epilepsy and cholera. The tincture is rubbed over the spine of children suffering for convulsion. Decoction of leaves when gargled relieves tooth-ache and gum boils. Flowers are cooling, used in bronchitis and eye sore. Fruits are cooling, tonic, laxative, good in colic and bleeding gum, diarrhoea and dysentery.

(Anonymous, 1946-1976)

Leaves are used as astringent for bowel troubles. Fruits are eaten preserved, made in jam, jellies and marmalade. Leaves also contain oil, used for flavouring. Wood is used for spear handles.

(Ambasta, 1986)

Leaves and bark are applied locally on blisters of mouth, also used for cough and cold, diarrhoea, dysentery, fever and head-ache. They are also used for gonorrhoea, stomach-ache and menstrual disorders. Fruit is employed as tonic and used in jaundice. Plant is used for carbuncle and sore.

(Jain, 1991)

Critical Notes

Uses of stem bark for dysentery and leaves for swelling and pains of eyes are not reported in past. It is cultivated for fruits. It is exotic native of tropical America.

Syzygium cumini (L.) Skeels

Local Name

Jambhul

Ethnobotanical Notes

Medicine

1. Paste of leaves is applied for a month to remove scar due to burning.
2. Decoction of leaves about half cup is drunk at morning to cure typhoid. It is followed till cure.

3. Leaves and seed powder is used to apply on spongy gums and bleeding gums.

Miscellaneous

1. ***Fodder***: Leaves are good fodder for cattle.
2. ***Wood***: Wood is used to make tool handles, and agricultural implements.

Notes from Classics

Decoction of bark is used as gargles in sore throat. Bark is astringent. Leaves are used in dysentery. They are used as platters or panch pallows and for pouring libations. They are also used for spongy and painful gums. Pulverized seed is taken as a antidote in nux-vomica poisoning. Powdered seeds are useful in the treatment of diabetes. Vinegar prepared from juice of unripe fruit is stomachic, diuretic and carminative. Wood is useful for house building, well-work and agricultural implements.

(Watt, 1889-1899)

Bark is astringent, used in preparation of gargles and mouth washes. Decoction of bark and powder seed is useful in diarrhoea and dysentery. Extract of bark, seeds, leaves and fruits are also useful for diabetes. Powder of seeds is used in diarrhoea and dysentery.

(Anonymous, 1946-1976)

Decoction of bark and powdered seeds are used in diabetes. Bark is also used in dyeing. Leaves are used as fodder for cattle, sheep and goats. Wood is useful for tool handles, agricultural implements, cart wheels, well curbs and furniture. Ripe fruits are edible, used in making jam, jellies and squash.

(Ambasta, 1986)

Bark is astringent, useful to cure piles. Leaves are remedy for blisters, dysentery and stomach-ache. Seeds are used for diabetes. Fruits are edible and employed for digestive complaints, stomach-ache. Fruits are tonic. Plants are used for anaemia, diarrhoea, cancer and colic. Leaves are used as vegetable, also used for thatching huts.

(Jain, 1991)

Critical Notes

Leaves are useful to remove scars due to burning. They are beneficial in typhoid. These uses are not reported earlier. It is generally planted for its edible fruits.

Family: Barringtoniaceae
Careya arborea Roxb.

Local Name

Kumbha, Kumbhi

Ethnobotanical Notes

Medicine

1. Extract of stem in water about one spoonful mixed in equal amount of cow ghee is consumed after every two hours for haemorrhage.
2. A cup of juice of leaves is given once a day specially at morning with empty stomach for five to seven days to treat cough and cold.
3. A spoonful of fruit powder with warm water is drunk to reduce cough till cure.
4. A spoonful of fruit powder is taken with warm water twice a day before meal. It helps to reduce over weight within a month or so.
5. Fruit pulp is applied to scalp regularly to promote hair growth.

Miscellaneous

1. Stem bark yields fibres for making cordage.
2. Root bark is sued as fish poison.
3. Wood is used for making agricultural implements.

Notes from Classics

The calyx with honey is used as demulcent in cough and cold. Flowers are used as tonic for ruptures caused by child-birth. Fruit is astringent and aromatic, used in the form of decoction to promote digestion.

(Watt, 1889-1899)

Stem bark is demulcent, given in cough and cold. It is antipyretic in eruptive fevers such as small-pox. Leaves are used in the treatment of ulcer and used for biddi wrapping. Calyx contain mucilage which is used as demulcent. Decoction of fruit promotes digestion. Seeds and leaves are used as fish poison.

(Anonymous, 1946-1976)

Stem bark is used for cordage. Timber is used for agricultural implements, for making cabinate and house post.

(Ambasta, 1986)

Bark is used against blood dysentery, cough, cold, diarrhoea and skin diseases. Bark is applied locally on cuts, wounds and swellings of face. Plant is used to cure fistula and swelling of face. Plant and bark are used as antidote on snake bite.

(Jain, 1991)

Critical Notes

Roots are employed for stupefying fishes. Bark is useful to treat haemorrhage. Leaves are beneficial for cold and cough, to promote hair growth and to reduce over body weight. All these uses are unrecorded in past literature.

Family: Lythraceae

Ammannia baccifera L. subsp. *aegyptiaca* (Willd.) Koehne

Local Name

Lalagio, Umarya, Bhan Jambhul

Ethnobotanical Notes

Medicine

1. A cup of leaf decoction is administered daily as a remedy against venereal diseases till cure.
2. Leaves are brushed for ring-worms.

Notes from Classics

Root juice is used for affections as a topical applicant. Leaves are sued for blisters and fever. Brushed leaves are applied as poultice to abscess. Juice of plant is given internally in spleen.

(Watt, 1889-1899)

Fresh leaves are used in skin diseases as a rubefacient. They are acrid.

(Anonymous, 1946-1976)

Herb is reported to possess antityphoid and antitubercular properties.

(Ambasta, 1986)

Leaves are used against rheumatism, skin diseases to expel ring-worm.

(Jain, 1991)

Critical Notes

To cure venereal disease use of leaves is a new finding.

Lawsonia inermis L.

Local Name

Mehandi

Ethnobotanical Notes

Medicine

1. Spoonful of bark powder is given at night with warm water to cure acidity till cure.
2. Decoction of leaves, about half cup, is advised twice a day for a week to cure allergic dermatitis.
3. Extract of leaves is applied locally to treat burns and scalds.
4. Spoonful of juice of leaf is taken per day as a oral contraceptive to prevent conception. It is followed till effect.
5. Paste of leaves is applied to sole of foot to stop burning due to summer heat.

Miscellaneous

1. Plant is used as a live hedge around houses.
2. Leaves yield dye which is used for colouring palms of hands and sole of feet and hair.

Notes from Classics

Bark is used in jaundice and enlargement in spleen inflammation, burns, head-ache and ulcer of mouth. Ointment made from leaves has healing properties. Decoction is astringent, used for gargle. Leaves and tender shoots relieve pain in leprosy. Leaf juice is used for spermatorrhoea. Infusion of flowers cure head-ache.

(Watt, 1889-1899)

Leaves are used as propylactic against skin disease and also used as astringent. Leaf decoction is used to treat boils, burns, bruises and skin inflammation. Decoction is used as gargle for sore throat.

(Anonymous, 1946-1976)

Leaves contain dye for colouring palm of hands, nails dying hair and beard. Leaves are used for colouring skin, leathers and fine wool. Leaves contain essential oil. Wood is used for making tool handles.

(Ambasta, 1986)

Root is employed for jaundice. Leaves are applied to stop burning of sole. Seeds are used against anti-fertility and used for weakness. Plant is useful for head-ache and pain in muscles. It is planted as hedge.

(Jain, 1991)

Critical Notes

Uses such as bark to reduce acidity, of leaves to treat dermatitis and as oral contraceptive have not been documented in past. It is usually planted near human habitation or as a live hedge. It is exotic, native of North America and Australia.

Woodfordia fruticosa (L.) Kurz.

Local Name

Dhaiti

Ethnobotanical Notes

Medicine

1. Flowers are powdered with some sugar. Two spoonful of it with milk for twice a day is consumed for a week to cure menorrhagia.
2. Flower powder is applied on wounds for quick healing.

Notes from Classics

Flowers are stimulant, astringent, cooling, prescribed in pregnancy, bilious and mucus disorders, heamorrhoids and bowel disorders. The powdered flowers and spinkled over ulcers. Decoction of flowers is given in female complaints.

(Watt, 1889-1899)

Flowers are edible. Dried flowers are stimulant and astringent. Dried flowers are sprinkled over ulcers and wounds. Paste of flowers is used for cough. Ointment of flowers is used on the pustules of small-pox. Fruits, flowers, buds or inflorescence are used in bowel complaints and haemorrhage. They are also administered for menorrhagia and feminal weakness. Extract of plant shows antipyretic properties.

(Anonymous, 1946-1976)

Ambasta (1986) gave uses are similar as those in Anonymous (*loc.cit.*).

Roots are used in rheumatism and skin disorders. Stem and stem bark cure wounds. Leaves are applied on veterinary sore and ulcer. Plant is considered useful for bleeding in pregnancy, bone fracture, carbuncle and dropsy. It also cures fever, menorrhoea, muscle pains, sore, spleen complaints and ulcer. Flowers are used to cure cancer, cough, dysentery, haemorrhage and injuries. They recover loss of appetite in pregnancy, menorrhoea, nausia, small pox and tonic. Flowers, fruits, stems are edible. Dye is obtained from flowers.

(Jain, 1991)

Family: Punicaceae

Punica granatum L.

Local Name

Dalimb

Ethnobotanical Notes

Medicine

1. Extract of leaves is applied on boils and scabies.

2. Half cup of fruit juice for twice a day given for two weeks to the patients suffering from jaundice.
3. Half spoon of rind powder homogenized in sugar syrup or honey is kept in mouth, for half an hour, for stomatitis. This is repeated three to four times a day till cure.
4. A spoonful of extract of rind prepared in water is taken orally with honey as a remedy against bleeding dysentery.

Notes from Classics

Juice of leaves and young fruits is used for dysentery, epistaxis and bleeding from nose. Juice of flower is snuffed in nostrils and flower buds are used for bronchitis. Flowers and buds are styptic and cicatrizing. Flower juice is cooling and refrigerant. Stem bark is anthelmintic, astringent and used for sore throat. Extract of stem-bark is used as wash for veterinary diseases. Root bark is given to uterine inertia. Decoction of root bark is a remedy for round-worms. Root bark is anthelmintic. Rind is astringent, used for diarrhoea and dysentery. Fruits are eaten to treat haemorrhage of tongue of a haemophilic child.

(Watt, 1889-1899)

Expressed juice of leaves, young fruits and decoction of bark are used in dysentery. Powdered flower buds are used in bronchitis. Fresh fruit juice is cooling, refrigerant and used to cure dyspepsia. Sweet fruits are mildly laxative and less sweet type are good for inflammation of stomach and heart pain. Seeds are stomachic and the pulp is cardiac and stomachic.

(Anonymous, 1948-1976)

Bark is used to expel tape-worms. Stem and roots are anthelmintic. Flower buds are used for bronchitis. Rind is astringent, used to cure diarrhoea and dysentery. Fleshy testa are edible. Seeds are used as condiments.

(Ambasta, 1986)

Roots and stems are anthelmintic. Leaves are used to check bleeding during delivery and miscarriage. They are also used to treat cholera and dysentery. Leaves are cooling in nature. Fruits are used in the treatment of cholera, cough, diarrhoea, dysentery and stomach-ache. Fruits are edible. Seeds are applied on pimples. Tanin and dye are obtained from fruit and rind.

(Jain, 1991)

Critical Notes

Leaves are useful to treat boils and scabies. Fruit juice is beneficial in jaundice. Fruit rind is a remedy against stomatitis. All these uses did not find place in earlier published literature. The plant is cultivated mainly for edible fruits. It is exotic and native of Iran.

Family: Caricaceae

Carica papaya L.

Local Name

Papayee, Popai

Ethnobotanical Notes

Medicine

1. Ripe pulp of fruit is mixed with small amount of cream of milk and applied regularly at night on cracks of foot sole.
2. Ripe pulp of papaya is mixed with leaf pulp of *Aloe vera* (L.) Burm. *f.* and the mixture is applied on face for some time to remove the black-spot and pimples from face.
3. Fresh pulp of ripe fruit is applied around the eyes to remove blackness surrounding the eyes and make the skin soft and wrinkleless.
4. Ten to eleven seeds are consumed at morning with empty stomach for three days to kill worms in child.
5. Milky latex and honey, three spoon each, are mixed and homogenized in a glass of warm water. It is drunk once a day to cure amaebic dysentery.
6. Latex of fruit is applied on the boils in nose regularly to cure early.
7. Latex is applied on gums to treat tooth-ache.

Notes from Classics

Leaves are used as poultice for bone fractures, and taken internally for constipation and digestion. Fruits are cooling, digestive and gastric. They control urine bladder complaints and are abortifacient. Latex from unripe fruits is abortifacient, also used to expel ring worms. It is applied externally for skin disease and tooth-ache. Latex of unripe fruits possess powerful anthelmintic properties. It gives relief in haemoptysis, bleeding piles, dyspepsia, diptheria and ulcer of urinary passage. Rubbing of latex two or three times cure ring-worm. Seeds are abortifacient. Fruits are edible, pickled and used as vegetable.

(Watt, 1889-1899)

Ripe fruits are stomachic, digestive and carminative. Roots cure yaws and piles. Poultice of leaves is applied for nervous pains and used for reducing elephantoid growth. Syrup of pulp of ripe fruit is expectrant, sedative and tonic. Milky juice of unripe fruit is anthelmintic and effective against expulsion of lumbrici. Green fruit is laxative, diuretic and possess ecobolic properties. Seeds are used to quech thirst. Fruits are tonic.

(Anonymous, 1946-1976)

Unripe fruits are used as vegetable. Mature ones are also eaten. Ripe fruits are employed in the preparation of candies, soft drinks and marmalades. Fruit pulp is ingredient in face creams and hair shampoos. Milky latex is used to remove freckles and blemishes from skin. It is anthelmintic. Dessicated latex is digestant. Plants yield blood anticoagulant. Seeds yield fatty oil.

(Ambasta, 1986)

Leaves are useful in constipation and applied locally on bone fracture. Fruits are cooling, digestive, useful in gastric and urine complaints. They are edible, also pickled and used in vegetable. Latex is applied to cure skin disease, tooth and gum-ache. Fruits, seeds, roots and latex are abortifacient.

(Jain, 1991)

Critical Notes

Usefulness of ripe pulp to cure foot cracks and to remove black spots, blackness around eyes is being reported as additional uses. To treat boils in nose, latex is useful. This is also unnoticed earlier. It is cultivated species for fruits. It is exotic, native of tropical America.

Family: Cucurbitaceae

Benincasa hispida (Thunb.) Cogn.

Local Name

Khohala

Ethnobotanical Notes

Medicine

1. A cup of fruit juice is drunk for about a week to promote urination.
2. Fruit is eaten along with meals to cure habitual constipation.
3. Half cup of fruit juice is mixed with equal amount fruit extract of *Emblica officinalis* Gaertn. One spoon of it is taken at morning to cure bleeding piles till cure.
4. Cotton balls soaked in fruit juice is kept on eyes in case of burning of eyes.
5. Fruit juice is applied on rough skin for softness.
6. Fruit paste is tied onto foot sole and hand palms for few hours to cure cracks.
7. Fruit juice about half litre is mixed with sesamum oil (*Sesamum orientale* L.) and also 200 ml of fruit juice of *Emblica offinalis*. A cup of this mixture is administered to stop bed dreams and talking when in sleep.
8. Seeds are crushed in milk or butter and applied on face to prevent pimples, rash and improve complexion.
9. Pulp of this fruit mixed with equal quantity of sugar are boiled. After cooling this jam is consumed every morning for healthyness and rheumatic pains.

Notes from Classics

Fruit pieces are candied called 'petha burfi'. Fruit possesses alternative and styptic properties. Fruits are cooling, tonic, nutritive, diuretic, specific for haemoptysis, useful in insanity, epilepsy and other nervous diseases. Ripe fruit is efficacious in phthisis and pulmonary tuberculosis. Fruit pulp is given in piles and dyspepsia. Fruits are used in vapour bath for syphilitic eruptions. Juice of fruit is purgative and alternative. Fruits are digestible and useful in diabetes.

(Watt, 1889-1899)

Decoction of fruits is laxative and styptic, given for respiratory troubles and internal haemorrhages. An Ayurvedic medicine, 'Kushmand leha' is used for consumption and as remedy against piles, dyspepsia and diabetes. Seeds are fried and eaten. They yield fatty oil which is considered anthelmintic.

(Ambasta, 1986)

Juice of fruit is considered useful in haemoptysis and other internal discharges and beneficial in phthisis. Seeds possess anthelmintic properties.

(Anonymous, 1946-1976)

Fruit juice is used in jaundice.

(Jain, 1991)

Critical Notes

The medicinal uses of fruits for habitual constipation, burning eyes, to cure cracks, rash, pimples and to reduce rheumatic pains are hitherto unrecorded. It is cultivated for fruits. It is exotic, native of Jawa.

Cucumis melo L. var. *melo*

Local Name

Kharbooz

Ethnobotanical Notes

Miscellaneous

1. ***Food***: Seeds are edible and nutritive.
2. ***Fodder***: Plant is relished by cattle, usually fed after fruit harvest.

Notes from Classics

It is extensively cultivated for fruits. Pulp of fruit is pleasant and tasty.

(Watt, 1889-1899)

Plant is extensively cultivated for its fruits valued as dessert.

(Anonymous, 1946-1976)

Fruits are eaten as dessert. Seeds are edible, nutritious and diuretic. Pulp is used in the treatment of eczema.

(Ambasta, 1986)

Fruits are edible.

(Jain, 1991)

Critical Notes

It is generally cultivated in rivers and also in fields for its edible fruits.

Cucumis setosus Cogn.

Local Name

Mekha

Ethnobotanical Notes

Miscellaneous

1. *Food*: Fruits are edible.

Notes from Classics

Fruits are edible. This species is cultivated in hot season and also in rainy season.

(Watt, 1889-1899)

Fruits are used in salads.

(Ambasta, 1986)

Fruits are edible.

(Jain, 1991)

Critical Notes

It is wild in our district, found amidst grasses in rainy season.

Coccinia grandis (L.) Voigt.

Local Name

Tondali, Tongala

Ethnobotanical Notes

Medicine

1. About ten gm of roots are pounded in water and it is taken orally thrice a day to cure dysuria.
2. About half cup of leaf juice daily in the morning is prescribed for three to four days to reduce acidity.
3. Fruit juice about a cup is taken twice a day for twenty consecutive days to control swelling of body.
4. Raw fruits are consumed twice a day to cure diarrhoea.

Miscellaneous

1. Plants are collected from hedges, for fodder of goats.

Notes from Classics

Ripe fruits are used as vegetable.

(Watt, 1889-1899)

Raw fruits are used as vegetable, ripe ones are eaten.

(Anonymous, 1946-1976)

Tender shoots are used as vegetable.

(Ambasta, 1986)

Fruits are used for cough and cold

(Jain, 1991)

Critical Notes

Medicinal uses of roots for dysuria, of leaves to reduce acidity and of fruits to control body swellings are all additional ones. Fodder value is also unnoted. It is cultivated species but has naturalized on hedges along roads.

Cucumis callosus (Rottl.) Cogn.

Local Name

Kakadi

Ethnobotanical Notes

Medicine

1. Extract of about 40 gm seeds with a cup of milk i administered twice a day for cough till cure.

2. One litre milk with about 30 gm of seeds and small amount of common salt is drunk daily as diuretic.

Notes from Classics

Oil of seeds along with the fruits is used for asthma.

(Watt, 1889-1899)

Seeds are cooling, tonic and diuretic. Seed kernels are edible and are used in confectionary.

(Anonymous, 1946-1976)

The properties reported in Anonymous (*loc.cit.*) are also documented by Ambasta (1986).

Fruits are used for skin diseases and wounds. They are used as vegetable. Leaves destroy lice if applied on head.

(Jain, 1991)

Critical Notes

Seeds are being reported for the first time to cure cough.

Citrullus colocynthis (L.) Schrad.

Local Name

Kadu-indrayan, Jangali tarbooj

Ethnobotanical Notes

Medicine

1. Fruit extract, about half cup, is taken orally for food poisoning.
2. Seed powder is mixed in cow ghee and tablets are prepared from it. Two or three tablets twice per day cure stomach swelling and ache. These are advised till cure.

Veterinary Medicine

1. About fifteen or twenty fruits are mixed in fodder of she-buffaloes to arouse sex.

Notes from Classics

Fruits are used in jaundice, ascities, rheumatism and urinary diseases. Fruits control the enlargement of abdominal viscera.

Leaves, fruits and seeds are used for jaundice, colic and to expel worms and for irritation of ulcers. Whole plant is used as hydrogogue, catharatic and purgative.

(Watt, 1889-1899)

Roots are purgative, used in jaundice, ascites and rheumatism. Roots also control urinary diseases. Dried pulp of unripe but full grown fruit forms a drug which is used for hydrogogue and catharatic.

(Anonymous, 1946-1976)

Seeds contain fixed oil.

(Ambasta, 1986)

Roots are used as abortifacient. Fruits can be used as galactogogue and purgative. They give relief in stomach-ache. Grains are edible.

(Jain, 1991)

Critical Notes

Fruit extract against food poisoning in case of human beings and the entire fruits to arouse sex in domestic animals are hitherto unreported. The uses of seeds for stomach-ache and swelling are also not reported earlier. It is denizen of barren lands. These lands are encroached for agriculture, hence their number is also being reduced in the district.

Diplocyclos palmatus (L.) Jeffery

Local Name

Shivlingi

Ethnobotanical Notes

Medicine

1. Five to six seeds powdered and given to a women in spoonful of honey for about six months, this promotes fertility in women.
2. Two or three fruits are crushed in a half cup of milk and advised daily to cure head-ache and fever till cure.
3. Plant extract about full of cup is given as antidote on snake poisoning.

Notes from Classics

Roots are applied to carbuncles and given with turmeric and myrobalans for gonorrhoea. Root is useful in inflammation of lung in cattle. Root bark boiled with gingelly oil and is used before bath for the relief from head-ache. Fruit is remedy for asthma.

(Watt, 1889-1899)

Paste of leaves is used on boils. Cooked young fruits are consumed as sajoor, a soup. Herb is used as bitter tonic and febrifuge.

(Ambasta, 1991)

Seed promotes fertility in women. Leaves are remedy in case of eye infection. Plant is considered to use ague, asthma, carbuncle, cholera, constipation, convulation and delirium. Plant cures fever, head-ache, paralysis, post natal complaints and sores. Plants control syphilis, tumour, sores and spleen complaints. It is used as antidote for snake bite, also a remedy on veterinary medicine.

(Jain, 1991)

Critical Notes

Fruits are beneficial to treat head-ache and fever, where as seeds are so far fertility in women. These uses are hitherto unrecorded.

Kedrostis rostrata (Rottl.) Cogn.

Local Name

Mirchi-kand.

Ethnobotanical Notes

Medicine

1. Ripe fruits are eaten as antidote for snake-bite.

Notes from Classics

Root is demulcent, used in piles and asthma. Ripe fruit and leaves are eaten.

(Anonymous, 1946-1976)

Roots are mucilaginous and demulcent, used in asthma and piles. Fruits and leaves are eaten.

(Ambasta, 1986)

Critical Notes

Use of ripe fruit as antidote for snake bite is still unreported.

Lagenaria siceraria (Melina) Standl.

Local Name

Dudhi, Dudhi-Bhopala

Ethnobotanical Notes

Medicine

1. Decoction of green fruit, about half cup, is taken at early morning to avoid heart attack. It is followed for one or two months.

Miscellaneous

1. Musical instruments of Pavara tribe called 'Pavri' and 'Gudgudi', a smoking tobacco pipe, are made from dried fruits.
2. Dried fruits are used as water bottle.
3. Seeds are edible.

Notes from Classics

Decoction of leaves is used for jaundice. They are purgative in nature. Seed is cooling and given internally for head-ache. Seed oil is applied externally for head-ache. Fruit pulp is purgative, antibilious, diuretic, cooling and anidote to certain poisons.

(Watt, 1889-1899)

Leaf juice is used for promote hair growth on bald. Fruit pulp is emetic, purgative, cooling, diuretic antibilious and applied externally for delirium. Juice of fruit mixed with Citrus juice is applied on pimples. Fruit pulp is boiled in oil and applied on rheumatic joints. Seeds are used in dropsy. It is anthelmintic. Seed oil is applied to cure head-ache.

(Anonymous, 1946-1976)

Fruits are used for making water bottles, bowls and snuff boxes. Fruits are used as vegetable. Seeds are edible and yield oil. Oil is applied on head to cure head-ache.

(Ambasta, 1986)

Leaves and roots are useful in jaundice. Fruits are emetic, purgative and remove cracks on sole of feet. Fruits are used for making bottles, bowls, containers, utensils and ornament as ear top.

(Jain, 1991)

Critical Notes

It is interesting to note that green fruit is used as controlling measure for heart attack. This is not noticed elsewhere. It is commonly cultivated for vegetable by the tribals near their abodes.

Luffa acutangula (L.) Roxb.

Local Name

Dodaka

Ethnobotanical Notes

Medicine

1. Extract of leaves mixed in turmeric powder (*Curcuma longa* L.). This paste is applied on urticaria.
2. Powdered fruits snuff is smelled as a snuff to stop head-ache due to migraine.

Notes from Classics

Fruit is cathartic and emetic. Dried fruit powder is used as snuff in jaundice. Juice of roasted fruit is applied for head-ache. Seeds are tonic and diuretic.

(Watt, 1889-1899)

Leaves are used for granular conjunctivitis, uremia and menorrhoea. Leaves are used as poultice in haemmorrhoids and leprosy. Seeds possess emetic and purgative properties. Cake is used as manure.

(Anonymous, 1946-1976)

Ambasta (1986) reported similar properties and uses as those in Anomymous (*loc.cit.*).

Fruits are used in fever and jaundice. Fruit powder is smelled as snuff. Tendrils act as antidote for snake-bite. Plant is used for madness, scabies, syphillis, tetanus, convulation and cramps.

(Jain, 1991)

Critical Notes

Leaves are employed in case of urticaria and fruit powder is used as snuff against migraine. Both uses are not noted earlier. It is generally grown near huts and houses, apart from crops.

Luffa cylindrica Mill.

Local Name

Gilke

Ethnobotanical Notes

Medicine

1. Juice of leaf, about half cup, is taken orally twice a day for a week in case of entire body swelling.
2. Seed oil is applied on scabies till cure.

Notes from Classics

Seeds are said to be emetic and cathartic. Fruits are edible, used in curries.

(Watt, 1889-1899)

Tender fruits are diuretic and galactogogue. Ripe fruits are bitter, pulverizing, carminative and anthelmintic. Juice of fruit is purgative. Seeds are bitter, emetic and cathartic. Seed oil is used on skin affections. Seeds can be used as mannure.

(Anonymous, 1946-1976)

Ambasta (1986) reported similar properties and uses as those use in Anonymous (*loc.cit.*).

Fruits cure diarrhoea, prickly heat and are used in veterinary medicine. Fruit is used in dyspepsia. Fruits are used as vegetable.

(Jain, 1991)

Critical Notes

Leaf juice is being noticed for the first time useful for body swellings. It is also grown near huts and houses, apart from crops.

Momordica dioica Roxb. ex Willd.

Local Name

Katirley

Ethnobotanical Notes

Medicine

1. Two fruits and spoonful of turmeric powder is pounded in *Aloe vera* (L.) Burm. *f.* jel. Pellets made from this mixture is taken daily for a week to cure asthma.
2. Half cup of fruit extract is drunk at morning daily for body swelling till cure.

Notes from Classics

Mucilaginous roots of female plants are used in terms of electuary in case of bleeding pile and bowel affection. Root of female plant is used as expectorant. Roots of male creeper is used in ulcer. The plant is cephalic, it is used as liniment. The plant relieves all pains due to head-ache.

(Watt, 1889-1899)

Root is astringent. Root paste is applied over the body as sedative in fever. Female plant root is applied on bleeding pile, bowel affection and urinary complaints.

(Anonymous, 1946-1976)

Fruits are used as vegetable. Kernels yield a semi-drying oil used in paints and varnish industries. Powder of infusion of dry fruit is introduced in nostrils produce errhine effect and provokes a copious discharge from schneiderian mucous membrance. Roots are used against fever and pile.

(Ambasta, 1986)

Fruits controls pains in breast due to swellings. Whole plant is useful for injuries. Plant is used against asthma, bronchitis, dysentery, delirium, spleen complaints, head-ache, phthis, post-natal complaints and swelling on head. Whole plant is used in veterinary medicine. Fruits are edible.

(Jain, 1991)

Critical Notes

Utility of fruits against asthma is noted newly.

Solena amplexicaulis (Lam.) Gandhi

Local Name

Gomathi, Tawguala

Ethnobotanical Notes

Medicine

1. Decoction of roots, about 20 ml, is administered twice a day for urinary troubles till cure.

Notes from Classics

Juice of roots with milk is given as a remedy for spermatorrhoea. Juice of leaves is applied to parts which are inflamed due to application of the marking nut juice. Ripe and unripe fruits are eaten and roots after boiling are also consumed.

(Watt, 1889-1899)

Roots are stimulant, invigorating and purgative, also used for gonorrhoea and dysuria. Juice of roots with cumin and sugar is given in cold milk as a remedy for spermtorrhoea. Juice of leaves is applied to the parts inflamed by application of marking nut juice.

(Anonymous, 1946-1976)

Roots are stimulant, invigorating and purgative, used in dysuria and spermatorrhoea. Juice of leaves is applied on parts inflammed by marking nut juice. Roots, leaves and fruits are eaten.

(Ambasta, 1986)

Roots are useful in antifertility. Roots are tonic, used for spermatorrhoea, and for veterinary sore of small-pox. Leaves are cooling, useful for spermatorrhoea, fever and diabetes. Fruits are useful in fever. Seeds are used as antidote for snake-bite. Plant is cooling, used in ear-ache, promote labour-pains, paralysis and applied on sores. Plant is used against stomach-ache syphilis, unconsciousness. Fruits are edible. Fruits and leaves are used as vegetable.

(Jain, 1991).

Critical Notes

Utility of roots to treat urinary troubles is not yet on record.

Trichosanthes tricuspidata Lour.

Local Name

Gawlan

Ethnobotanical Notes

Medicine

1. Half spoon of pulp of fruit is given early morning to expel cough.
2. Half fruit is crushed and homogenised with few leaves of *Piper betel* L. It is consumed for dog-bite for human being as well as buffaloes.
3. Two seeds wrapped in *Piper betel* L. leaf is consumed orally twice a day to cure asthma.

Notes from Classics

Roots are used in the inflammation of lungs of cattle. Roots with colocynth roots are rubbed into paste and applied to carbuncles, also given internally for gonorrhoea.

(Watt, 1889-1899)

Roots with equal part of colocynth roots made into paste and applied to carbuncles. Roots are used as veterinary medicine for inflammation of lungs. Tubers mixed with equal part of roots of *Strychnos potatorum, Symplocos racemosa* and *Terminalia chebula* and rubbed in cow-urine were applied to tumours caused by plague. Fruits paste in coconut oil is applied on sore.

(Anonymous, 1946-1976)

Roots are useful on carbuncle. Roots are also used in veterinary medicine for inflammation of lungs.

(Ambasta, 1986)

Roots are useful for diarrhoea, fever, pneumonia rheumatism, vomiting and antidote of snake-bite. Leaves are applied on burns. Fruits are edible. Seeds are used for dysentery.

(Jain, 1991)

Critical Notes

Fruits are useful to expel cough and on dog-bite. Seeds are especially beneficial in case of asthma. These utilities have not been found elsewhere. The mature red fruits are sold in village markets for medicine especially in tribal areas.

Family: Cactaceae

Opuntia elatior Mill.

Local Name

Nagphan

Ethnobotanical Notes

Miscellaneous

1. Rigid spiny bushes planted on borders of cultivated fields or farmyard to prevent entry of cattle.
2. ***Food***: Ripe fruits are eaten by the cowherds and goatherds.

Notes from Classics

The prickly pear is much used in India as a hedge plant around fields and fences around the home steads.

(Watt, 1889-1899)

Fruits are eaten. Plant yields cordage fibre. Plant is fed to cattle during scarcity after burning of the spines.

(Anonymous, 1946-1976)

Ambasta (1986) reports similar properties and uses as those in Anonymous (*loc.cit.*).

(Ambasta, 1986)

Fruits are edible.

(Jain, 1991)

Critical Notes

It is planted as live hedge. It is exotic, native of tropical America.

Family: Apiaceae

Centella asiatica L.

Local Name

Bhrahmi

Ethnobotanical Notes

Medicine

1. Decoction of leaves is taken orally, one spoonful per day, to reduce fever due to sun-stroke.

Notes from Classics

Leaves are stimulant to the cutaneous circulation in skin disease, also used in tubercular leprosy and eczema. Leaves are also used in syphilis, chronic callos ulcer and rheumatism. Leaves are employed in diarrhoea, also used as emmenagogue in case of amenorrhea. The plant is largely used in brain affection and insanity.

(Watt, 1889-1899)

Leaves are used as nervice tonic, also used against epilepsy, insanity and hysteria. Dried leaves are useful for nervous break down. It is a cardiac tonic.

(Anonymous, 1946-1976)

Leaves are used as a diuretic and aperiant. Plant is diuretic and tonic used in leprosy.

(Ambasta, 1986)

Leaves are useful for blood dysentery, diarrhoea, cough, cold and gastric diseases. Leaves are used in the treatment of head-ache, insanity, stomach-ache, tuberculosis, liver nervine and respiratory diseases. Fruits are used in respiratory, nervine and skin diseases. Whole plant is anthelmintic, used for nervine, skin and urinary tract diseases. Plant is good brain tonic. Plant is edible, used in preparing local drinks.

(Jain, 1991)

Critical Notes

Leaves are found beneficial to treat sun-stroke and fever caused due to it. This use is not yet tapped.

Coriandrum sativum L.

Local Name

Kothambir, Kothmir

Ethnobotanical Notes

Medicine

1. Few drops of leaf juice are dropped in nose after every three hours to treat migraine.
2. Paste of leaves and common salt is used as a poultice on eyes for conjunctivitis and burning of eyes.

3. About half cup of decoction of leaves and cinnamon (*Cinnamonum zeylanica* Blum.) is given for about three months to cure menorrhagia.
4. Juice of leaves about half cup is advised to take orally twice a day for hysteria.
5. About 25 gm of green fruits are soaked in a glass of water, which is drunk at morning daily for fever till cure.

Notes from Classics

Fruits are carminative, diuretic, stomachic and antibilious in action. They are also refrigerant and aphrodisiac. Leaves are used in dyspepsia.

(Watt, 1989-99)

Extract of leaves is used as a eye wash for conjunctivitis. Fruits are carminative, refrigerant, diuretic and are good tonic. Fruits are used as aphrodisiac. Strong decoction of seeds with milk and sugar is given in bleeding piles. Bruished plant is a cooling application for head-ache.

(Anonymous, 1946-1976)

Leaves are aromatic and used as flavouring agent in coco, chocolate. Oil cake is used as fodder.

(Ambasta, 1986)

Leaves are useful for bleeding piles, colic and flatulence. Seeds are remedy on chicken-pox, stomach-ache and are tonic. Leaves and seeds are edible. Fruits are used as condiment.

(Jain, 1991)

Critical Notes

For the first time, the bark is observed useful to treat migraine, menorrhagia and hysteria. It is cultivated species for condiment. It it exotic, native of Mediterranean regions.

Daucus carota L.

Local Name

Gajar

Ethnobotanical Notes

Medicine

1. Leaf juice, full of cup, is given twice a day for bleeding piles till cure.
2. Juice of tubers is drunk for diarrhoea. About a cup of juice is taken 2-3 times a day till cure.
3. A glass full of juice of fruit is taken twice a day for controlling polyurea.
4. Leaf paste is consumed two gm per day for blood vomiting till cure.

Notes from Classics

Root is used as tonic. Ointment of root is used to applied on burns and scalds. Poultice of root is used in chronic ulcers, also used against dysentery and enlargement of spleen. Decoction of leaves and seeds stimulate uterus. Fruits are aromatic, stimulant and carminative. They are used in diseases of kidney, flatulence, colic pains and dropsy. Seeds are nervine tonic and bring about abortion.

(Watt, 1889-1899)

Infusion of root is considered diuretic, deobstruent and stimulant. Fruits yield both volatile and fatty oils.

(Ambasta, 1986)

Tubers are folk remedy for thread worms. It helps to increase quantity of urine and elimination of uric acid. Tubers are boiled and given to cattle for making them fat. Leaves and seeds are useful to facilitate child-birth. They are efficacious in tooth-ache.

(Jain, 1991)

Critical Notes

Medicinal use of leaves for bleeding piles and vomiting are hitherto unrecorded. Utility of tubers for controlling diarrhoea and polyurea is likewise unrecorded. It is cultivated species. It is exotic, native of Europe, N.Africa and Temperate Asia.

Trachyspermum ammi (L.) Spr.

Local Name

Ajwan, Owa

Ethnobotanical Notes

Medicine

1. Fruits are chewed as remedy for better digestion and cleaning mouth.
2. Fruits are smoked after wrapping in paper for cold and cough.
3. Mother chews the fruits thoroughly. It is given along with saliva to her kid for better digestion.

Notes from Classics

Aqueous solutions left after separation of essential oil is called 'omum water'. It is used in flatulence and grip.

(Watt, 1889-1899)

Hot and dry fomentation of fruit on the chest is a common remedy for asthma.

(Anonymous, 1946-1976)

Roots are carminative, diuretic, used in febrile condations and stomach disorders. Fruits are stimulant, antispasmodic, tonic, carminative, administered in flatulence, atonic dyspesia, diarrhoea and cholera. Fruits are used for relaxed sore throat and bronchitis and cough. Paste of crushed fruits is applied to the abdomen externally for colic. It shows antibiotic activity and is used in lotions, applied for checking chronic discharge. Ajawan oil is an aromatic, carminative and antiseptic, expectorant in emphysema, bronchial pneumonia and other respirator ailments. Fruit yield fatty oil which is used externally on rheumatic swellings. It is also used in soaps. Oil cake is used as seed and fertilizer.

(Ambasta, 1986)

Critical Notes

Ajawan seeds (or fruits) are available in local markets. It is also cultivated in this district. They are used for better digestion of adults and kids as well. Smoking them to treat cough and cold is a common practice especially in rural areas. It is exotic and native of Africa.

Family: Alangiaceae

Alangium salvifolium (L.*f.*) Wang.

Local Name

Ankok, Akali

Ethnobotanical Notes

Miscellaneous

1. Fruits are edible.
2. Dried branches are used for fencing farms.
3. Fresh stem pieces are used as tooth brushes.
4. Wood is used for house construction and agricultural implements.

Notes from Classics

Roots are laxative and useful in worms, colic, inflammations and poisonous bites. Root bark is useful in leprosy. Oil of root bark is useful in external applications in acute rheumatism. Root bark is anthelmintic, purgative, febrifuge and emetic. It is useful in leprosy and skin disease and also employed in dropsical cases. Root bark is reputed antidote in snake bites. It is diaphoretic, alterative, tonic and antipyretic. It is used in cases of bites from rabid animals. Fruit is said to be cooling, tonic, nutritive, useful in burning of body and in haemorrhages.

(Watt, 1889-1899)

It is suitable for inlaying and carving wood is also useful as fuel. Fruits are edible.

(Anonymous, 1946-1976)

Fruits are edible, relished by children. Seeds yield fatty oil which is used as an illuminant. Wood used for pestles, for oil mills, agricultural implements, house construction, musical instruments, for ornamental and cabinet work.

(Ambasta, 1986)

Roots are used to cure fever, skin diseases, poisonous bites of dogs and jackals. Stem bark is useful to control blood pressure. Stem is utilized for abortions and to cure menstrual complaints. Leaves are applied externally for boils, rheumatism and skin diseases. It is also used as antidote for poisons. Fruits are useful in eye and stomach complaints. Fruits are eaten and used on skin diseases. Plant is used to cure anasarca, asthma, bronchitis, cholera, colic, dropsy, fistula, phthisis, snake bites and skin complaints. Wood is used for making musical instruments.

(Jain, 1991)

Family: Rubiaceae

Catunaregam spinosa (Thunb.) Tirvengadum

Local Name

Gela, Gel

Ethnobotanical Notes

Medicine

1. Half cup of fruit extract is boiled and then mixed in spoonful ground nut oil once daily to cure fever till cure.
2. Spoonful of fruit extract is mixed in ground-nut oil. Few drops are dropped in nose at morning to control asthma.
3. Powder of dry fruit is mixed in oil and applied to legs for massage in case of paralysis till cure.
4. Spoonful of fruit pulp is consumed for vomiting in case of food poisoning.

Miscellaneous

1. Fruit pulp is used as fishing poison.
2. Leaves are placed inside food grains containers.
3. ***Fodder***: Plant is browsed by goats and is lopped for fodder.
4. Dried branches are used for making fence.

Notes from Classics

Bark is given internally and applied externally for the bone-ache during fever. Bark is mixed in cow-dung and applied to bruises to releave the pain. Bark is astringent, used in diarrhoea and dysentery. Rind and fruit pulp is used as emetic. Pulp of fruit is emetic in small dose. It is nauseant, expectorant, diaphoretic, anthelmintic and abortifacient. Powdered pulp is applied to tongue and poultice for fever, also for incidental ailments for children while teething. Externally, fruit pulp is applied to rheumatic swellings, abscess and other cutaneous affections. In colic, fruit is rubbed into paste with rice water over navel to relieve pains. It is nervine calamative and antispasmodic.

(Watt, 1889-1899)

Bark is used in dysentery and diarrhoed. Infusion of bark is emetic, abortifacient, sedative and externally applied in rheumatism. Fruit pulp is emetic in small doses. It is nauseant expectotorant, diaphoretic, anthelmintic and abortifacient. Fruit is used as nervine calamative and antispasmodic. Unripe fruits and roots are used as fish poison. Unripe fruits are used in soaps. Fruits extract exhibit insecticidal and insect repellent properties. Seeds are used as tonic.

(Anonymous, 1946-1976)

Roots are bitter tonic, used against cancer and scrofula. They are externally applied to ulcer, boils and abscess. Leaves are astringent antisyphilitic, used in scrofula and herpes. Flower buds are tonic, diuretic and sedative. Fruits are considered cooling, demulcent, used in the treatment of urinogenital diseases. Herb is diaphoretic, diuretic, emolliant and sedative. Decoction of plant is used in chronic malaria, leucorrhoea and urinary diseases. Seeds are used for inflammatory swellings. Seed oil is used for herpes. It is used for making soaps and paints.

(Ambasta, 1986)

Roots used in the treatment of ear-diseases, gonorrhoea and malaria. Stem bark is prescribed for rheumatism and wounds. Fruits are used for amebiasis, boils and dysentery. They are emetic. Seeds and fruits are edible. Fruit and leaves are used as vegetable. Fruit powder is applied to store grain to protect from pest. Seeds are used for stupefying fish. Wood is used for making combs and smoking pipes.

(Jain, 1991)

Critical Notes

Fruits are employed in asthma, paralysis and food poisoning. These are unreported earlier. The miscellaneous uses except fish-poison are also not reported.

Haldina cordifolia (Roxb.) Ridsd.

Local Name

Haldu, Haldwan.

Ethnobotanical Notes

Medicine

1. A spoonful of leaf juice is taken orally twice a day to cure cough till cure.

Miscellaneous

1. Wood is valuable timber, used for house construction, especially important for door and poles.
2. Leaves are used as fodder for cattle.

Notes from Classics

Roots are used as medicine. Small bud ground with pepper are snuffed in to nose for head-ache.

(Watt, 1889-1899)

The bark is febrifuge and antiseptic. Wood is best timber suitable for flooring and for paneling railway carriages, also for cigar boxes, toys, combs and agricultural implements.

(Anonymous, 1946-1976)

Bark is febrifuge. Leaves are used as a fodder. Timber is easy to saw, seasons well and durable. Wood takes good polish. Wood pulp is used for making wrapping, writing and printing paper. Spent bark is used for boards. Wood is accepted as grade I commercial and moisture proof plywood timber. Leaves are used as fodder.

(Ambasta, 1986)

Bark is antiseptic, used against fever. It is used to remove scars on skin and yellowing of body, also prescribed for urine complaints. Bark is applied on sores to expels worms. Stem is a remedy on eye complaints. Stem is also used as a veterinary medicine. Leaves are used in the treatment of cold, cough and head-ache, also applied on wounds. All plant parts are used to cure night blindness, carbuncle and cholera. Wood is used for making combs and tools. Women wear leaves in friendship ceremony.

(Jain, 1991)

Critical Notes

The tree is a source of important timber. It is largely used in construction of houses and huts especially in tribal areas. The number of trees have been considerably reduced. Leaves are also useful as cattle fodder. It should find place in afforestation program of forest department.

Ixora pavetta Andrew

Local Name

Lokhandi

Ethnobotanical Notes

Miscellaneous

1. ***Fodder***: Leaves are used as fodder for cattle.
2. Wood is used as fuel. Wood is used in construction of huts.

Notes from Classics

Santhals eat ripe fruits. Leaves are used as fodder for buffaloes. Wood is used as furniture and building purposes.

(Watt, 1889-1899)

Leaves are used as fodder. Fruits are eaten by santhals. Wood is used as fuel as well as furniture and building purposes.

(Anonymous, 1946-1976)

Decoction of bark is given in debility and anaemia.

(Ambasta, 1986)

Fruits are edible. Wood is used for basketry and walking sticks.

(Jain, 1991)

Meyna laxifora Robins

Local Name

Awli

Ethnobotanical Notes

Miscellaneous

1. ***Fodder***: Leaves are used as fodder for domestic animals.

2. *Food*: Fruits are edible.
3. Stem twigs are used as tooth-brush.

Notes from Classics

Leaves offer poor quality of fodder. Powdered leaves are used in diptheria. Fruits and leaves are eaten. Dry fruits are narcotic, also used for dysentery and boils. Seed yields fats.

(Anonymous, 1946-1976)

Ambasta (1986) documented similar properties as those in Anonymous (*loc. cit.*).

Fruits are edible.

(Jain, 1991)

Critical Notes

Stem pieces are used to clean tooth. This use is yet to be documented.

Mitragyna parvifolia (Roxb.) Korth.

Local Name

Kalam, Jangli hirmundi

Ethnobotanical Notes

Medicine

1. Paste of bark is applied on tumour regularly to suppress it.
2. Leaves are applied as poultice on rheumatic swellings.

Miscellaneous

1. Wood is used for the agricultural implements, employed for construction of houses and huts and also for cradle of ghost.
2. Leaves are fed to domestic animals.

Notes from Classics

Bark and roots are used for fever and colic. Timber is used to make planks, rafters, furniture match-box etc. Bark yields fibre.

(Anonymous, 1946-1976)

Trees are lopped for fodder.

(Ambasta, 1986)

Bark is used as contraceptive. It cures fever, ear-ache, leucorrhea, muscle pains, stomach-ache and syphilis. It is used for cholera, carbuncle, fever and sores. From bark fibre is obtained.

(Jain, 1991)

Critical Notes

Uses of bark to suppress tumours on body and application of leaves specially for rheumatic swellings are noteworthy.

Morinda pubescens J.E.Sm.

Local Name

Bartondi, Jangli Ramphal

Ethnobotanical Notes

Miscellaneous

1. Ripe fruits are edible.

Notes from Classics

Green fruits are picked up by Hindus and eaten with curries. Fruits are edible.

(Watt, 1889-1899)

Root bark yields dye. Wood is used for making plates and dishes.

(Anonymous, 1946-1976)

Ambasta (1886) reported similar uses as those in Anonymous (*loc.cit.*)

From wood musical instruments are made.

(Jain, 1991)

Spermacoce hispida L.

Local Name

Phangoda

Ethnobotanical Notes

Medicine

1. Fresh flowers are collected and made into garland. It is placed around the neck of a patient's maternal uncle to improve suppurating body tumours.

Notes from Classics

Roots are alternative. Leaves are eaten as vegetable.

(Watt, 1889-1899)

Leaves are used as fodder, also consumed as a vegetable in times of scarcity. Extract of leaves is given for haemorrhoids and gall stones. Seeds are demulcent, used in diarrhoea and dysentery.

(Anonymous, 1946-1976)

Ambasta (1986) reported similar uses and properties as those in Anonymous (*loc.cit.*).

Roots are used as an antidote to snake-bite, for antifertility, to prevent bleeding during child-birth, to check miscarriage, to treat gum trouble and tooth ache. Leaves are useful in scabies, eye and stomach complaints, applied on ulcer and wounds. Leaves are used as vegetable. Whole plant is used for body-ache and swelling. Seeds are stimulant.

(Jain, 1991)

Critical Notes

Flowers are beneficial in treating body tumours. This is not reported in classical literature.

Spermacoce pussila Wall.

Local Name

Rumne

Ethnobotanical Notes

Medicine

1. A spoonful of root decoction is administered daily for three months to cure asthma.

Critical Notes

The plant is unrecorded as useful. For the first time, utility of roots to cure asthma is being reported.

Family: Asteraceae

Caesulia axillaris Roxb.

Ethnobotanical Notes

Miscellaneous

1. *Food*: Leaves are used as vegetable

Notes from Classics

Whole plant is used as vegetable.

(Jain, 1991)

Echinops echinatus Roxb.

Local Name

Udkata

Ethnobotanical Notes

Medicine

1. Paste of roots is applied onto joints for rheumatic swellings.
2. Paste of roots is applied locally on itch and eczema.
3. A cup of decoction of entire plant, about 10-15 ml, is drunk twice daily for cough and cold till cure.

Veterinary Medicine

1. Fumes obtained from the entire plant are useful for cattle, in case they are reluctant to eat fodder.

Notes from Classics

Roots are applied to destroy maggots on wounds of cattle. Roots are mixed with *Acacia* gum and applied to hair to destroy lice. Plant is bitter tonic, diuretic, used in cough, hysteria, dyspepsia, scrofula and opthalmic disorders.

(Anonymous, 1946-1976)

Ambasta (1986) recorded uses similar to those in Anonymous (*loc.cit.*)

(Ambasta, 1986)

Roots are prescribed for bone-ache and skin diseases. Root paste is applied on wounds of animal for healing. Roots also facilitate child-birth. Whole plant controls fever and malaria.

(Jain, 1991)

Critical Notes

Medicinal utility of roots for rheumatic swellings and eczema have been recorded for the first time. Likewise, its application in the form of fumes for pet animals when they are reluctant to eat fodder is unrecorded in past literature.

Emilia sonchifolia (L.) DC. ex Wight

Local Name

Pathri

Ethnobotanical Notes

Miscellaneous

1. ***Fodder***: Entire plant are used as fodder for all domestic animals.

Notes from Classics

Cauline leaves are cooked and eaten as vegetable.

(Anonymous, 1946-1976)

Ambasta (1986) reported uses similar to those of Anonymous (*loc. cit.*).

Leaves are used in salad.

(Jain, 1991)

Critical Notes

Being a weed of cropped fields and also waste lands and bandhs of fields is removed manually by farmers. It is cherished by domestic animals. Thus it is a useful weed as fodder and being noted as such for the first time. It is Afro-Asian element, naturalized in this district.

Glossocardia bosvellea (L.*f.*) DC.

Local Name

Mirgi

Ethnobotanical Notes

Medicine

1. Few drops of juice of leaves are dropped in ears for ear-ache.
2. A spoonful of root extract is administered twice daily for five days in the treatment of epileptic fits.

Notes from Classics

Plant is employed for female complaints. Leaves are eaten as green vegetable.

(Watt, 1889-1899)

Leaves are used as emmernogogue. They are bitter in taste, eaten as a vegetable.

(Anonymous, 1946-1976)

Ambasta (1986) documented similar uses those in Anonymous (*loc. cit.*).

Critical Notes

Utility of leaves for ear-ache and of roots to treat epileptic fits is not noted in past. It is exotic, native of East Indies.

Sonchus oleraceus L.

Local Name

Mhatara, Pathari

Ethnobotanical Notes

Miscellaneous

1. Entire plant is used as fodder for pet animals.

Notes from Classics

Cattle are very fond of it.

(Watt, 1889-1899)

Plant is used as fodder for cattle, sheep and goats. Plant is used as vegetable or as a salad.

(Anonymous, 1946-1976)

Herb is consumed as a salad and vegetable, also used as fodder for sheep, goats and cattle.

(Ambasta, 1986)

Leaves are galactogogue. It is a tonic.

(Jain, 1991)

Critical Notes

It is a exotic weed of waste and cultivated lands. It is native of Europe.

Sphaeranthus senegalensis DC.

Local Name

Gorakhmundi

Ethnobotanical Notes

Miscellaneous

1. Leaves are used as vegetable after boiling.

Notes from Classics

Fibre is obtained from bark, used for making ropes and bow string. Seed oil is used in cooking and anointing. Decoction of roots is used in chest pain.

(Watt, 1889-1899)

Bark is mixed in whey and used on piles. Leaf juice is boiled with milk and given in cough. Leaves are also eaten as pot herbs. Flower is alternative, depurative and tonic. Seeds are anthelmintic. Plant is styptic, used in liver, gastric disorders and applied on itch. Plant is used as fish poison. It is also stuffed into holes of crab to kill them.

(Anonymous, 1946-1976)

Herb is used as fish poison. It is also stuffed into hole of crab to kill them.

(Ambasta, 1986)

Flowers are used as vegetable. Fruits are used as fish poison. Whole plant is edible.

(Jain, 1991)

Spilanthus calva DC.

Local Name

Akkalkara

Ethnobotanical Notes

Medicine

1. Leaf paste is applied on itching.
2. Flowers heads are chewed for cleaning of mouth and throat infection.

Notes from Classics

Flower head and whole plant cause salivation. They are powerful stimulant and sialogogue, used in head-ache paralysis of tongue, affection of throat, gums and tooth-ache.

(Watt, 1889-1899)

A tincture of herb is used against tooth-ache, scurvy and gum trouble. Internally tincture and flower head are used for gout and bladder pains.

(Anonymous, 1946-1976)

Roots are purgative. Pungent flower are used for relief in throat affection and paralysis of tongue. Tincture of capitula act as a sialogogue and stimulant, used in caries and inflammation of jaw bones. Herbs are used in dysentery, scabies and psoriasis. Decoction of herb is diuretic. Plant is used as fish-poison.

(Ambasta, 1986)

Roots are used for cough and diarrhoea. Leaf powder is applied on scabies. Flowers are as anaesthetic and as remedy of tooth-ache.

(Jain, 1991)

Critical Notes

Leaves are applied to treat itching. This is not reported as such earlier. The plant is sometimes cultivated in house gardens.

Tridax procumbens L.

Local Name

Ed-dandi

Ethnobotanical Notes

Medicine

1. Juice of leaves is applied on fresh cuts immediately to avoid infection.
2. Paste of leaves is applied in case of eczema and boils.
3. Spoonful of leaf extract is given at every morning for fifteen days to remove urinary bladder stone.

Notes from Classics

Leaves are employed in bronchial catarrah, dyspepsia, diarrhoea and for restoring hair. Leaf juice possess antiseptic, insecticidal and parasitcidal properties. It is used to check bleeding from cuts and wounds.

(Anonymous, 1946-1976)

Ambasta (1986) reported similar uses as those in Anonymous (*loc. cit.*).

(Ambasta, 1986)

Roots are employed to cure fever and skin diseases. Leaves are useful in bronchial catarrah, diarrhoea, dysentery, eye diseases, stomach-ache, to remove urine stone and veterinary wound. Leaves are applied externally for leprosy, scorpion bite, sores and tooth-ache.

(Jain, 1991)

Critical Notes

It is a common weed of waste lands and cultivated fields. It is exotic, native of South America.

Vernonia anthelmintica (L.) Willd.

Local Name

Kala-jeera

Ethnobotanical Notes

Medicine

1. Paste of achenes prepared in lemon juice is applied for scabies.
2. Spoonful of leaf extract is taken regularly once a day for habitual constipation.

Notes from Classics

Achenes are remedy for leprosy and skin diseases. They are anthelmintic. Powdered seeds with sesamum oil (*Sesamum orientale* L.) are used for perspiration. Seed poultice is used for dispersing tumours. Infusion of seeds is used for cough, flatulence and anasarca. It is also used as antidote for snake bite. Seeds are used to remove worms in intestine.

(Watt, 1889-1899)

Achenes are diuretic, antiseptic and stimulant. Effect of achenes on the skin is described in indigenous medical literature.

(Anonymous, 1946-1976)

Achenes are anthelmintic and effective against worms.

(Ambasta, 1986)

Roots give relief in pains. Leaves are anthelmintic and cure paralysis and swellings of body. Seeds reduce acidity. It is given in the treatment of asthma, cough and lecoderma. They are diuretic and febrifuge. Plant is used for cholera, constipation and syphilis.

(Jain, 1991)

Critical Notes

Leaves are being reported as additional use especially for habitual constipation.

Vernonia cinerea Less

Local Name

Sahdevi

Ethnobotanical Notes

Medicine

1. Decoction of roots, about half cup, is given once a day for diarrhoea.

2. Roots are wrapped in black cloth and tied on wrist for seven days to cure high-temperature in fever.

Notes from Classics

Decoction of flowers promote perspiration in febrile affection. They are also useful for conjunctivitis. Seeds are alexipharmic and anthelmintic. Whole plant is given as remedy for spasma of bladder and strangury. Leaves are useful as a pot herb.

(Watt, 1889-1899)

Roots are anthelmintic, used for diarrhoea and stomach-ache cough and cold. Leaf juice is given in amoebiasis. Poultice of leaves is used against herpes, eczema, ring-worm and guinea-worm. The juice is boiled with oil and used for elephantiasis Seeds are anthelmintic and alexipharmic, effective against round-worm and thread-worm. Seeds are given in cough, flatulence, intestinal colic, dysuria, leucoderma, psoriasis and skin diseases. Seeds with lime juice destroy pediculi. Flowers are used in conjunctivitis, fever and rheumatism. Decoction of plant is used in malaria, promote perspiration in febrile condition. Juice of plant is given to children for incontinence of urine, to cattle for swollen throat and indigestion.

(Anonymous, 1948-76)

Ambasta (1986) documented similar properties as those given in Anonymous (*loc.cit.*).

Roots are used for leucorrhoea. Leaves are used in constipation and malaria. Seeds are used against the treatment of skin diseases and thread worms. Plant are employed in cholera, dysentery, fever, impotency, lactation, night-blindness, piles, spleen-complaints and wounds.

(Jain, 1991)

Critical Notes

Roots are being reported for the first time useful to treat fever.

Xanthium strumarium L.

Local Name

Landga, Kutri

Ethnobotanical Notes

Medicine

1. Half cup of decoction of roots is drunk once a day for a week to cure body-ache.
2. Few drops of leaf juice are dropped in eyes to cure cataract.

Notes from Classics

Prickly involucre is applied to the ears to cure hemicrania. Fruit is efficacious in small-pox, irritability of bladder, gleet, leucorrhoea and menorrhagia. Fruit is burnt and applied to sore on the lips and mucous membrane of the mouth.

(Watt, 1889-1899)

Bark is astringent, given in diarrhoea and dysentery. Infusion of bark is emetic and abortifacient. Bark is sedative and given to relieve pain due to bruises and bone ache. It is useful during fever and used for rheumatism. Seeds are used to increase appetite. Fruits are cooling and emetic. In small doses, fruits are expectorant and diaphoretic. Unripe fruits and roots are used as fish poison. Wood is used for making walking sticks and umbrella handle. Plant is broused by goat and lopped for fodder.

(Anonymous, 1946-1976)

Ambasta (1986) reported similar properties as those in Anonymous (1946-1976).

Roots cure boils and ulcers. Leaves are used to treat herpes, malaria and helps in expelling worms. They also cure scrofula. Fruits are cooling and cure eye diseases. Fruits are useful for head-ache. Garland of fruits cure night blindness. Seeds are remedy on cancer, wound and head-ache. Fumes of seeds cure piles. Seed oil is applied for rheumatism.

(Jain, 1991)

Critical Notes

Roots are medicinally important for body-ache, and leaves are so to cataract. These uses have remained unreported elsewhere. The species is exotic, noxius weed of waste and cultivated lands. However, it is used as stated above. It is native of south America.

Family: Primulaceae

Anagallis arvensis Hook. *f.*

Local Name

Jokmari

Ethnobotanical Notes

Miscellaneous

1. Leaf extract is diluted in water and then spread over crops in fields to repel insects.

Notes from Classics

Plant is used to intoxicate fish and to expel leeches from the nostrils of cattle. It is used in cerebral affections, leprosy, hydrophobia, dropsy, epilepsy and mania. Plant is also used for inflammation of stomach and said to be poisonous to dogs.

(Watt, 1889-1899)

Plant intoxicate fish and expels leeches from nostrilis of cattle. It is poisonous to dog.

(Anonymous, 1946-1976)

Ambasta (1986) reported uses similarly as in Anonymous (*loc.cit.*).

Critical Notes

To protect crops from pests, the plant is being reported useful for the first time. The uses documented also suggests its poisonous nature. It may come up as a good biological control if used judiciously.

Family: Sapotaceae

Madhuca longifolia (Koen.) Macbr. var.*latitolia* (Roxb.) Che.

Local Name

Mahu

Ethnobotanical Notes

Medicine

1. Half spoon of seed powder in a cup of milk if drunk twice a day cures throat infection.

2. Pinch of seed powder is snuffed once a day for a week to cure hysteria.
3. Cup of decoction of flowers is consumed once a day for a week to cure dysuria.
4. Seed oil is applied daily on sole of feet to remove cracks.

Miscellaneous

1. ***Food***: Oil is extracted from seeds and used as edible oil.
2. Flowers are crushed in Sorghum flour and bread is prepared from it.
3. Dining plates and bowls are made from the leaves.

Notes from Classics

Bark is an astringent and tonic. Bark is also used for rheumatic affection. Flowers are used in cough. Flowers are mixed in milk and used in impotency. Spirits distilled from the flowers is astringent, tonic and appetizing. Plants are stimulant, demulcent and emollient. Seed oil is used in the treatment of skin diseases. The oil cake is employed as an emetic.

(Watt, 1889-1899)

Wood is used for building purposes, beams, doors window frame, post, furniture, boat, bridge well construction and agricultural implements. Flowers are eaten raw or cooked. Flowers are used in preparation of distilled liquor. They are cooling, demulcent and tonic, used in cough, cold and bronchitis and show the antibacterial properties. Honey of flowers is edible and used for eyes diseases. Seed are galactogogue. Seed oil has emollient properties. It is used in skin diseases, rheumatism, head-ache. Seed oil is laxative emetic and used in habitual constipation, pile and haemorrhoids. Oil cake has insecticidal and pesticidal properties, applied to lawns and golf grasses. It is used with 'shikekia' as a hair-wash.

(Anonymous, 1946-1976)

Ambasta (1986) described uses similar as in Anonymous (*loc.cit.*).

Bark is antidote to snake bite and is astringent, used for diabetes, gum trouble and itch. Bath in bark-added water is useful to treat jaundice. This water is prescribed for rheumatism. Bark is used in tooth-ache and it is a tonic also used for ulcer. Leaves in form of paste is applied on burns. Flowers are administered for bronchitis, colic and cough and antidote for scorpion bite. Flowers are edible, used for making alcoholic drink. Seeds are used as antidote on snake bite. Oil is employed for pneumonia and chest pain. Seeds are also edible. Seed oil is edible. It is also used for lumination.

(Jain, 1991)

Critical Notes

Use of seed powder for throat infection and hysteria, seed oil to treat crack on feet and flowers for curing dysuria are hitherto unreported uses. The leaves are also found for the first time useful as for making dining plates.

Manilkara hexandra (Roxb.) Dub.

Local Name

Khirni

Ethnobotanical Notes

Miscellaneous

1. ***Food***: Ripe fruits are edible.
2. ***Fodder***: Leaves are used for cattle fodder.
3. ***Wood***: Wood is used for agricultural implements.

Notes from Classics

Dried fruits are eaten by Hindus on fast day. Poor class eat it mixed with whey.

(Watt, 1889-1899)

Fruits are eaten fresh. They are sweet and astringent. Leaves are used as cattle fodder. Tree yield gum used in fever and as tonic. Wood is used for sugar mills, oil press, post beams, cart rollers, furniture, agricultural implements, tool handle etc.

(Anonymous, 1946-1976)

Fruits are eaten fresh or dried. Leaves are used as cattle fodder. Timber is used for agricultural implements, railway sleepers, tool handles, turnery, pens, pencils etc. Seed yield edible oil.

(Ambasta, 1986)

Fruit are edible.

(Jain, 1991)

Family: Ebenaceae

Diospyros melanoxylon Roxb.

Local Name

Tembhurni, Tendu

Ethnobotanical Notes

Medicine

1. Ripe fruits along with fruits of *Ficus religosa* L. and seeds of *Syzygium cumini* (L.) Skeels are boiled in milk. This mixture, about 10 ml daily at morning, is taken orally in the treatment of back-ache and leucorrhoea till cure.
2. Extract of fruit is applied over fistula.
3. Fruit paste is used as face mask to improve skin fairness.
4. Flower extract, about half cup, is given for eight days for anurri.

Miscellaneous

1. ***Food***: Ripe fruits are edible.
2. Leaves are used for bidi-wrapping by the tribals.

Notes from Classics

Bark is astringent, used in diarrhoea and dysentery. Bark is a good tonic, used as lotion of eyes. Powder of bark and pepper is given in dysentery.

(Watt, 1889-1899)

Bark is astringent, used in diarrhoea and dyspepsia. Dilute extract of bark is used in lotion for eyes.

(Anonymous, 1946-1976)

Bark and fruits are edible. Fruits are germicidal and applied as paste on skin diseases. Fruits also cure urine complaints. It is a source of coromandal ebony.

(Jain, 1991)

Leaves are esteemed for wrapping bidies. Fruits are edible, laxative, carminative and styptic. Dried flowers are used in urinary and skin troubles.

(Ambasta, 1986)

Critical Notes

Fruits are medicinally useful for back-ache, leucorrhoea, fistula and annuria. These utilities have not been yet noticed elsewhere. Due to illicit felling of trees and over-exploitation for various purposes healthy trees are rarely observed. Moreover, in afforestation programmes it did not find place to date. To raise its captive plantation is necessary to boost the economy of tribals.

Family: Oleaceae

Nyctanthes arbor-tristis L.

Local Name

Parijat

Ethnobotanical Notes

Medicine

1. Spoonful of stem bark powder is consumed twice as day in the treatment of asthma till cure.
2. Spoonful of paste of young leaves is consumed once a day febrifuge for three days.

Notes from Classics

Leaves are used in fever and rheumatism. Fresh juice of leaves is given with honey in fever. Decoction of leaves is given for obstinate sciatica. Six to seven leaves are given for obstinate fever. Seed powder cures scurfy. Seeds are antibilious and expectorant.

(Watt, 1889-1899)

Leaf extract is cholagogue, laxative, diaphoretic and diuretic. It is given to kids for expulsion of round and thread worms. Powdered seeds are useful for scurfy affection of the scalp.

(Anonymous, 1946-1976)

Corolla tubes were formerly used for dying silk. Flowers yield essential oil.

(Ambasta, 1986)

Bark is used for bone fracture and internal injuries. Leaves are anthelmintic, employed for cough, fever, hiccup, malaria, rheumatism, sciatica and snake-bite. Leaves are used in curries. Inflorescence is used for cough. Seeds are edible and seed oil is applied on scalp for scurfy affection. Plant is useful in the treatment on cancer, dysentery, indigestion, manorrhoea and snake-bite. Tannin is yellow dye obtained from bark. Flowers are used, an hairdo as ornament. Twigs are used in basketary.

(Jain, 1991)

Critical Notes

Stem bark is beneficial in treating asthma, a use await publication.

Schrebera swietenoides Roxb.

Local Name

Mekha, Nakti

Ethnobotanical Notes

Medicine

1. Bark powder is applied externally for boils.
2. Fruit paste is applied externally around ears for ear-ache.

Miscellaneous

1. ***Food***: Leaves are used as vegetable.
2. ***Wood***: Wood is used as fuel.

Notes from Classics

Leaves are used as vegetable. Wood is used for making combs, cart wheels and mathematical instruments.

(Watt, 1889-1899)

Roots are used in the treatment of leprosy. Bark is used for burns and boils. Leaves possess stomachic properties and used in the treatment of urinary discharge and enlargement of spleen. Leaves are crushed and employed for healing worms in wounds of cattle. Leaves are edible. Fruits curves hydrocele.

(Anonymous, 1946-1976)

Roots are employed in leprosy, also useful to expel worms from animals. Bark is administered for cough and mental depression.

Leaves are used in the treatment of tooth-ache. Fruit is employed for hydrocele.

(Jain, 1991)

Leaves are used to cure enlargement of spleen and urinary discharge. Wood is used for making rail shuttles, utensils, poles, wheels and mathematical instruments, fuel and cabinet work.

(Ambasta, 1986)

Critical Notes

Generally foliage of very few trees species have been reported useful as vegetable in past. The observation in Jalgaon district and those noted in classical literature by Watt (*loc.cit.*) and Anonymous (*loc.cit.*) appear warranting for scrutiny of leaves of this tree species from nutritional point of view. Use of fruit against ear-ache is not reported.

Family: Salvadoraceae

Salvadora persica L.

Local Name

Pelu, Pilu

Ethnobotanical Notes

Medicine

1. Stem pieces are used as tooth-brushes.
2. Powder of stem bark is used to strengthen gums and to remove foul smell.

Notes from Classics

Roots and berries are used as vesicant. Decoction of leaves is used as purgative to horses. Fruits are aphrodisiac and edible. Seed oil is applied in painful rheumatic affection and also after child-birth.

(Watt, 1889-1899)

Root bark is vesicant. Decoction of root is used against gonorrhoea and vesical catarrah. Extract of the roots relieves pain due to spleen troubles. Decoction of root-bark is tonic, used in fever and as an emmenogogue. Stem-bark is used an ascarifuge and for gastric troubles. Leaves are antiscrobutic and astringent. Decoction of leaves is used in asthma and cough. Poultice of leaves is applied to painful tumours, piles and in case of rheumatism. Fruits possess deobstruent, carminative, diuretic, lithontriptic and stomachic properties, are used in biliousness and rheumatism. Seeds are purgative, diuretic and tonic. Seed oil is applied on the skin in rheumatism.

(Anonymous, 1946-1976)

Leaves are eaten as vegetable, also used as fodder for domestic animals. Fruits are edible, used in sweets and in preparation of fermented drinks.

(Ambasta, 1986)

Roots are used to expel ring-worms. Stem is vermifuge. Bark is used for pain in back and limb. Leaves are used to treat asthma and cough. It is purgative. Fruits are used in constipation. Fruits are edible. Seeds are used for rheumatism.

(Jain, 1991)

Critical Notes

Stem is useful as tooth-brush. Bark is employed to strengthen gums and to remove foul smell. These uses are not documented earlier.

Family: Apocynaceae

Carissa carandas L.

Local Name

Karwand, Dongrachi Kali-Maina

Ethnobotanical Notes

Medicine

1. Emulsion of roots is mixed with citrus juice [*Citrus aurantifolia* (Christm.) SW.] and camphor and applied on boils of kids.
2. Decoction of spines is drunk, about half cup twice a day for a week, in case of excessive bleeding in menstrual period.

3. Unripe fruits are used in dysentery. Five to ten fruits are eaten daily till cure.
4. In fever unripe fruit paste is applied on body to lower down the body temperature.

Miscellaneous

1. ***Food***: Ripe fruits are edible. Tribals collect them and sold in local market.
2. Pickles are prepared from fruits.
3. Shrub is spiny so used as live hedge around farms and farmyards.

Notes from Classics

Roots are stomachic. Unripe fruits are astringent, used in bilious complaints. Ripe fruits are good expectorant and edible.

(Watt, 1889-1899)

Roots are stomachic and anthelmintic. Decoction of leaves is used in fever. Whole plant is used for fencing. Wood is used for spoons and combs.

(Anonymous, 1946-1976)

Fruits are sweet, edible, suitable for tarts, pudding and jellies. Unripe fruits are pickled. Plant is planted for fence.

(Ambasta, 1986)

Root paste is applied on wounds for healing. Roots are stomachic and anthelmintic. Root bark is stomachic and anthelmintic. Root bark is useful to expel worms from wounds of animals. Unripe fruits are astringent. Fruits are useful in bilious complaints. They are good expectorant and edible. Plant is used for anasarca, bite of rabid dog, for convulsions, dropsy and epilepsy. It is good remedy on madness, post-natal care, rheumatism and sores.

(Jain, 1991)

Critical Notes

Roots are important for boils, spines are so for controlling excessive bleeding at the time of menstruation. Unripe fruits are medicinally useful for dysentery and to bring down body temperature during fever. These uses have not been recorded earlier. Ripe fruits are collected by the local people and sold in villages and towns of this district.

Holarrhena pubescens (Buch.-Ham.) Wall. ex G. Don

Local Name

Dahi kadi, Kuwadi, Kuda

Ethnobotanical Notes

Medicine

1. Decoction of bark is used to gargle in case of tooth-ache.
2. Spoonful of bark powder is consumed twice a day for five to six days for malarial fever.
3. Spoonful of leaf extract is drunk once a day for two days to stop dysentery.

Miscellaneous

1. Wood is used as fuel.
2. Latex is used to curdle milk.

Notes from Classics

Root is used in dysentery, to stop bleeding after child-birth and also bleeding from nose. Stem bark is astringent, anti-dysenteric and anthelmintic. It has stomachic, febrifugal and tonic properties. Bark is also used in the treatment of amoebic dysentery, diarrhoea, piles, dyspepsia, chest affection and diuresis. Bark is used in pessaries for promoting conception and for toning up vaginal tissue after delivery, also used for veterinary purposes. Leaves are used in chronic bronchitis, boils, and ulcer. Flowers are edible and employed in blood diseases. Wood is used for making combs, picture frames, toys, cup, paper knives, sticks, plough, plateforms etc.

(Watt, 1889-1899)

Bark of stem and roots are powerful antidysenteric. Bark powder is applied over body in dropsy. Fruits are used as antidote in snake-bite, to allay swelling and irritation. Seeds are astringent, febrifuge and anthelmintic. They are also carminative, astringent and used in pulmonary affections. Seeds are tonic aphrodisiac.

(Anonymous, 1946-1976)

Root is used as veterinary medicine. Bark is astringent, anthelmintic, stomachic, antipyretic, tonic and antidysenteric. Bark is used for amoebic dysentery and diarrhoea. Bark oil is anthelmintic. Wood is used for making articles like combs, spoon, boxes, bobbins, pen holdes, brushes etc.

(Ambasta, 1986)

Bark is antidote of snake-bite, cures bronchital fever, cold and cough, used also in the treatment of diarrhoea, dysentery, fever and gastric diseases. Bark is prescribed for head-ache, malaria, menstrual complaints, paralysis and piles. It revives test and cure stomach-ache. Flowers are edible and used as a vegetable. Flowers are appetizer, cure dysentery and help to expel worm from intestine. Latex is efficacious in eczema, also used to curdle milk. Leaves are used as appetizer. They are used in the treatment of dysentery and leucoderma. Plant is used against colic, constipation, fever, epilepsy hematuria, spermatorrhea, to facilitate child-birth, spleen and stomach complaints. Wood is used for making combs.

(Jain, 1991)

Critical Notes

This taxon is a reputed medicinal species and finds place in classic systems of medicine as well. The plant is used widely as fuel wood by the tribals, although it is of medicinal significance. Awareness in this regard is a dire necessity.

Nerium indicum Mill.

Local Name

Kanher

Ethnobotanical Notes

Medicine

1. Stem powder is applied for gum trouble.
2. Spoonful of leaf powder is taken with honey twice a day for one month to cure asthma.
3. Paste of leaves in butter milk is applied for ventilago.
4. Paste of flowers is applied on piles.

Notes from Classics

Root paste is applied on skin diseases. Fresh juice of root is dropped in eyes for cataract. Oil of root bark is used for application on skin diseases of scaly nature, leprosy and itch. Decoction of leaves is recommended to reduce swelling.

(Watt, 1889-1899)

Bark and flowers possess cardiotonic properties.

(Anonymous, 1946-1976)

Roots are resolvent and attenuant. Oil of root bark is used in skin disease of scaly nature.

(Ambasta, 1986)

Roots are helpful in expelling worms in veterinary medicine. Stem and roots are abortifacient. Latex is applied on blisters and boils. Flowers and latex are applied on eyes for eye diseases. Plants are used for asthma, bronchitis, dysentery, gum trouble and menorrhoea.

(Jain, 1991)

Critical Notes

Stem is useful to treat gum trouble. Leaves are beneficial for ventilago. Both uses are not published earlier. It is planted in gardens as ornamental. It is exotic, native of Meditterranian region.

Tabernaemontana divaricata (L.) Br.ex Roem and Schult.

Local Name

Chandani, Taggar

Ethnobotanical Notes

Medicine

1. Root powder is applied on wounds of snake bite as a antidote on snake bite.

Notes from Classics

Wood is employed medicinally as refrigerant.

(Watt, 1889-1899)

Root is employed as local anodyne, chewed for relief from tooth-ache. It is employed as vermicide. Red pulp surrounding the seeds is employed for dyeing fabrics. Wood is used as incense and in perfumery.

(Anonymous, 1946-1976)

Root is employed as a local anodyne, chewed for relief from tooth-ache. It is also employed as a vermicide. Red pulp surrounding the seed is employed for colouring fabrics, wood and perfumery.

(Ambasta, 1986)

Leaves are used to give relief in boils, fever, head-ache and sores. Flowerbuds are used for eye complaints. Red dye is obtained from pulp of follicle.

(Jain, 1991)

Critical Notes

It is usually planted as ornamental. Use of roots as antidote for snake-bite is still unreported.

Wrightia arborea (Dennst.) Mabb.

Local Name

Kuda

Ethnobotanical Notes

Medicine

1. Spoonful of bark powder mixed in spoonful of curd is taken twice a day for about a week to treat urine stone.

Notes from Classics

Bark is given in menstrual and renal complaints. Stem bark and root bark is useful in snake bite and scorpion sting. Leaves are eaten at pot-herb.

(Watt, 1889-1899)

Roots and leaves are used for fever. Leaves are administered with salt for relief from tooth-ache. Milky-juice is used to stop haemorrhage. Leaves and tender twigs are edible.

(Anonymous, 1946-1976)

Leaves, flowers buds and inflorescences are used in bowl complaints, haemorrhage, menorrhagia and seminal weakness. Dried flowers are stimulant and astringent, also used to prepare cooling drink. Flowers, twigs and leaves yield red dye. Wood is suitable for axe handle and also used as fuel.

(Ambasta, 1986)

Roots are used to control fever of kids. Bark is useful for menstrual complaints and for healing wounds. Leaves are useful for the ear-ache. Plant is used in the treatment of dysentery, tumours and stomach-ache.

(Jain, 1991)

Critical Notes

Bark is useful to treat urine stone. This use is not noted earlier.

Family: Periplocaceae

Cryptolepis buchanani Roem. and Schult.

Local Name

Kawalicha wel

Ethnobotanical Notes

Miscellaneous

1. Plant is used as cordage to tie bundles of fuel-wood and fodder.

Notes from Classics

Hill people of Vizianagram make cordage and kind of cloth from the fibre derived from this plant.

(Watt, 1889-1899)

Plants yield fibres which are used by hill tribes for cordage and for making a kind of cloth.

(Anonymous, 1946-1976)

Plant yields fibres which are used by tribals for cordage and for making a kind of cloth.

(Ambasta, 1986)

Stem juice is used as drink.

(Jain, 1991)

Hemidesmus indicus L.

Local Name

Anant-mul

Ethnobotanical Notes

Medicine

1. A spoonful of powder of roots mixed in cow milk is drunk for a week to stop polyuria.
2. Equal amount of powder of roots of *H.indicus* L. and *Asparagus recemosa* Willd. are mixed. Spoonful of it with curd is administered to cure ventilago.
3. Paste of roots is regularly applied on eczema.
4. A spoonful of root pounded in honey is advised to take daily for a month to cure leucorrhoea.
5. Powder of roots, about spoonful, is taken orally with a glass of water for conjunctivitis till cure.
6. Spoonful of powder of roots is mixed in lemon juice (*Citrus aurantifolia* (Christm) SW. and taken daily to purify blood till cure.
7. Paste of roots is applied on to scalp at night regularly to promote hair growth.

Notes from Classics

Root is alterative, tonic, prescribed for dyspepsia, syphilis, leucorrhoea, strangury, piles, skin problems and loss of appetite. Decoction of roots with cumin seeds is used to purify blood. Milky juice is applied for conjunctiva and other eye problems. Milky juice is mixed in root powder for genito-urinary disease. Milky juice is applied externally on swelling, syphilis, rheumatic affection, cough and cold.

(Watt, 1889-1899)

Roots have reputation as tonic, alterative, demulcent, diaphoretic, duretic and blood purifier. Dried roots are employed for nutritional disorders, syphilis, rheumatism, gravel, urinary and skin diseases. Leaves are chewed as mouth refresher. Milky latex of plant is used for relieving inflammation of eyes.

(Anonymous, 1946-1976)

Ambasta (1986) also reported similar uses as those in Anonymous (*loc.cit.*).

Roots are used as an antidote to snake-bite. These are appetizer, blood purifier, tonic and galactogogue and cooling. Roots are useful to cure debility, diarrhoea, dysentery and digestive diseases. Roots are used in the treatment of hypertension, impotency, laziness, menstrual complaints, nervous disorders, skin diseases, snake-bite, spermatorrhoea, stomach-ache, tonsillitis, urine and veneral complaints. Roots are used as snuff. Roots are eaten or worn to ward off evil spirits. Plant is used for tooth-ache and as an antidote of snake-bite.

(Jain, 1991)

Critical Notes

Uses of roots in case of polyuria, ventilago, leucorrhoea and hair growth are additional.

Family: Asclepiadaceae

Calotropis gigantea (L.) R.Br.

Local Name

Safed Rui, Mhatari, Rui

Ethnobotanical Notes

Medicine

1. Infusion of roots, about half cup, is given as antidote on snake poisoning immediately after snake bite.
2. Flowers and roots in equal amount are fried in oil or ghee and are consumed twice a day for gastro-intestinal disorders.
3. Starting consumption with single flower and betel leaf (*Piper betle* L.) upto five days with increasing number of flowers reduces body swellings.
4. A spoonful of powder of flowers with honey is consumed twice a day for a month to removal urine stone.
5. Latex is applied locally on wounds caused by scorpion sting.

6. Spoonful of latex mixed homogeneously with oil of *Brassica juncea* (L.) Czern et Cass and applied on skin locally for scabies and to treat itching.
7. A spoonful of latex is mixed in oil of *Brassica juncea* (L.) Czern. et Coss. and a pinch of powder of black pepper (*Piper nigrum* L.). It is applied locally to cure ventilago.
8. Latex mixed in fruit powder of *Emblica officinalis* Gaertn. is applied on scalp to prevent hair-fall.
9. Latex is mixed in clay soil and applied on tumour to retard its growth.
10. A thorn inserted in foot sole is easily removed on application of latex.

Notes from Classics

Root bark promotes milk secretion and useful in secondary syphilis and skin diseases. Root bark control enlargement of abdominal viscera, expels intestinal worms and cough. Root bark act on ascities ad on skin diseases. Root bark powder is emetic and expectorant. Leaf juice is dropped in ear for otorrhoea. Fresh juice with salt cure sprain. Powdered roots with leaves and black pepper can cure jaundice. Flowers are considered digestive, stomachic and nervine tonic. Flowers are useful in asthma, catarrha, leprosy, loss of appetite, hysteria and epilepsy.

(Watt, 1889-1899)

Root bark is a good remedy on dysentery. In small dose, it is diaphoretic and expectorant. Paste of root bark is applied in elephantiasis. A tincture of leaves is used in the treatment of intermittent fever. Powdered flowers are useful in cold, cough, asthma and indigestion.

(Anonymous, 1948-1976)

Bark and fibres are used for fishing nets. Seed yield fixed oil. The floss obtained from feeds is used for stuffing pillows.

(Ambasta, 1986)

Roots are used to expel guinea worms. They act as an antidote on scorpion and snake-bite, and useful to cure leprosy. Leaves lessen body ache, are applied on burns, skin diseases and swellings. Leaf juice is dropped in ears for ear-ache and also dropped in eyes to expel ring-worms and intestinal worms. Latex is abortifacient, used for asthma, leprosy, hydrophobia, rheumatism, sprain, post-natal complaints and stomach-ache. It is applied on skin disease, wound tooth-ache and ring-worms. It is a remedy on rabid dog bite.

(Jain, 1991)

Critical Notes

Usefulness of flowers for urine stone, of latex for ventilago, to prevent hair fall, to remove thorn easily from foot sole and to cure tumour are interesting and additional to the literature of ethnobotany.

Calotropis procera (Ait.) R.Br.

Local Name

Ruchkin, Rui

Ethnobotanical Notes

Medicine

1. Root powder, about spoonful, is drunk with water for vomiting after poisoning.
2. Tender leaf extract, about few drops, is dropped in nose as an antidote on scorpion sting.
3. Paste of leaves is warmed and then applied on rheumatic joints.
4. Few drops of extract of leaves and garlic bulblets (*Allium sativum* L.) are dropped in ears to cure ear-ache.
5. Leaves of this plant, *Datura metal* L. and common salt are burnt together. Ash obtained, about one spoonful with honey, is given to cure asthma till cure.
6. Equal amount of leaf powder of this plant and (*Piper betle* L.) mixed in honey. One spoonful of it is drunk with the treatment of fever till cure.
7. Decoction of leaves, about a half cup, is taken at night to cure paralysis of legs.
8. Spoonful latex is mixed homogeneously in cow-ghee, dropped in nose to cure migraine.
9. About half cup of latex is mixed in about 400 ml of milk and boiled constantly. The condensed milk about half cup

is drunk once a day when there is no fever. It cures malaria. The patient never suffers from malaria in life again.

10. Few drops of latex are dropped on teeth for tooth-ache.

Notes from Classics

Roots are used as tooth brush to treat tooth-ache.

(Watt, 1889-1899)

Root bark is used for dropsy. Stem yield fibres similar to that of *Calotropis gigantea*. Wood yield gum powder and charcol.

(Ambasta, 1986)

Roots are good remedy on asthma, fever and to reduce swelling. Leaves give relief in cough and cold. Leaves are applied on boils, eczema and skin diseases. Leaves control stomach diseases and spleen complaints. Latex is abortifacient. It act as antidote on scorpion sting. Latex is antiseptic and used against cornea, dropsy, leposy, taenia and rheumatism. Latex helps to expel ring-worms. Flowers are useful for cholera and stomach disorders. Plant is a cure against lepsory, small-pox and sores. Plant is also employed for insect-bite.

(Jain, 1991)

Critical Notes

Roots are being reported useful for the first time, against poisoning. Leaves are medicinal to treat, rheumatic joints, migraine, ear-ache asthma, fever, malaria and paralysis. Leaves are used to cure ear-ache. All these have not been recorded in past.

Caralluma adscendens Haw.

Local Name

Molya, Vanjari Bhaji

Ethnobotanical Notes

Miscellaneous

1. ***Food***: Plant is used as vegetable.
2. Flowers are used to design doors and windows, hand sticks of cow-herds.

Notes from Classics

Fleshy plant is eaten by natives in the form of pickle. Chatney is also prepared.

(Watt, 1889-1899)

Plant is used as vegetable.

(Anonymous, 1946-1976)

Ambasta (1986) documented similar uses as in Anonymous (*loc.cit.*).

Stem is used as vegetable.

(Jain, 1991)

Critical Notes

Very few plants are observed in the district. In past, these have been over-exploited. There is an urgent need to stop its collection from this region.

Wattakaka volubilis (L.*f.*) Stapf

Local Name

Morsa

Ethnobotanical Notes

Medicine

1. Leaves are powdered and applied on boils and itch.

Miscellaneous

1. *Food*: Flowers are used as vegetable.
2. *Fodder*: Leaves and fruits are good feed for goats.

Notes from Classics

Roots and tender stalks are emetic and expectorant. Leaves are applied to boils and abscesses, eaten as green vegetable. Tender end of the creeper with its juice if touched to nose causes excessive sneezing. Roots and tender stalks are emetic and purgative.

(Watt, 1889-1899)

Leaves are used in application for boils and abscesses. Seeds are edible. Plant juice is used as sternutatory. Leaves, flowers and the wall of unripe fruits are boiled and eaten as a vegetable or used in curries.

(Anonymous, 1946-1976)

Uses similar to those in Anonymous (*loc.cit.*) are mentioned by Ambasta (1986).

Roots are anthelmintic and used for fever, regurgitation. Leaves and roots are used for gastric ulcer and urine complaints. Leaves are applied on abscess, boils, skin diseases and for healing wounds. Plant is useful for dropsy, failure of lactation, sores and urticaria.

(Jain, 1991)

Critical Notes

Medicinal uses of leaves for itch, as also the leaves and green fruits as goat feed are new reports awaiting documentation. Use of flowers should be avoided for vegetable purpose. This checks its wild populations. Feeding goats on green fruits, further aggravates this problem.

Sarcostemma intermedium Decne.

Local Name

Dudhikand

Ethnobotanical Notes

Miscellaneous

1. *Fodder*: Stems are fed to milching cattle to increase lactation.

Notes from Classics

Plant is used to entirpate white ants from sugar-cane fields.

(Watt, 1889-1899)

Plant is used to entirpate white ant from sugar cane fields.

(Anonymous, 1946-1976)

Dried stem is emetic. Infusion of root is given to person bitten by rabid dogs.

(Ambasta, 1986)

Roots are used for antifertility. Leaves and flowers are used as antidote for snake-bite. Latex is applied on boils. Plant paste is applied on bone fractures.

(Jain, 1991)

Critical Notes

To increase lactation in case of milching pet animals, stems are fed to them. This is not documented in past.

Pentatropis spiralis (Forsk.) Decne

Local Name

Popat-chochi

Ethnobotanical Notes

Miscellaneous

1. ***Food***: Young seeds are eaten raw.
2. ***Fodder***: Climbers are used as fodder for goats.

Notes from Classics

Small tubers are peeled and eaten in Punjab.

(Watt, 1889-1899)

Tubers are eaten.

(Anonymous, 1946-1976)

Sweet tubers are peeled and eaten. Roots are refrigerant and also used in gonorrhoea.

(Ambasta, 1986)

Plant is used to treat diarrhoea.

(Jain, 1991)

Critical Notes

Edibility of young seeds is a additional use. Climbers are cut and removed from hedges, along roads and bandhs of fields by the goatherds as fodder for goats. Now-a-days, few plants are found growing in the region. This trend will obviously affect its very existence in near future, if not checked. Seeds are collected by the local people being edible. This further aggravates the problem.

Pergularia daemia (Forsk.) Chiov.

Local Name

Utran

Ethnobotanical Notes

Medicine

1. Paste of roots is applied on forehead for head-ache.
2. Juice of leaf, about full of cup, is mixed in floter of *Eleusine coracana* (L.) Gaertn. flour and bread is prepared from it. It is given for a week on dog-bite.
3. About half cup of decoction of leaves is given twice to cure cough till cure.
4. About a cup of decoction of fruit is warmed and drunk twice a day to get relief in sprain.
5. About 30 gm leaves of this plant alongwith equal quantity leaves of *Ocimum sanctum* L. and roots of *Cynodon dactylon* (L.) Pers. are crushed. This extract, about half cup, is taken orally once a day for burning pain of body. Rice flour is soaked in extract and poultice of it is applied for burning body pains till cure.

Miscellaneous

1. ***Fodder***: It is good fodder for goats.

Notes from Classics

Roots are purgative and used for rheumatism. Leaves are employed for diseases of children. Decoction of leaf is used as anthelmintic. Leaf juice is prescribed for asthma and applied externally for rheumatic swellings of limbs. It controls intestinal diarrhoea.

(Watt, 1889-1899)

Decoction of leaves is used in asthma. Leaf juice is applied to rheumatic joints. Pulped leaves are applied to carbuncles. Plant is emetic. Juice of leaves is used in the treatment of catarrhal affection. Combined with lime juice, leaf juice is applied to rheumatic swellings. Leaf juice is given in rheumatism, amenorrhoea and dysmenorrhoea. Root bark is used as purgative in rheumatic cases. Plant extract is used for uterine and menstrual trouble and to facilitate parturition.

(Anonymous, 1946-1976)

Ambasta (1986) reported similar uses as given in Anonymous (*loc.cit.*).

Roots are used for head-ache. Leaves are used for asthma, cataract, eye disease, diarrhoea, head-ache, urine complaints and wounds. Latex and fruits are used for calf-laying, which is a veterinary medicine. Flowers are used to treat cough and used as vegetable. Whole plant is administered for purifying blood, in gastrointestinal disorders, ulcer, leucoderma, rheumatism and snake-bite.

(Jain, 1991)

Critical Notes

Roots are important for head-ache. Leaves are useful in case of dog-bite and for cough. Fruits are beneficial in sprain. Its fodder value is still unnoticed. All these uses are not documented till date.

Tylophora dalzelli Hook. *f.*

Local Name

Umethicha

Ethnobotanical Notes

Medicine

1. Extract of roots, about spoonful, is given daily alongwith tea to check fever till cure.
2. A spoonful leaf extract with mother's milk is given to kids for three days in case of abdominal swellings.

Notes from Classics

Roots and leaves are remedy for dysentery and catarrah. Liniment of root is applied on the head for cephalagia and neuralgia. Roots are emetic. Plant is cathartic, expectorant, diaphoretic and useful to stop vomiting in children.

(Watt, 1889-1899)

Leaves are used to cure asthma.

(Jain, 1991)

Critical Notes

Leaves help in treating abdominal swelling. Root are helpful to check fever. These local medicinal uses are not noticed elsewhere.

Family: Loganiaceae

Strychnos potatorum L.*f.*

Local Name

Nirmali

Ethnobotanical Notes

Medicine

1. Powdered seeds are applied on lower eyelids to prevent constant watering of eyes till cure.
2. Powder of seeds is taken orally, about one spoonful, with honey. It is advised twice a day with honey for a week to cure trachoma of eyes.
3. Paste of seeds is applied on wound of scorpion bite to reduce effect of its poison.

Miscellaneous

1. Ripe fruits are eaten.
2. Seeds are used for fish killing.

Notes from Classics

Seeds are used in eye diseases. They are rubbed with water and rock-salt, applied on chemosis in conjuctival and corneal ulcer. Seeds are cold and dry, useful for dysentery, irritation of urinary organs and gonorrhoea. Seeds are used in colic, applied to eyes for strengthening eye sight.

(Watt, 1889-1899)

Fruit pulp is edible. Powdered bark is mixed in lime juice and given to treat cholera. Leaves are used as poultice over ulcer infested with maggots. Fruits are substitute for 'ipecae' in the treatment of dysentery and bronchitis. Wood is used for building huts and carts.

(Anonymous, 1946-1976)

Seeds are tonic, stomachic, demulcent, emetic and for eye troubles. Fruits are eaten and used in preserves.

(Ambasta, 1986)

Roots are prescribed for syphilis, fever and night wetting. Bark is used to relieve muscle pains, stomach-ache and to accelerate labour pains. Fruits are edible and yield brown dye. Fruits are used as fish poison. Fruits are used for eating.

(Jain, 1991)

Critical Notes

Seed paste is useful to reduce poison of scorpion sting. This has remained unnoticed elsewhere.

Family: Gentianaceae

Enicostemma axillare (Lam.) Rajnal

Local Name

Naikai

Ethnobotanical Notes

Medicine

1. A cup of root extract is taken orally immediately after scorpion sting as antidote.
2. About 20 ml of decoction of whole plant twice a day for eight days is advised for dropsy.
3. Plant extract is useful to stop hiccup.

Notes from Classics

Plant is used as stomachic. It is also tonic and laxative.

(Watt, 1889-1899)

Plant is tonic, stomachic and laxative. Whole plant is dried, powdered and given with honey as blood purifier, and in the treatment of dropsy, rheumatism, abdominal ulcer, hernia, swelling, itches and insect poisoning. It is lso effective against malaria.

(Anonymous, 1946-1976)

Ambasta (1986) noted uses similar to those of Anonymous (*loc.cit.*).

Leaves are used as a febrifuge. Whole plant is administered for malaria and veneral disease. Whole plant is used as antidote for snake bite.

(Jain, 1991)

Critical Notes

The plant is already reported useful for snake bite. The roots are particularly reported as antidote for scorpion sting. Entire plant is useful for hiccup. These uses have not been available in past literature.

Family: Boraginaceae

Coldenia procumbens L.

Local Name

Tripunkhi

Ethnobotanical Notes

Medicine

1. Leaves are levigated in water and applied onto joints to reduce rheumatic pains.

Notes from Classics

Equal part of dry plant and fenugreek are rubbed to fine powder and applied warm on boils to bring its suppuration. Fresh leaves are ground and applied to rheumatic swellings.

(Watt, 1889-1899)

Fresh leaves are ground and applied for rheumatic swellings. Whole plant is used on boils. Roots are prescribed for menstrual complaints. Leaves are useful in rheumatic complaints.

(Anonymous, 1946-1976)

Ambasta (1986) reports similar properties as in Anonymous (*loc.cit.*).

Whole plant is used for boils. Roots are prescribed for the menstrual complaints. Leaves are used in rheumatic complaints.

(Jain, 1991)

Cordia dichotoma Forst.*f.*

Local Name

Bhokar

Ethnobotanical Notes

Medicine

1. Leaf extract, about spoonful, is taken twice a day for a week to cure urethritis.
2. Decoction of leaves, about half cup, is given at morning to child for expelling intestinal worms.
3. Fruits are eaten raw at morning and evening to cure stomach ache.
4. About five gm of gum is dissolved in water. It is taken orally as expectorant till cure.

Miscellaneous

1. ***Fodder***: Leaves are used as fodder for cattle.
2. ***Wood***: Wood is used for making agricultural implements.
3. ***Food***: Fruits are edible and unripe fruits are pickled.

Notes from Classics

Decoction of bark is used in dyspepsia and fever. Kernels are used for external application for ring-worms. Fruits are edible and astringent. They are anthelmintic, diuretic and demulcent. Fruits are also used in diseases of chest and urinary passage. They are good expectorant. Wood is used for boat building, well curbs and agricultural implements.

(Watt, 1889-1899)

Fruits are anthelmintic, diuretic and demulcent.

(Anonymous, 1946-1976)

Ambasta (1986) records similar properties as those of Anonymous (*loc.cit.*).

Leaves are used for chest infection and lung diseases. It gives relief in cough, fever, head-ache, jaundice, ulcer, lung and colic diseases. Fruits are good expectorant and edible. Seeds are anthelmintic. Plant is used for anasarca, cholera and dyspepsia.

(Jain, 1991)

Critical Notes

Leaves are beneficial for urethritis and against intestinal worms. Fruits are useful for stomach-ache. These do not find place in past documentation.

Cordia gharaf (Forsk.) Ehrenb and Asch.

Local Name

Bhokar, Ghondani

Ethnobotanical Notes

Miscellaneous

1. ***Food***: Ripe fruits are edible.

Notes from Classics

Fruits are eaten by poorer classes and also pickled. Flowers and fruits are eaten during famine in 1877-78.

(Watt, 1889-1899)

Fruits are edible. Bark fibres are used for ropes and caulking boats.

(Anonymous, 1946-1976)

Ambasta (1986) mentioned similar properties as in Anonymous (*loc.cit.*).

Fruits are edible.

(Jain, 1991)

Ehretia aspera Willd.

Local Name

Tambulya, Datarangi

Ethnobotanical Notes

Medicine

1. Young twigs are chewed to treat mouth ulcer and for better gums.

Miscellaneous

1. Stem pieces are used as tooth brushes.
2. ***Fodder***: Leaves are used as fodder.
3. Fruits are eaten at the time of scarcity.

Notes from Classics

Fruits are tasteless but eaten during famine. Leaves are used at cattle fodder.

(Watt, 1889-1899)

Decoction of fresh root is given in veneral diseases.

(Anonymous, 1946-1976)

Wood is suitable for brush backs, shoe lasis, match boxes and splints. Inner ark and fruits are edible. Leaves are used as cattle fodder.

(Ambasta, 1986)

Leaves are applied on muscle pains. Decoction of plant is drunk for syphilis. Stems are used for thatching huts. Fruits and inner bark are edible.

(Jain, 1991)

Critical Notes

Young twigs are used to treat mouth ulcer and for strengthening gums are not documented in the past literature. Stem is also reported for the first time as tooth brush.

Tricodesma indicum (L.) Lehm. var. *indicum*

Local Name

Kataini

Ethnobotanical Notes

Medicine

1. Leaf juice is given one spoonful per day at morning for three days to cure acidity.

Notes from Classics

Roots are pounded and to paste is applied to reduce swellings of joints. Leaves are cooling and dupurtive and cure snake bite. They are an emollient poultice.

(Watt, 1889-1899)

Root paste is applied on swelling of joints. Roots are used to cure dysentery and fever. Infusion of leaves is depurtive. Flowers are sudorific and pectoral. Herb is diuretic and also used as emollient poultice. Herb is used for expulsion of dead feotus. Roots are used in dysentery and also applied to swelling.

(Anonymous, 1946-1976)

Infusion of leaves is depurtive. Flowers are sudorific and pectoral. Herb is emollient, diuretic and used for expulsion of deal feotus. Leaves and flowers are eaten.

(Ambasta, 1986)

Roots are employed for stomach ache, also used as vermifuge. Leaves are edible and cure eczema. Fruits are used for swellings. Whole plant is tonic for brain

(Jain, 1991)

Family: Convolvulaceae

Convolvulus arvensis L.

Local Name

Chandvel

Ethnobotanical Notes

Medicine

1. Roots 25 gm are pounded in honey and given one spoon per day as purgative.

Notes from Classics

Roots possess cathartic properties.

(Watt, 1889-1899)

Roots are cathartic.

(Anonymous, 1946-1976)

Seeds contain fixed oil.

(Ambasta, 1986)

Roots are purgative. Whole plant is used as detergent.

(Jain, 1991)

Critical Notes

Usefulness of roots as purgative although noted elsewhere it is also important from the view-point of cross country observation. Present notice increases its authenticity.

Ipomoea aquatica Forsk.

Local Name

Naipala, Nai

Ethnobotanical Notes

Medicine

1. Half cup of decoction of leaves is given orally twice a day to treat snake-bite.
2. Infusion of leaves, about half cup, is drunk twice daily to cure malarial fever.

Miscellaneous

1. ***Food***: Young terminal shoots and leaves are used as vegetable.
2. ***Fodder***: Leaves are used as fodder.

Notes from Classics

Plant is used as fish food.

(Watt, 1889-1899)

Stem and leaves are cooling. Flower buds are used in the treatment of ring-worm. Plant is used for nervous and general debility and piles. Juice of plant is purgative. It is used as emetic in case of opium and arsenical poisoning.

(Anonymous, 1946-1976)

Roots are eaten at the time of scarcity. Young terminal shoots and leaves are used as vegetable and also used as green fodder. Juice of plant is emetic in case of opium and arsenical poisoning. Plant is also used for women suffering from nervousness and general debility.

(Ambasta, 1986)

Stem paste is applied on boils. Leaves and stems are used for stomach disorders. Stem is employed for gonorrhoea. Whole plant is purgative and used to treat liver disease. Whole plant, shoot and leaves are used as vegetable.

(Jain, 1991)

Critical Notes

Decoction and infusion of leaves are useful in case of snake-bite and malarial fever respectively. These have not been yet documented in botanical literature.

Ipomoea carnea Jacq. subsp.*fistulosa* (Mart. ex Choisy) D.Austin.

Local Name

Besharam, Jahari-Nagin

Ethnobotanical Notes

Medicine

1. A woven matt called 'Tatta' is made from the stem for bullock-cart.
2. It is also used for partition wall for huts and bathrooms in rural and tribal region of the district.
3. It is planted as live hedge around fields.

Notes from Classics

The plant is sometimes grown at hedge plant. It is mild purgative.

(Anonymous, 1946-1976)

Ambasta (1986) reported similar uses as those in Anonymous (*loc.cit.*)

Leaves are used as vegetable.

(Jain, 1991)

Critical Notes

The plant was introduced in Jalgaon district as live hedge prior to 1970. It has widely spread and perfectly naturalized in the region. It is now troublesome, especially in the places like rivers, nalas or in marshy places. Jain(1991) reports leaves useful as vegetable. However, a word about caution is necessary. We observed contradictory reports from the rural people of this district. The plant is reported as poisonous. If the leaves are placed near foodstuffs, the later are avoided for consumption due to poisoning by the local people. It is, therefore, locally called 'Jahari-Nagin' meaning poisonous female cobra. However, the use of stem is reported in this district is interesting. The troublesome plant is cut by the local people for good purpose. It is exotic, native of South America.

Ipomoea nil (L.) Roth

Local Name

Nila Wel

Ethnobotanical Notes

Miscellaneous

1. *Fodder*: Plants are used as fodder for goats.

Notes from Classics

Seeds are used as safe purgative.

(Watt, 1889-1899)

Seeds are used as purgative and substituted for jalap.

(Ambasta, 1986)

Whole plant is used as vegetable.

(Jain, 1991)

Critical Notes

The use of plants as fodder for goats forms a first report.

Ipomoea obscura (L.) Ker.Gawl.

Local Name

Pungalicha-pala

Ethnobotanical Notes

Miscellaneous

1. *Fodder*: Leaves are used as fodder for goats and cattle.

Notes from Classics

Leaves are valuable application for apthous affection.

(Watt, 1889-1899)

Leaves are used in apthous affection. In admixture with the leaves of *Argyreia mollis*, they are used for sores.

(Anonymous, 1946-1976)

Leaves are aromatic and mucilaginous. Roasted, powdered and boiled in ghee are useful in apathous affection.

(Ambasta, 1986)

Stems are employed for cold and cough.

(Jain, 1991)

Critical Notes

Fodder value of this species is being reported for the first time.

Ipomoea pes-tigridis L.

Local Name

Panja

Ethnobotanical Notes

Medicine

1. Powder of leaves is smoked to relieve bronchial spasm.

Miscellaneous

1. ***Fodder***: Plant is used as cattle fodder.

Notes from Classics

Plant is said to be used medicinally.

(Watt, 1889-1899)

Roots are purgative. Leaves are used as a poultice for boils, pimples and sores. Plant is used in the treatment of boils and carbuncles. It is antidote to dog-bite.

(Anonymous, 1946-1976)

Plants are used as cattle fodder, both as green and hey.

(Ambasta, 1986)

Roots are used as laxative and antidote on dog-bite. Leaves are used in case of fever, head-ache and eye diseases. Seeds are edible.

(Jain, 1991)

Critical Notes

Leaves are useful to treat bronchital spasm, a use hitherto unnoticed.

Ipomoea turbinata Lagasca

Local Name

Bhavara

Ethnobotanical Notes

Miscellaneous

1. *Food*: These are used as vegetable.
2. *Fodder*: It is collected by tribals as fodder for milching goats.

Notes from Classics

Plant is eaten at the time of scarcity. Swollen pedicels are edible.

(Watt, 1889-1899)

Seeds are cathartic.

(Anonymous, 1946-1976)

Calyx is edible.

(Jain, 1991)

Critical Notes

In Jalgaon district, only leaves are found in use as vegetable. However, entire plants are served specially to the milching goats. The latter use is not on record.

Merremia dissecta (Jacq.) Hall. *f.*

Local Name

Jangli Bhowra

Ethnobotanical Notes

Medicine

1. Leaf juice is applied on tumours of any part of the body.

Notes from Classics

Seeds possess odor of oil of bitter almonds and used in preparation of liquor.

(Ambasta, 1986)

Critical Notes

To treat tumours, leaves are being reported for the first time.

Merremia emarginata (Burm.*f.*) Hall.*f.*

Local Name

Undirkani, Chuha-kani, Bandhchari

Ethnobotanical Notes

Medicine

1. About half cup of root infusion is drunk after dinner once a day for a week to cure epileptic fit.
2. Root powder is snuffed to prevent nose bleeding.
3. Equal amount of leaves of this plant and *Withania somnifera* (L.) Dunaj. are crushed thoroughly. Two pellets of this powder are consumed daily to leucorrhoea til cure.

Notes from Classics

Plant is considered deobstruent, diuretic and alterative. It is used in rheumatism and neuralgia also controls cough. Juice of plant is dropped in ears to cure sore.

(Anonymous, 1946-1976)

The plant is deobstruent and diuretic. The plant juice is given in rat bite and dropped in ears to cure sores in ears, also used at pot herb.

(Ambasta, 1986)

Critical Notes

Roots are useful for epileptic fits and nose bleeding. Leaves are also important medicinally to treat leucorrhoea. All these uses still await for documentation. This species is generally mistaken for Bhramhi (*Centella asiatica* Linn.) by the local people.

Rivea hypocrateriformis (Desr.) Choisy.

Local Name

Fangola

Ethnobotanical Notes

Medicine

1. Paste of roots is applied on belly to reduce inflammation of abdominal region.
2. Paste of leaves is applied on abdomen for stomach-ache.
3. Paste of two leaves in cow-ghee is taken orally for a week. It is administered twice a day for lochia.
4. Spoonful of fruit pulp prepared in water and is applied around eyes to prevent constant watering of eyes.

Miscellaneous

1. Flour of gram (*Cicer arietinum* L.) is applied on leaves and then steamed. A local food article 'Funke' is prepared.
2. ***Fodder***: These vines are good fodder for goats. It is sold sometimes in local market.

Notes from Classics

Leaves and young shoots are boiled by natives with salt and chilies and eaten young shoots are eaten as vegetable.

(Watt, 1889-1899)

Root is given in Bihar to women after child-birth. Leaves and young shoots are eaten as vegetable and used in preparing bread.

(Anonymous, 1946-1976)

Young shoots and leaves are eaten as vegetable. Roots are given after parturition.

(Ambasta, 1986)

Roots are used against fit and antidote to snake-bite. Leaves are useful for sex-weakness and used as vegetable. Plant is useful for gum trouble and tooth-ache.

(Jain, 1991)

Critical Notes

Roots are useful for inflammation of abdomen. Leaves are important in case of stomach-ache and lochia. Fruits are employed to treat constant watering of eyes. Plants constitute good fodder for goats. It is locally used as vegetable. It is worth to scrutinize its active principle especially from nutritional point of view.

Family: Cuscutaceae

Cuscuta chinensis Lam.

Local Name

Amarvel

Ethnobotanical Notes

Medicine

1. Stem paste is applied on the scalp at night to remove dandruff.
2. Two to three drops of plant juice are dropped in ears for convulsion.
3. One spoonful of stem powder with honey is taken at night for three days to cure dysentery.
4. Equal amount of powder of stems of this plant and sandal wood (*Santalum album* L.) is mixed in whey and taken spoonful for to times to cure ventilago.
5. Paste of stem is lemon juice [*Citrus aurantifolia* (Christrn.) SW.] is applied on skin to cure eczema.
6. Decoction of stem, about half cup, is taken at morning for three days to expel worms.

Notes from Classics

Whole plant is purgative, used internally in protracted fevers, retention of wind and induration of liver. Decoction of plant is used to wash sores, also the paste is applied externally on itch. Seeds are tonic, diaphoretic, demulcent and carminative seeds are alterative and used to purify blood.

(Watt, 1889-1899)

Plant is used in flatulence and liver complaints. Seeds are used as purgtive.

(Anonymous, 1946-1976)

Ambasta (1986) mention ed similar uses as in Anonymous (*loc.cit.*). Plant is used for anasarca, bone fracture, cancer and dropsy. Plants also help to reduce labour complaints.

(Jain, 1991)

Critical Notes

Stem is useful to remove dandruff, to cure for ear complaints, to treat dysentery and to expel worms and ventilago. All uses are new to science.

Family: Solanaceae

Capsicum annuum L.

Local Name

Mirchi

Ethnobotanical Notes

Medicine

1. Few drops of leaf extract are dropped in nostrils for relief in case of migraine.
2. Extract of ripe fruits, two or three drops are dropped in ears it is then wiped out. After twenty minutes, few drops of coconut oil (*Cocos nucifera* L.) are dropped to cure pain in gums.
3. Half spoon of seed paste is consumed only once with jaggery to cure stomach-ache.

Notes from Classics

Fruits are used externally as plaster. They are stomachic and stimulant. Fruits are good remedy for hoarseness and dyspepsia.

(Watt, 1889-1899)

Fruits are used as counter-irritant in lumbago and neuralgia. They are used in rheumatic disorders and atonic dyspepsia. Fruits are used as tonic and carminative.

(Anonymous, 1946-1976)

Ambasta (1986) noted similar uses as in Anonymous (*loc.cit.*).

Fruits are edible and used as condiment and vegetable.

(Jain, 1991)

Critical Notes

Leaf extract is useful for migraine. Fruits are serviceable to cure pains in gum. Seeds are also useful against stomach-ache. All these uses are not noted by earlier authors. It is cultivated species. It is exotic, native of tropical America.

Datura innoxia Mill.

Local Name

Dhotra

Ethnobotanical Notes

Medicine

1. Fresh latex is applied around eyes for eye injuries.
2. Leaf paste is applied on foot sole for burning in summer.
3. Fresh leaves are fried and tied on mumps till cure.
4. Fresh leaves are crushed and applied the suppurating tumours.
5. Leaf paste is applied on pimples and boils.

Notes from Classics

Dried leaves are smoked to relief in spasmodic asthma, emphysema of lungs and used inchronic catarrh. Leaf extract is sedative and narcotic. Poultice of leaves is used in rheumatic joints, tumours and piles. Juice of leaves is used in epilepsy and mania. It is useful in neuralgia, rheumatism, dysmenorrhoea, cancerous sores and syphilitic pains. Juice of leaves is an antiperiadic, used in intermittent fever. It is a strong poison and causes delirium. Paultice of leaves is applied on breast for arresting secretion of milk. It is also used foı expulsion of guinea-worm. A pill of seed is used for tooth-ache. Pulp of leaves is applied on sweating of feet.

(Watt, 1889-1899)

Whole plant is used as pre-anesthetic in surgery. It helps also in child-birth. Plant cures eye disease and controls motion of sickness.

(Ambasta, 1986)

Leaves are used for asthma and boils. Seeds are used in country liquor and controls the pyorrhea. Whole plant is antihydrophobic in nature.

(Jain, 1991)

Plant is source of scopolamine and cerebral depresent and useful in manical conditions.

(Ambasta, 1986)

Critical Notes

Leaves are remedial on foot sole burning in summer period on mumps and pimples. These are not documented in classical literature. However, it is to be noted that medicinal utility of leaves is reported for many other human ailments. Medicinal utility of latex for eye injuries is although being reported for the first time, its use of such may be cautioned for obvious reasons. It is common weed of waste lands. It is exotic, native of America.

Datura metel L.

Local Name

Kala-dhotra

Ethnobotanical Notes

Medicine

1. Massage with leaf juice on back. It cures tetanus. This is followed for 8-10 days twice a day.
2. Tender leaves are warmed and tied on knees for arthritis.
3. Leaf juice, about 10 ml, is consumed mixed with curd for fifteen days to cure manorrhagia.
4. Leaf paste is applied on forehead at night in case of head-ache.
5. Leaf powder is used as snuff for asthma.
6. Leaves of this plant and *Calotropis gigantea* (L.) R.Br.mixed with common salt honey are made into paste. One spoon of it is taken orally twice a day to cure asthma.
7. Leaves of this plant, stem bark of *Nerium indicum* Mill. and *Moringa oleifera* Lamk. and extracted in the mixture of sesamum (*Sesamum orientale* L.) oil and water. This mixture is useful in massage for paralysis.

8. Flowers are crushed and homogenised in honey. Three to four spoonful of it are consumed regularly to overcome impotency in men.
9. Crushed flowers are boiled in sesumum oil. It is filtered. The filtrate is applied on body and massaged to relieve, body-pain.

Notes from Classics

Dried leaves are smoked for relief in spasmodic asthma. They cure emphysema of lugs and chronic catarrah. Leaves are also useful in epilepsy, syphilitic pains and dysmenorrhoea. Juice of leaves is used to dilate the pupils. It also gives effect in intermittent fever. Pounded seed pill is placed in decayed teeth to relieve tooth-ache.

(Watt, 1889-1899)

Roots, fruits, leaves and seeds are used for the treatment of boils and blisters. Dried leaves are used as belladonna leaves.

(Anonymous, 1946-1976)

Leaves are narcotic and antipasmodic. Seeds contain a fixed oil.

(Ambasta, 1986)

Corolla, leaves and roots are used for asthma. Leaves are used for bronchitis, cough, gonorrhoea and skin diseases. These are used as an antidote on snake bite, expelling ring-worms, guinea-worms and wound healing. Fruits are prescribed for cerebral complaints, fever and skin diseases. Seeds are good remedy on piles, fistula, leprosy, tooth decay and tumour. The plant is considered useful for anasarca, cramps and dropsy. It is also advised for epilepsy, hydrocele, head-ache, rheumatic pain, mumps, small-pox and syphilis. Madness of dog is controlled by seeds if given internally.

(Jain, 1991)

Critical Notes

Medicinal utility of leaves for the treatment of tetanus, arthritic, menorrhagia, and paralysis have not been noticed elsewhere. Usefulness of corolla is noted for asthma (*cf* Jain, 1991). However, flowers are useful in case of body pains and impotency in and men are being reported for the first time. It is a weed of waste places. It is exotic, native of tropical America.

Solanum anguvi Lamk.

Local Name

Ubhi-ringani, Dorli

Ethnobotanical Notes

Veterinary Medicine

1. Leaf or root juice is used to kill maggots infested on the abscess of cattle wound.

Notes from Classics

Root is one of the ingredients of 'Dash-mula'. Roots are expectorant, useful in cough and catarrhal affection. They are cardial, aphrodisiacal, astringent, resolvant, used in asthma, cough, chronic febrile affection, colic flatulence, worms and tooth-ache. Roots are used for dyuria, is churia.

(Watt, 1889-1899)

Roots are expectorant, useful in cough, and catarrahal affection dysurea and colic. A decoction of root is tonic and used in difficult parturition. Roots used for nasal ulcer. Fruits are tonic, laxative and digestive. Smoke from burning seeds relieve tooth-ache. Half ripe fruits are used in carries chutneys and preserves. Leaves are used as vegetable and fodder for goats.

(Anonymous, 1946-1976)

Roots are carminative and expectorant, useful in cough and catarrhal affection, dysuria and colic. These are applied to nasal ulcer. Half ripe fruits are used in curries, chutneys and preserves. Leaves are eaten as vegetable and fodder. Fruits are laxative and digestive.

(Ambasta, 1986)

Critical Notes

Maggots of cattle are killed by using leaf or root juice. This use is not noted earlier.

Solanum melongena L.

Local Name

Wange

Ethnobotanical Notes

Medicine

1. Extract of leaves, about half cup, is drunk with sugar during menstrual period for three days as oral contraceptive.
2. Fruit is roasted and eaten with honey at night for deep and sound sleep.

Notes from Classics

Leaves are said to have narcotic properties. Seeds are stimulant.

(Watt, 1889-1899)

Decoction of twigs is given in chest troubles. Unripe berries are poisonous, used as fish poison. Plant is antiseptic, used for tooth-ache and sore throat.

(Anonymous, 1946-1976)

Root is antiseptic and stimulant. Paste of roots is applied on ulcer of nose. Leaves are sialagogue, used in bronchitis, asthma and dysuria. Fruits are given in liver complaints. Seeds yield fatty oil.

(Ambasta, 1986)

Fruits and stalks are used for fistula and piles. Fruits are edible.

(Jain, 1991)

Critical Notes

Use of leaves as contraceptive and roasted fruit with honey for sound sleep are not documented in past literature. It is cultivated as a vegetable crop and also grown near huts and houses. It has different varieties also.

Solanum virginianum L.

Local Name

Kateri-ringni, Bhui-rigni, Ran-wangi.

Ethnobotanical Notes

Medicine

1. Root powder, about half spoon, is mixed in a cup of butter milk, and drunk daily for a week to treat urine stone.

2. Stem powder, about half spoon, is consumed in early morning once a day for asthma till cure.
3. Stem extract is applied to gums to cure tooth-ache.
4. Fumes of fruits are inhaled to check bacterial disease of teeth.

Notes from Classics

Roots are ingredient of 'Dashamula'. It is useful for cough, asthma, catarrah and pain in chest. Roots are beneficial in consumptive complaints in humoral asthma. Stems, flowers and fruits are carminative, useful in burning of chest. Fumigation of seeds to tooth-ache act as sialogogue. Decoction of seeds is diuretic, useful in chronic diarrhoea and as a febrifuge. It is stomachic and antbillious. Seeds are rubbed with water and applied for local inflammation and swellings. Decoction of seeds is given in cough, cola and chronic malarial fever. It promotes secretion of urine.

(Watt, 1889-1899)

Juice of leaves mixed with black pepper is prescribed in rheumatism. Plant is diuretic and used to cure dropsy. Juice is mixed with whey, chiretra and ginger, and given in fever. Plant is useful in dengue fever, acute bronchitis and chest affections. Fruits are used in carries. Seeds are eaten. Plant is fodder for goats.

(Anonymous, 1946-1976)

Root is expectorant, used in asthma and for pain in chest. Stems, flowers and fruits are carminative, used in burning sensation of feet. Juice of leaves with black pepper is prescribed in rheumatism. Juice of berries is used for sore throat. Fruits are used in caries. Seeds are administered as expectorant, in a asthma and cough. Extract of whole plant show antiviral activities against 'Ranikhet' viral diseases. Roots are used for cold, cough, fever, scorpion-bite, tooth-ache, urine complaints. Roots and fruits are useful for bronchitis, cough, ear complaints, gum troubles, paralysis, piles, sores, tooth-ache and tooth decay. Seeds are useful in the treatment of migraine, pains in jaws and pyorrhea. They are smoked in asthma. Plant is used for dropsy, gonorrhoea, heart diseases and swelling.

(Jain, 1991)

Critical Notes

Stem is useful for asthma and tooth-ache. This plant part in unreported as such.

Nicotiana tabacum L.

Local Name

Tambaku

Ethnobotanical Notes

Medicine

1. Powder of young leaves is applied locally for scabies.
2. Extract of leaves of this plant and *Azadirachta indica* A.Juss. is used for cleaning wounds caused due to diabetes.
3. Leaf powder is burnt and applied for tooth-ache.

Notes from Classics

Dried leaves are officinal, sedative and antispasmodic. They are useful in curing tetanus, dropsy, abdominal problems and for retention of urine. Dried leaves are advised for strangulated hemi dislocation, relieving pains, irritation in rheumatic swelling, syphilitic sore and skin diseases.

(Watt, 1889-1899)

Nicotine extraction from leaf is used in varnish and paints. Seed cake is used as feed for cattle.

(Anonymous, 1946-1976)

Leaves are employed in medicine as a sedative, antispasmodic and vermifuge. They are used in skin troubles and gastro-intestinal troubles. Cigarettes, cheroot, beedi and chewing tobacco is made from leaves.

(Ambasta, 1986)

Leaf powder is applied on eczema and wounds, also as a remedy on snake-bite and caterpiller sting. Leaves are applied for tooth-ache and spongy gums. Seeds are anthelmintic.

(Jain, 1991)

Critical Notes

Medicinal use of leaves for scabies is not reported earlier. Few plants are grown near abodes by some tribal people. It is exotic, native tropical America.

Physalis minima L.

Local Name

Phat phangora, Phuga

Ethnobotanical Notes

Miscellaneous

1. *Food*: Ripe fruits are edible.
2. *Fodder*: Entire plants are used as fodder for domestic animals.

Notes from Classics

Fruits of the cultivated species is eaten by natives.

(Watt, 1889-1899)

Fruits and leaves are edible.

(Anonymous, 1946-1976)

Ambasta (1986) recorded similar uses as those in Anonymous (*loc.cit.*).

(Ambasta, 1986)

Fruits and leaves are edible.

(Jain, 1991)

Critical Notes

It is exotic weed, native of South America. It is observed in the waste lands and also in cultivated ones. Although, it is a weed species, it is useful as reported presently.

Withania somnifera (L.) Dunal

Local Name

Dhor-kamhuni, Ashwagandha, Askand

Ethnobotanical Notes

Medicine

1. Decoction of roots, about half cup, is taken orally to reduce rheumatic pains of joints till cure.
2. A cup of extract of roots is drunk for fortnight to arouse sexual desire.
3. Infusion of roots with milk, about half cup, is taken orally after 4th day of menses for pregnancy.
4. Equal amount of roots and leaf powder are mixed homogenously in cow-ghee and sugar. It is consumed, one spoonful regularly, in case of male sterility.
5. Powder of plant fried in cow-ghee is advised to women after delivery for back-ache and weakness. A spoonful of it is taken orally.

Notes from Classics

Roots are alexipharmic, narcotic, tonic and diuretic. Leaves are used in the treatment of rheumatism, dyspepsia, diuretic, tonic, alterative, aphrodisiac, also employed in the emaciation and debility. Leaves are browsed by goats in Punjab. Seeds are diuretic and hypnotic in properties.

(Watt, 1886-96)

Roots are used for hiccup, female disorders, cough, rheumatism, dropsy, as sedative, in senile debility, dyspepsia and lumbar pains. Leaves are given in fever. They are brushed and applied to lesions, painful swellings, sore eyes and syphillitic sores. The plant is useful in inflammation, ulcer and scabies when locally applied.

(Anonymous, 1946-1976)

Roots are used for hiccup, cough, dropsy, rheumatism and internal disorders. Tender shoots used as vegetable and fodder for goats.

(Ambasta, 1986)

Roots are prescribed for asthma, bronchitis, cough, dropsy, dyspepsia, epilepsy, insanity and wounds. Leaves are used for boils, eye complaints, skin diseases, swellings of hands and feet. Seeds are employed for chest pain. Plant is used for lumbago.

(Jain, 1991)

Critical Notes

Roots are used to arouse sexual desire, to overcome male sterility and also used for conception. These have not been reported earlier. It is a common weed of waste places.

Family: Scrophulariaceae

Striga gesneroides (Willd.) Vatke

Local Name

Chikhalya, Chikas

Ethnobotanical Notes

Medicine

1. The injuries caused in between fingers of foot due to continuous contact with mud is cured by paste of root stock.

Notes from Classics

The plant is used in diabetes, but recent laboratory trials with animals show that it has no hypoglycaemic effect.

(Anonymous, 1946-1976)

Ambasta (1986) reports similar uses as those in Anonymous (*loc.cit.*).

Critical Notes

Root stock is useful to treat injuries caused due to constant touch of mud or muddy water. This use is not on earlier records. It is a parasitic plant.

Veronica anagallis-aquatica Linn.

Local Name

Feshi

Ethnobotanical Notes

Miscellaneous

1. Leaves are crushed and paste is spread in water to stupefy fishes.

Notes from Classics

Herb is credited with antiscorbutic properties.

(Anonymous, 1946-1976)

Roots are used for gargle. Herbs are used for salad. Herb is antiscorbutic.

(Ambasta, 1986)

Plant is blood purified and also used in skin disease.

(Jain, 1991)

Critical Notes

Use of leaves as fish poison is not reported in past.

Verbascum chinense (L.) Sant.

Local Name

Meno

Ethnobotanical Notes

Miscellaneous

1. Crushed leaves are employed to stupefy fishes by tribal people.

Notes from Classics

Leaves are eaten by camels and goats.

(Watt, 1889-1899)

Juice of leaves is sedative and astringent, useful in diarrhoea and dysentery. Plant juice is used in fever and skin eruptions. Leaves are eaten by camels and goats.

(Ambasta, 1986)

Critical Notes

Leaves as a fish poison is not reported in past.

Family: Bignoniaceae

Dolichandrone falcata Seem.

Local Name

Mendha-shingi, Kakad-shingi

Ethnobotanical Notes

Medicine

1. Paste of leaves is applied on swellings.
2. Fruit powder is mixed in "Triphala Churna" and turmeric powder (*Curcuma longa* L.). It is taken orally once a day for three days. It removes unpleasant smell of mother's milk.
3. Whole plant is powdered in cow urine and applied locally on sores.

Veterinary Medicine

1. Seed powder is dusted on wounds of domestic animals to heal and reduce swelling.

Notes from Classics

Decoction of fruit is given medicinally.

(Watt, 1889-1899)

Bark is used as fish poison.

(Anonymous, 1946-1976)

Bark yield coarse fibre. Wood used for building construction and agricultural implements.

(Ambasta, 1986)

Leaves are used for body-ache and snake bite. The plant is used as abortifacient.

(Jain, 1991)

Critical Notes

Medicinal efficacy of leaves for swelling, of fruit to improve mother's milk, of entire plant for sore are not documented yet.

Family: Pedaliaceae

Pedalium murex L.

Local Name

Motha gokhru, Ubha-gokharu

Ethnobotanical Notes

Medicine

1. Decoction of leaves, half cup per day, is drunk for two weeks in case of menolepsis.
2. A spoonful of powder of fruits is consumed with honey twice a day for paralysis of legs to cure.

Notes from Classics

Decoction of roots is antibilious. Young twigs, leaves, fruits and seeds are used as demulcent, diuretic, tonic for diseases of urinogenital system, gonorrhoea and dysuria.

(Watt, 1889-1899)

Leaves are useful in ulcer. Decoction of root is antibilious. Leaves are used in curry for the treatment of enlarged spleen. Mucilaginous water obtained from leaves is useful in unnatural heat of kidneys. It gives relief in strangury, gonorrhoea, lochial discharge and puerperal diseases. It is also used for spermatorrhoea, impotence and applied locally on apthae.

(Anonymous, 1946-1976)

Fruits are used as aphrodisiac, and given as decoction for incontinence in urine, nocturnal emission, spermatorrhoea and impotence.

(Ambasta, 1986)

Fruits are antispasmodic, aphrodisiac and used for dysmenorrhoea, leucorrhoea, and urine complaints. Fruits are also tonic for sex vigour.

(Jain, 1991)

Critical Notes

Leaves are beneficial in case of menolepsis. This local use has not been noted elsewhere in our country. It is exotic, native of Africa.

Sesamum orientale L.

Local Name

Til, Kothum-rie

Ethnobotanical Notes

Medicine

1. Seed powder, about half spoon, is given twice a day for a week to minimize the pains in piles and helps to overcome weakness.
2. Seeds are crushed and applied on skin for smooth and fair complexion.
3. Pinch of seeds are crushed in water, cotton ball is soaked in it and kept in ears to treat worms.

Notes from Classics

Roots are employed for hair wash. Mucilage from leaves is used as demulcent, cures cholera, diarrhoea, dysentery, catarrah and diseases of urinary passage. Leaves also serve for preparation of emollient cataplasm. Flowers are panacea for all eye diseases. Seeds are considered emollient, nourishing, tonic, diuretic and lactogogue. They regulate bowels and piles. Poultice of seeds is applied to ulcer. Seeds and oil are used as demulcent in dysentery and urinary diseaes. Seed oil is used for rheumatism and boils. Seeds are powerful emmenogogue, capable of producing abresion. Decoction of seeds is prescribed for cough. Compound decoction with linseed is employed as aphrodisiac. Ground seeds are applied to scald and burns.

(Watt, 1889-1899)

Seeds are emollient, diuretic, lactogogue, nourshing, tonic and helpful in piles. Decoction of seeds is emmengogue, given in cough-combined with linseed. The decoction of seeds is used as aphrodisiac. Plaster of ground seeds is applied to burns and scalds. Poultice of seeds is applied to ulcer. Powdered seeds are used in amenorrhoea and dysmenorrhoea.

(Anonymous, 1946-1976)

Sesamum oil is ingredient of confectionary and is used in making soaps, consmetics perfumery, insecticide, hair oil etc. Fresh leaves are used for affection of kidney and bladder. Externally it is applied for ophthalmic and cutaneous complaints.

(Ambasta, 1986)

Seeds are abortifacient, used for boils and ulcer. Seeds promotes lactation. Seeds are edible. Plants are given in dysentery, leprosy and spleen complaints.

(Jain, 1991)

Critical Notes

Seeds are useful for fair complexion and to treat worms in ears. These uses have not been recorded in past. The plant species is cultivated mainly for oil-seeds.

Family: Martyniaceae

Martynia annua L.

Local Name

Wagh-nakhi, Winchu

Ethnobotanical Notes

Medicine

1. Fruit paste is applied on scorpion bite.

Miscellaneous

1. Leaves are spread around bed while sleeping to repel bed bugs.

Notes from Classics

Fruits are used as antidote to scorpion bite, venomous bite and sting.

(Watt, 1889-1899)

Leaves are eaten at the time of scarcity. They are used in epilepsy and applied on tuberculous gland. Juice of leaves is used as gargle for sore throat. Fruit is considered alexiteric and used for inflammation.

(Anonymous, 1946-1976)

Ambasta (1986) mentioned similar properties as in Anonymous (*loc.cit.*).

Roots cure tonsillitis. Leaves cure epilepsy, sore throat and pain in neck, also applied in veterinary wound. Fruits give relief in itch and rheumatism. Paste of young twigs is applied on boils. Plant is useful for carbuncle, scabies and hair fall.

(Jain, 1991)

Critical Notes

Leaves are bed bug repellant. Probably, it is due to the secretion from glandular hairs present on leaves. It is exotic and native of tropical America. It has naturalized on waste land.

Family: Acanthaceae

Adhatoda zeylanica Medic.

Local Name

Adulsa

Ethnobotanical Notes

Medicine

1. Stem bark of this plant and leaves of *Melia azadirachta* L. are cooked and this cooked paste is applied on to rheumatic joints for relief.
2. Leaves of this plant and *Tribulus terristris* L. are extracted in milk. About half cup of it is administered twice a day to cure leucorrhoea. This is followed till cure.
3. Leaves are crushed in alcohol or wine. This paste is fried in 'Deshi ghee'. It is consumed regularly for curing piles.
4. Leaf juice is mixed with equal quantity of honey. About 20 ml of it is advised per day to cure tuberculosis.
5. About two spoonful of leaf decoction is drunk regularly to treat cough till cure.
6. Six young leaves and turmeric powder (*Curcuma longa* L.) are crushed in cow-urine and then applied over scabies.

Notes from Classics

Roots and leaves are important remedy for all sorts of cough and also used in phthisis. Roots, leaves and flowers are antispasmodic. Juice of leaves is employed for diarrhoea, dysentery and also bleeding during dysentery and malarial fever.

(Watt, 1889-1899)

Roots and leaves are antispasmodic and used in cough. Leaves are used for bronchitis, asthma and fever. These constitute good remedy against jaundice.

(Anonymous, 1946-1976)

Fresh and dried leaves constitute drug used in bronchital troubles and consumption. Leaf juice is also used in diarrhoea, dysentery and glandular tumours. Powdered leaves are used for skin eruptions.

(Ambasta, 1986)

Roots are anthelmintic, antipyretic, antiseptic and antispasmal. They are used against asthma, bronchitis and eczema. It is reported expectorant and also facilitates child-birth. Leaves are also used for asthma, bronchitis cough, cold and applied on dislocation of joints, rheumatic pains and swellings. It is good febrifuge and used against malaria. Leaves are used as antispasmal in cough and cold. Seeds are used in boils. Plant is useful for veneral complaints.

(Jain, 1991)

Critical Notes

Leaves are found useful in treating scabies, leucorrhoea, piles and tuberculosis. Stem bark is an important medicine against rheumatic pains. These uses have not been recorded earlier. The plants run wild in waste places. It is also planted as medicinal plant.

Barleria cristata L.

Local Name

Nili-koranti

Ethnobotanical Notes

Medicine

1. Root powder is mixed in water. After three to four hours, a cup of this infusion is advised to a patient till cure suffering from leucorrhoea.
2. Fruit of this plant and that of *Gmelina arborea* Roxb. are boiled together in water. Three spoons of this mixture are administered for a week to cure swelling of body.

Notes from Classics

Roots and leaves are used to reduce swellings and their infusion in given in cough. Seeds are reported useful as antidote for snake bite.

(Watt, 1889-1899)

Roots are used for inflammation, cough and cold. Leaves are also used for inflammation and cough.

(Anonymous, 1946-1976)

Similar uses are noted by Ambasta (1986) as in Anonymous (*loc. cit.*).

Roots find place in the treatment against anemia and headache. Leaves and roots are chewed for tooth-ache. Plant is used for body pain and swellings of legs.

(Jain, 1991)

Critical Notes

Infusion of roots is beneficial for leucorrhoea. This is unrecorded earlier. Roots and leaves are already reported as remedy against body swellings. Present notice of its fruits for similar purpose is being reported for the first time.

Barleria prionitis Sant.

Local Name

Kate karvanti, Piwli koranti

Ethnobotanical Notes

Medicine

1. Leaf paste is applied on gums for tooth-ache and gum complaints.
2. Leaf extract, about two spoons, is taken orally for cough and cold for three to four times a day till cure.

Notes from Classics

Juice of leaves is a favourite medicine in catarrhal affection and fever. Native apply leaf juice to their feet in rainy season to harden them preventing lacreation and cracking of sole. Leaves are beneficial in syphilitic affection as an alterative. They are used in cough and infantile diarrhoea. They are used as diaphoretic and expectorant.

(Watt, 1889-1899)

Roots are used for catarrahal affection of children, boils and glandular swellings. Roots are febrifuge. Leaves are employed for tooth-ache.

(Anonymous, 1946-1976)

Similar uses are noted by Ambasta (1986) as in Anonymous (*loc.cit.*).

Roots are used for boils. Bark is useful on anasarca, gland swelling and dropsy. Leaves are employed for catarrhal affections, rheumatism and ear complaints. It controls laceration and cracking of feet. It is also used to make feet tolerant to heal. Leaves are also used against piles and tooth-ache. Plant is used to cure tuberculosis.

(Jain, 1991)

Critical Notes

Medicinal use for gum complaints is the only additional one.

Blepharis repens (Vahl.) Roth

Local Name

Hadsan, Had-Sandhi

Ethnobotanical Notes

Medicine

1. Stem powder about 3 gm mixed in spoonful of honey and taken twice a day till the bone fractures cure.
2. Leaf paste is applied on forehead to stop head-ache.
3. Plant extract, about a cup, is drunk regularly at early morning to cure fever.

Notes from Classics

Leaves are applied on head for head-ache.

(Watt, 1889-1899)

Critical Notes

Utility of stem powder to treat bone fracture and plant extract for curing fever is not reported in past.

Hygrophila schulli (Buch.-Ham.) M.R. and S.M.Almeida

Local Name

Talimkhana

Ethnobotanical Notes

Medicine

1. Half cup of decoction of leaves is administered for stomach-ache till cure.
2. Paste of whole plant is applied on herpes.

Notes from Classics

Seeds are aphrodisiac and tonic. Whole plant is cooling, diuretic, used in dropsy, rheumatism and diseases of genito-urinary tract.

(Watt, 1889-1899)

Roots, leaves and seeds are diuretic, used in jaundice dropsy, rheumatism, anasarca and urinogenital complaints. Seeds yield semidrying oil.

(Ambasta, 1986)

Roots are employed for dropsy and rheumatism. Stem is aphrodisiac. Leaves are applied locally for body pains and to cure jaundice. Seeds are tonic and used to treat impotency. Plant is used to cure malaria, swelling of face and body, cancer and fistula. Plant is diuretic. Leaves, stem and twigs are used as vegetable.

(Jain, 1991)

Critical Notes

Leaves for stomach-ache and all part for herpes are beneficial. These are not yet reported.

Lepidagathis cristata Willd.

Local Name

Bendgawat

Ethnobotanical Notes

Medicine

1. Powder of flowering heads is filled in a smoking pipe called 'Chillam' it is smoked. Then cotton dipped in sesamum oil (*Sesamum orientale* L.) is applied inside the mouth to cure mouth ulcer.

2. Spoonful of powder of flowering head of this species and *Calotropis procera* (Ait) R.Br. stem in equal amount is given with water for dog-bite. It is followed for few days.

Notes from Classics

Ash of plant is applied on sores, itch, and for other skin diseases. Plant is administered in fever. Plant is a tonic. Whole plant is used as bitter tonic, and useful in fever. It is also applied on itchy affections of the skin.

(Anonymous, 1946-1976)

Ambasta (1986) reported similar properties as in Anonymous (*loc.cit.*).

Leaves are febrifuge. Whole plant is antiallergic, used for boils and wounds. Plant is useful to cure scabies, sores and syphilis.

(Jain, 1991)

Critical Notes

Uses of inflorescence as a remedy against mouth ulcer and also against dog-bite are new reports.

Peristrophe paniculata (Forsk.) Burm. *f.*

Ethnobotanical Notes

Medicine

1. The root powder in the form of poultice is applied over eyes for insomnia.
2. Ash of plant is mixed with coconut oil. This paste is applied over the swollen part of body.

Notes from Classics

Plant is macerated in infusion of rice and it is a useful remedy in case of poisonous snake-bite.

(Watt, 1889-1899)

It is given as fodder for horses and used as green manure.

(Anonymous, 1946-1976)

Ambasta (1986) reported similar uses as those in Anonymous (*loc.cit.*).

Leaves and fruits are used for eye diseases. Roots are used for gout and rheumatism. Whole plant is used for fractures and sprain.

(Jain, 1991)

Critical Notes

Use of roots for insomnia and plant ash for swollen body are previously not reported.

Rungia repens (L.) Nees

Ethnobotanical Notes

Miscellaneous

1. *Fodder*: Plant is relished by domestic cattle.

Notes from Classics

Leaves are applied to scalp for tinea capitis. Plant is used for cough and fever, also considered as vermifuge.

(Watt, 1889-1899)

Herb is used for cough and fever, also it is vermifuge and diuretic. Leaves are applied on scalp for tinea capitis in children.

(Anonymous, 1946-1976)

Ambasta (1986) reported similar uses as in Anonymous (*loc.cit.*).

Critical Notes

It is a common weed of cultivated fields. It is removed from fields while weeding out and fed to pet animals.

Family: Verbenaceae

Clerodendrum phlomidis Linn.*f.*

Local Name

Arni

Ethnobotanical Notes

Veterinary Medicine

1. Extract of leaves is applied on body of domestic animals to kill lices.

Miscellaneous

1. ***Fodder***: Leaves are good fodder for goats.

Notes from Classics

Plant is given to cattle to cure diarrhoea worms and in case of stomach swelling.

(Watt, 1889-1899)

Critical Notes

Leaves constitute good fodder by goats. Leaf extract is useful to kill lices in case of animals. Both uses are not noted in past. Interestingly, the author has observed that the cuttings of young branches sprout in wet soil. If cultivated by vegetative propagation *i.e.* cuttings, it may be a good fodder plant especially for goats.

Lantana camara L.

Local Name

Ghaneri

Ethnobotanical Notes

Miscellaneous

1. Stems are used for thatching roof and for making partition walls of huts.
2. It is planted as live hedge around crop fields and farmyards.

Notes from Classics

Leaves and twigs are occasionally used as green manure in forest areas. They are also composted along with other materials. Leaves yield oil which is used as antiseptic. Fruits are edible. Stems are used as tooth-brush.

(Anonymous, 1946-1976)

Fruits are edible. Leaves and flowers yield essential oil.

(Ambasta, 1986)

Fruits are edible.

(Jain, 1991)

Critical Notes

The plant species is very useful for the purposes reported above. It is usually planted as live hedge, also found running wild in waste places. It is exotic, native of Tropical America.

Gmelina arborea Roxb.

Local Name

Shivan

Ethnobotanical Notes

Medicine

1. Paste of roots is mixed with a pinch of common salt and turmeric powder (*Curcuma longa* L.). It is applied locally on herpes.
2. Paste of leaves is applied locally on scorpion bite for quick relief.
3. Fruit powder, about three spoons, is mixed in a cup of milk or in cow ghee. This is administered for two times during 3rd month of pregnancy for better settlement of foetus.
4. Few ripe fruits are crushed in a glass of milk and given to children daily as a tonic and to improve body weight.

Miscellaneous

1. *Fodder*: Leaves are used as fodder for cattle. Green fruits are used as fodder for goats.
2. *Wood*: Wood is used for making doors, windows and poles.

Notes from Classics

Roots are bitter, tonic, stomachic, laxative and used in anasarca. They are used in fever, indigestion and to increase secretion of milk. Leaves are used as refrigerant, anthelmintic, to cure gonorrhoea, cough and cold. They are demulcent. Leaves are used as fodder. Fruits are cooling, used as decoction in fever and bilious disorders.

(Watt, 1889-1899)

Roots are tonic, stomachic, laxative and galactogogue. Tender leaves are demulcent, applied on head for head-ache. Leaf juice is used to wash ulcer. Flowers are used in blood diseases. Drupes are used as refrigerant. Decoction of these is used in fever and bilious affections. Leaves are used as fodder.

(Anonymous, 1946-1976)

Wood is used for furniture, carriages, printing blocks, carving musical instruments, frames, axles, bridges, well work, match boxes and ship buildings. Decoction of fruits is used in fever and bilious affection.

(Ambasta, 1986)

Roots are blood purifier, laxative, tonic and control stomach diseases. Bark is antidote to poisonous insects. It is applied locally in bone fracture. Bark is useful to cure catarrah of bladder and stomach disorders. Leaves are carminative, used for cough, gonorrhoea, head-ache and wound. Plant is useful for bronchitis, cholera, colic, disorders, diarrhoea, dyspepsia, dropsy, epilepsy, and fever. It givers relief in phthisis, rheumatism, small-pox, sores and spleen complaints. Plant is useful for syphilis, throat swelling and urticaria. Fruits are edible. Wood is used for drums house buildings etc.

(Jain, 1991)

Critical Notes

Uses of roots for herpes, of leaves against scorpion bite, of fruits for settling foetus, as tonic to increase body weight are additional. The number of plants in forest areas are very few. It has been also planted by the Department of Social Forestry along the roads at some places. From fodder and timber point of view, it appears worth trying for captive plantation near villages on barren or waste lands in this district.

Tectona grandis L.*f.*

Local Name

Sag, Sagwan

Ethnobotanical Notes

Medicine

1. Blackness of skin near nose due to acidity is removed by regular application of leaf juice.

2. Spoonful of seed powder in cold water is taken daily to cure strangury.
3. Spoonful of seed powder mixed with honey is taken orally for few days to cure urine complaints.

Miscellaneous

1. Wood is used as timber for making furniture, agricultural implements mortar and many other articles.
2. Leaves are used for thatching huts.

Notes from Classics

Bark is astringent, applied on skin for relieving itching. Bark is tonic and used for leucorrhoea. Plaster of powdered wood is used in biliousness, inflammatory swellings and dyspepsia. Wood quenched in poppy juice reduced to a paste is applied to swelling of eyelids and strengthen eye sight.

(Watt, 1889-1899)

Powder of teakwood is said to be useful in allaying skin inflammation caused by *Melanorrhoea vsitlatissima*. Wood ash is applied on swollen eyelids. Oily product obtained by distillation of wood chips is useful for the treatment of eczema and ringworms. Oil is mixed with pounded seeds of *Derris oblonga* and used as ointment to cure itching. Maggots breeding in sores of cattle can be presented by applying wood tar.

(Anonymous, 1946-1976)

Leaves are used for thatching huts. Flowers are used in biliousness, bronchitis and urinary discharge. It is diuretic. Oil is used for scabies and to promote growth of hair. Root bark is used for colouring matter. Wood is used for poles, bims, columns roofs, window-frames, planks, staircases, carriages and ship buildings. Kernels yield oil which is used in scabies.

(Ambasta, 1986)

Stem bark is utilized for the treatment of diarrhoea and stomach-ache. Wood is employed for eye diseases, head-ache, swellings and also used for making musical instruments. Wood oil is applied externally for eczema, ring-worm and in sores of animals. Seeds are used in urine complaints. Oil of seeds is used as hair tonic. Red dye is used to paint furniture and leaves are used as umbrella.

(Jain, 1991)

Critical Notes

Blackness of skin is reduced by applying leaf juice; seed powder cures strangury and urinary complaints. These are additional uses.

Vitex negundo L.

Local Name

Nirgudi

Ethnobotanical Notes

Medicine

1. Root paste is applied externally on swellings.
2. Root paste is applied locally on abdomen to help easy delivery.
3. Decoction of root, about half cup, is given twice a day for urethritis.
4. Spoonful of stem powder is consumed a week for abortion.
5. Extract of roots and leaves in equal amount is mixed sasamum oil (*Seasamum orientale* L.) and boiled. This mixture is applied on skin for leprosy and scabies.
6. The aforesaid mixture about a spoonful is consumed at night for seven days to cure thyroid.
7. Leaf paste is applied on umbilicus to cure rheumatism.
8. Half cup of leaf juice is drunk with few drops of castor oil (*Ricinus communis* L.) to treat for back-ache.
9. About 25 gm of leaves pounded in 10 gm of ginger is boiled in 100 gm of water. Half cup of this filtrate is drunk twice a day to cure sciatica.
10. Stem is used as tooth brush.

Miscellaneous

1. Branches are used for partition wall of huts.

Notes from Classics

Roots are tonic, febrifuge and expectorant. Leaves are aromatic, tonic and febrifuge. Decoction of leaves is used in catrrahal fever with heaviness of head and dullness of hearing. Juice of leaves is said to remove foetid discharges and worms from ulcer. Leaf oil is applied to sinuses and scrofulous sores. Leaves are applied on swellings of joints and for gonorrhea. Leaves are used in warm bath for women after delivery. Powdered leaf is taken with water to cure intermittent fever. Fruits are nervine, cephalic and emmenogogue. The leaves are baked and applied to the head, while warm gives relief in head-ache. Plant is used in the treatment of cholera, stomach-ache and syphilis.

(Watt, 1889-1899)

Leaves are tonic and vermifuge. Decoction of leaves with long pepper is given in catarrhal fever with heaviness of the head and dullness of head. Leaves are smoked for head-ache and catarrah. Decoction of leaves and vapours are employed in bath for the treatment of febrile, catarrhal and rheumatic affections. Leaves possess discutient properties and are applied to rheumatic swellings of joints and sprains. Leaf juice is used in the treatment of foetid discharge. Ointment of leaves is applied as hair tonic.

(Anonymous, 1946-1976)

Roots are tonic, febrifuge and diuretic, used in rheumatism and dyspepsia. They are anthelmintic and used as demulcent in dysentery and pile. Flowers are astringent, used in diarrhoea, fever and liver complaints. Seeds are eaten after boiling. Young shoots are employed for basketry.

(Ambasta, 1986)

Roots are diuretic and expectorant. Stems are abortifacient. Stem is useful on bone fracture, gout, fever and veterinary diseases. Leaves are prescribed for blisters, boils, body-ache, cold, cough, dropsy and fever. They are also administered for gum trouble, head-ache, itch, paralysis, pile and rheumatism. They reduce sex-desire. Leaves are used in the treatment of swellings, tooth-ache and ulcer. They are vermifuge and sterilizer and applied on wounds. Flowers are the remedy on diarrhoea, dysentery, leprosy, eczema, pneumonia, diarrhoea and dysentery. Flowers are also applied locally on skin diseases and syphilis, head-ache, piles, mental disease, post-natal problems and swellings. Twigs are used for basketary and as insect repellent.

(Jain, 1991)

Critical Notes

Roots are useful for reducing swellings, easy delivery and urethritis. Stems are used as tooth brush. Leaves are employed for migriaine, leprosy, scabies, to treat typhoid, back-ache and sciatica. These uses have not been recorded so far. The species is wild and indigenous. It is to be noted that, it has been also planted near some human habitation for easy availability of the plant as it is widely used for different human aliments or diseases.

Family: Lamiaceae

Anisomeles indica (Linn.) O.Ktze.

Local Name

Rantil

Ethnobotanical Notes

Medicine

1. A half glass of decoction of plant is given at morning and evening to the patient suffering from rheumatism.

Notes from Classics

Herb is used as an astringent and carminative. Essential oil is useful in uterine affection.

(Anonymous, 1946-1976)

Roots are used as antidote to poisonous bites. Leaves are used for swelling neck and is also applied on wounds. Whole plant is astringent, carminative, used for fever muscular pains and nervous complaints. The whole plant is refrigerant, also used in veterinary medicine. Plant oil is useful in uterine affections.

(Jain, 1991)

Critical Notes

The plant is found useful for treating rheumatism. This is not noted in past literature.

Lavandulla bipinnata (L.) O.Ktze.

Ethnobotanical Notes

Medicine

1. Paste of leaves is applied on gums to relieve pains of tooth-ache.

Notes from Classics

Oil and tincture of plant is stimulant, antispasmodic and used in nervous and hysterical cases, flatulence and flatulent colic

(Watt, 1889-1899)

Paste of root is applied externally for stings and bite of poisonous animals.

(Anonymous, 1946-1976)

Flowers and leaves yield essential oil.

(Ambasta, 1986)

Plant is used to remove skin outgrowth.

(Jain, 1991)

Critical Notes

To get rid off from pain due to tooth-ache, leaves are not noted as such in past literature.

Leonotis nepetifolia (L.)R. Br.

Local Name

Khata, Dipmala

Ethnobotanical Notes

Medicine

1. Paste of seeds is applied on forehead at the time of delivery for easy delivery.
2. Paste of plant is applied on swelling of body.

Notes from Classics

Ash of flower buds is applied on burns and scalds. Ash of flower buds with curd is applied on ring worm and itching of skin.

(Watt, 1889-1899)

Decoction of leaves is used as tonic and febrifuge. Leaves are also used in the treatment of rheumatic affection. Ash of flowers is applied to scalds and burns. Plant is considered depurtive, emmenogogue, febrifuge, laxative and narcotic. Plant is also used in skin diseases amenorrhoea and fever.

(Anonymous, 1946-1976)

Ambasta (1986) reported similar uses as those in Anonymous (*loc.cit.*).

Stem paste is applied on swelling of body. Ash of inflorescence is useful for eye complaints. Calyx is used to remove stomach pains. Flowers are applied on burns, scalds, eczema and also helps in expelling worms. Seeds are used as remedy on dropsy, malaria, seed paste is applied on breasts. Seeds are diuretic. Plants control convulsions and cramps.

(Jain, 1991)

Critical Notes

Seed paste is useful for easy delivery and paste of entire plant is beneficial for body swellings. Both uses are not published anywhere.

Mentha spicata L.

Local Name

Pudina

Ethnobotanical Notes

Medicine

1. Leaf juice, about half cup, is given at morning and evening to cure jaundice till cure.
2. Decoction of leaves, about two spoons, is given twice a day for gastric troubles.

Notes from Classics

Oil of plant is stomachic, carminative and expectorant. Leaves are useful in fever, bronchitis and in apthe. Plant is carminative.

(Watt, 1889-1899)

Green leaves are used for making chutney and as a flavouring culinary preparations, vinegar, jellies and iced drinks. An alcoholic beverage is prepared from leaves and also used as an antidote for poison. Leaves are used in fever and bronchitis. Herb is considered stimulant, carminative and antispasmodic. A sweetned infusion of herb is given as a remedy for infantile troubles, vomiting in pregnancy and hysteria.

(Anonymous, 1946-1976)

Ambasta (1986) mentioned similar properties as those given in Anonymous (*loc.cit.*).

Leaves and shoots are edible.

(Jain, 1991)

Critical Notes

Use of leaves to treat jaundice is still untapped. It is cultivated especially for its use in culinary preparation. It is exotic, native of Europe.

Ocimum gratissimum L.

Local Name

Jangali tulsi, Rantulsi

Ethnobotanical Notes

Medicine

1. Leaf juice, about half cup, is taken as antidote on snake-bite.
2. Decoction of leaves, about two spoonful, is taken twice a day for a week for stomach-ache.

Notes from Classics

Decoction of leaves is valued in case of seminal weakness. Juice of leaves are used in gonorrhea. Leaves are used in mercurial rheumatism and mercurial salivation. Seeds are given in head-ache and neuralgia.

(Watt, 1889-1899)

Oil is used for external application on inflammed joints tooth-ache and ear-ache. It is taken internally for abdominal colic. Seeds are given for head-ache, neuralgia and dysentery. Decoction of seeds is used in urinary disorder. Decoction of seeds is given in seminal weakness.

(Anonymous, 1946-1976)

Leaves are mosquito-repellent and antibacterial. Herb is used in aromatic bath of fumigation for rheumatism and paralysis.

(Ambasta, 1986)

Leaves are used for intestinal disorder and gonorrhoea.

(Jain, 1991)

Critical Notes

Leaf juice is administered as an antidote for snake bite. This is still untapped.

Plectranthus mollis (Ait.) Spreng.

Local Name

Rantil

Ethnobotanical Notes

Medicine

1. Leaves are used as mosquito-repellent and bed-bug repellent. Plants are placed beneath the bed or cot by the tribals.

Notes from Classics

Leaves are used to stop bleeding and to cure fever. Leaves are used as mosquito-repellent.

(Anonymous, 1946-1976)

Leaves are mosquito-repellent. Leaves and flower tips yield essential oil which act as a cardiac depressent vaso constrictor.

(Ambasta, 1986)

Seeds are tonic, used in rheumatism.

(Jain, 1991)

Family: Nyctaginaceae

Boerhavia diffusa L.

Local Name

Punarva, Ghotali

Ethnobotanical Notes

Medicine

1. Roots of this plant and that of *Ricinus communis* L. are extracted in water. One cup of this extract is taken at morning and evening to cure body-ache till cure.
2. Root powder about two gm is homogenized in spoonful of honey and consumed at morning for six months as blood purifier.
3. About one cup of decoction of leaves is given twice a day to cure jaundice till cure.
4. Half cup of leaf extracts administered for eye health, daily for one or two months.
5. Leaves about 10 gm mixed with one spoonful of turmeric powder and few leaves of *Tinospora cordifolia* (Willd.) Miers. are crushed in about one liter water. It is boiled till the volume reduces to ¼ the of the total volume. A cup of this concentrate is taken orally at morning to cure stomach-ache.
6. Leaf paste is applied regularly on itch till cure.
7. Stems, roots and leaves in equal proportion are made into paste. It is applied locally around eyes to cure conjunctivitis.

Notes from Classics

Roots are laxative, diuretic, cooling, anthelmintic and used as expectorant. In large doses, they are emetic. Externally, root paste is used for dropsical swellings.

(Watt, 1889-1899)

Roots have diuretic, laxative and expectorant properties. They are useful in asthma on large doses, they are emetic.

(Anonymous, 1946-1976)

Roots are more powerful in action as compared to other diuretics on certain types of ascites, liver and chronic peritonitis.

(Ambasta, 1986)

Roots are abortifacient and facilitate child-birth. They are blood purifier. They reduce body heat and decrease temperature during fever. Roots are used for cough, inflammation, heart disease, kidney, stomach and urine complaints. They are also employed in piles, jaundice, wounds and dysentery. Leaves are useful in eczema, itch and rheumtism. Whole plant is useful for cold, epilepsy, fistula, eye liver and menstrual complaints. It cures pains in abdomen.

(Jain, 1991)

Critical Notes

The medicinal utility of roots for body-ache, of leaves for eye health and of entire plant for conjunctivitis are not documented in past.

Mirabilis jalapa L.

Local Name

Gulbakshi

Ethnobotanical Notes

Medicine

1. Leaves are crushed and applied on injuries and disinfectant.

Notes from Classics

Dried root powder is given in ulcer. Leaves are applied for abscesses and oils, also used as stimulant to ulcer. Dried root powder is given in milk as 'poultice'. Root powder is applied as 'lep' on contusion.

(Watt, 1889-1899)

Roots are purgative. Leaves employed as poultice for abscesses and boils. Juice from leaves is applied on wounds and for allaying itching in urticaria.

(Anonymous, 1946-1976)

Ambasta (1986) reported similar uses as those in Anonymous (*loc.cit.*).

Roots are useful on piles, also used on local cosmetics. Leaves are applied on boils and blisters. Tubers are used as vegetable, used at times of child-birth. Seeds are used as ornament for decoration.

(Jain, 1991)

Critical Notes

It is grown as ornamental but also naturalized on waste places. To avoid infection after injury, leaf paste is applied. This is not noted earlier. It is exotic native of tropical America.

Family: Amaranthaceae

Achyranthes aspera L.

Local Name

Aghada, Aghadla, Chirchira

Ethnobotanical Notes

Medicine

1. Root paste is applied externally on abdomen of a women, it helps for quick delivery.
2. Root paste is applied on snake bite for quick healing.
3. Half cup of infusion of roots is taken orally as a remedy against indigestion.
4. Leaf juice is applied on wound caused due to thorns. It helps to kill pains.
5. Poultice of leaves is tied on body part to remove easily thorn or spine from it.
6. Few drops of leaf juice are dropped in nose to stop head-ache.
7. Juice of leaves is also dropped in ears to reduce ear-ache.
8. Leaf extract, about 20 ml, is taken orally to avoid vomiting.
9. A spoonful of extract of leaves is drunk to strop excessive sweating.
10. Paste of entire plant is applied on swelling and for itching of legs and hands.
11. A pinch of ash of entire plant is mixed in cold water to cure strangury.
12. Seeds alongwith some sugar is powdered finely, about spoonful of it is advised in water daily to cure jaundice.

Notes from Classics

Fresh leaves are crushed into paste and mixed with black pepper and garlic and used as antiperiodic. Dried leaves are smoked in asthma and useful in dysentery. Leaves relieve the pains of scorpion bite. Leaf juice is applied on primary syphilic sores. They are emetic, useful in hydrophobia and irritations of boils. Plant is astringent, diuretic, useful in the treatment of menorrhagia and diarrhoea. Plant-juice is applied to relieve tooth-ache. Juice of plant if injected in uterus, labour pains are set up rapidly. Plant is used in ascites, anasarca, dropsy, boils, piles and eruptions of skin. Powdered plant is given to children for colic. Flower spikes are regarded protective against a scorpion, insect and snake bites. Seeds are emetic and useful in hydrophobia.

(Watt, 1889-1899)

Whole plant is astringent and diuretic. Leaves, flowering spikes and seeds are protective against scorpion bite. Decoction of entire plant is diuretic and useful in renal dropsy.

(Anonymous, 1946-1976)

Ambasta (1986) recorded uses similar to those of Anonymous (*loc.cit.*).

Roots are used as abortifacient and cause antifertility in women. It also stops bleeding in delivery and facilitate child-birth. Roots are used against asthma, bronchitis and pains in ribs, stomachache, whooping cough and wounds. Leaves cure colic, cold and cough, ear and eye complaints. Leaf juice is applied on wound of insect bite and also controls liver complaints and is used as a febrifuge. It is an antidote on scorpion bite. Leaves are chewed for stomach ache and sores. Inflorescence is used in cough and cold. Seeds are employed for hydrophobia, cough and reduced hunger. They are good tonic. Whole plant is used in the treatment of cough, dropsy, dog bite, liver complaints and piles. It cures syphilis, rheumatism, scabies and renal complaints. Whole plant is used to expel placenta in veterinary medicine.

(Jain, 1991)

Critical Notes

The uses such as root for snake bite, leaves to kill pains and to remove spine and thorn, for head-ache, ear-ache, to avoid vomiting, to stop excessive sweating etc. are being noticed for the first time. Likewise, ash of entire plant for strangury and seeds for jaundice are additional medicinal uses.

Alternanthera sesillis (L.) R.Br. ex DC.

Local Name

Khadak-shepu

Ethnobotanical Notes

Miscellaneous

1. ***Food***: Young leaves are used as vegetable.
2. ***Fodder***: Young shoots and leaves are good fodder for pet animals.

Notes from Classics

Plant is largely eaten as a vegetable especially by mothers to increase flow of milk. It is also used as a wash for eyes.

(Watt, 1889-1899)

Leaves are accredited with galactogogue properties. Leaves are used in soaps.

(Anonymous, 1946-1976)

Properties similar to that of Anonymous (*loc.cit.*) are reported by Ambasta (1986).

Leaves are used as vegetable. Root juice spray repels bees.

(Jain, 1991)

Critical Notes

The plant is also favoured by pet animals, a fact still remained unnoticed.

Alternanthera tenella Colla

Local Name

Bhaji

Ethnobotanical Notes

Miscellaneous

1. ***Food***: Leaves are used as vegetable.
2. Leaves and young shoots are favoured by live-stocks.

Critical Notes

This species is being reported for the first time for its ethnobotanical use. It is exotic weed, fast naturalized in this region. It's another form is used as border in gardens, but it reproduces vegetatively or by cuttings. However, the weedy form noted above produces seeds in large numbers.

Amaranthus polygonoides L.

Local Name

Bhaji

Ethnobotanical Notes

Miscellaneous

1. ***Food***: Leaves are used as vegetable.

Notes from Classics

Plant is used as pot herb. It is considered useful drug for convulsant.

(Watt, 1889-1899)

Critical Notes

It is a fact that many of the amaranthaceous species are useful as vegetable, especially the foliage. This species can be added in this list. It is a weed of waste and cropped lands.

Amaranthus viridis L.

Local Name

Bhaji

Ethnobotanical Notes

Miscellaneous

1. ***Food***: Leaves are used as vegetable.
2. ***Fodder***: Entire plant is useful as fodder for cows, bullocks and buffaloes.

Notes from Classics

Tender tops are eaten by the natives, though not so much esteemed as a cultivated sorts.

(Watt, 1889-1899)

Plants are used as a pot herb and also as cattle fodder.

(Anonymous, 1946-1976)

Plants are used as a pot herb. It tastes like spinach when boiled. It is also used as cattle fodder.

(Ambasta, 1986)

Leaves are used as vegetable.

(Jain, 1991)

Critical Notes

This plant species is weedy in nature. Apart from waste places, it is found in cropped land as well. It is manually removed and then used as fodder for cattle. Farm labours sometimes use the foliage as vegetable.

Celosia argentea L.

Local Name

Kurdu, Kombada, Shahu-medha

Ethnobotanical Notes

Medicine

1. A spoonful of root extract is mixed with five seeds of *Diospyros melanoxylon* Clarke. It is homogenized thoroughly and given to a woman orally as contraceptive every fifteen days before menses.
2. Root paste is applied locally on abdomen at night for removal of urine stone.
3. Root powder is mixed with few drops of Rosha oil [*Cymopogon martinii* (Roxb.) Wats.] and juice of lemon [*Citrus aurantifolia* (Christm.) SW]. It is applied on face for facial beauty.
4. Leaf extract, about half cup, is taken orally once a day for fifteen days in the treatment of urine stone.
5. A cup of leaf extract is drunk regularly to cure paralysis of legs.

Miscellaneous

1. *Fodder*: Plant is used as fodder for cattle.

Notes from Classics

Seeds are used as a remedy in diarrhoea. Plant is used as pot-herb in scarcity. It is also used as fodder for cattle.

(Watt, 1889-1899)

Seeds are used for diarrhoea and blood dysentery. It is a remedy on mouth sores. Seeds are useful for clearing vision and diseases of eyes.

(Anonymous, 1946-1976)

Plant is eaten as pot-herb in times of scarcity.

(Ambasta, 1986)

Seeds are prescribed as a blood purifier and in diarrhoea. It is also used in sores of mouth. Plant is useful to control blood dysentery. Leaves and shoots are used as vegetable.

(Jain, 1991)

Critical Notes

Utility of roots as contraceptive for women, to treat urine stone and facial beauty have not been found in published literature. Leaves are also used for urine stone and paralysis of legs. These are also not recorded earlier. It is noxious weed of crops and also found in grasslands.

Digera muricata (L.) Mart.

Local Name

Kundra, Gotaya-kirdu.

Ethnobotanical Notes

Miscellaneous

1. ***Food***: Leaves and shoots are used as vegetable.
2. Leaves also constitute a good fodder for domestic animals.

Notes from Classics

Plant serves as a pot-herb. Leaves and tender tops are used in the curries. Plant is used as fodder.

(Watt, 1889-1899)

Plant is relished by cattle. Leaves and tender shoots are eaten as pot herbs.

(Anonymous, 1948-1976)

Tender shoots are eaten as pot-herb. Plant is relished by cattle.

(Ambasta, 1986)

Whole plant and leaves are used as vegetable.

(Jain, 1991)

Critical Notes

The food and fodder value is again proved by the present author's observations. The plant is common weed of cultivated fields. It is removed manually and fed to cattle. Leaves are also used as vegetable by poor field labours. Thus it is a useful weed.

Pupalia lappacea (L.) Juss.

Local Name

Lapeti

Ethnobotanical Notes

Miscellaneous

1. *Food*: Leaves are used as vegetable by the tribal people.
2. *Fodder*: It is used as fodder for goats.

Notes from Classics

Fruits are applied to cuts and ingredients of enema preparation with palm oil, it is applied to boils. Fruit soup is given for cough and fever. Plant ash is applied to leprous sores after making them bleed.

(Anonymous, 1946-1976)

Ambasta (1986) gives similar properties as those are given in Anonymous (*loc.cit.*).

Critical Notes

Leaves are being reported for the first time useful as vegetable and fodder for goats.

Family: Chenopodiaceae

Chenopodium murale L.

Local Name

Chil, Chiwal-ni-bhaji

Ethnobotanical Notes

Miscellaneous

1. *Food*: Leaves and tender shoots are used as vegetable.

Notes from Classics

Plant is used as pot herb.

(Watt, 1889-1899)

Plant is used as pot herb.

(Anonymous, 1946-1976)

Plant is used as pot herb.

(Ambasta, 1986)

Leaves and stems are used as vegetable.

(Jain, 1991)

Critical Notes

The food value of this species is proved and can be reiterated.

Family: Aristolochiaceae

Aristolochia bracteata Retz.

Local Name

Kidamari

Ethnobotanical Notes

Veterinary Medicine

1. Decoction of leaves is applied on neck swelling or wound of cattle. It is practiced till cure.

Notes from Classics

Leaves are given internally to children for colic. They are supposed to be a emmenogogue. Leaf juice is applied to sore. Whole plant is anthelmintic, purgative and antiperiodic in fever. Whole plant is used in case of bowel complaints and on worms. It increases uterine contractions during labour pains. It is a good tonic.

(Watt, 1889-1899)

Decoction of roots was found efficacious in expulsion of round worms. Bruished leaves are mixed with castor oil and applied to control eczema. Seeds yield non drying fatty oil. Plant is purgative and anthelmintic.

(Anonymous, 1946-1976)

Uses similar to those of Anonymous (*loc.cit.*) are reported by Ambasta (1986).

Roots are used on fever. Leaves are vermicide and hasten delivery. Leaf paste is applied locally on eczema. Seeds are purgative. Whole plant is anthelmintic.

(Jain, 1991)

Critical Notes

It is a common weed of waste places and cropped fields. The observation from this district and those from earlier literature is suggestive of its chemical scrutiny especially for active principle, if not evaluated.

Family: Loranthaceae

Dendrophthoe falcata (L.*f.*) Etting

Local Name

Bandgul

Ethnobotanical Notes

Medicine

1. Ripe fruits are crush and mixed in cow ghee. Two spoonful of it is taken orally for 10 to 15 days in the third month of pregnancy for settlement of foetus.
2. Half cup of stem infusion alongwith cow ghee in 1:1 ratio is taken orally early in the morning for a fortnight to cure eye complaints.

Miscellaneous

1. Ripe fruits are edible.

Notes from Classics

Bark is used s a substitute for betel nut.

(Watt, 1889-1899)

Bark is astringent and narcotic. It is used in wounds and menstrual complaints. It is a good remedy for constipation, asthma and mania. Bark is used as substitute of betel.

(Anonymous, 1946-1976)

Roots are useful in skin diseases. Stem and whole plant is used as abortifacient. Stem, stem bark and whole plant is used against antifertility. Stem is good brain tonic and stops head-ache. Stem and leaves are useful for quick healing of wounds. Whole plant is used to recover paralysis and impotency.

Properties written in wealth of India (Anonymous *loc.cit.*) also reported by Ambasta (1986).

Critical Notes

The medicinal utilities of roots for curing eye complaints and leaves for settlement of foetus in case of human beings have not been noted elsewhere.

Viscum nepalense Spreng.

Local Name

Bhalki

Ethnobotanical Notes

Veterinary Medicine

1. Extract of plant is given when the cattle are reluctant to eat fodder.

Notes from Classics

Plant is given for fever attended with aching limbs.

(Watt, 1889-1899)

Plant is used as an aphrodisiac and febrifuge. Its paste is applied to cuts.

(Anonymous, 1946-1976)

Ambasta (1886) documented similar properties as those given in Anonymous (*loc.cit.*).

Stems are used against head-ache and antifertility. Leaves are used for mending bone fracture and bruises. Whole plant is used for fever, itches, pain in legs, rheumatism, syphilis, hemiplegia, torn pinna of ear etc.

(Jain, 1991)

Critical Notes

The aforesaid ethno-veterinary use is unrecorded.

Family: Santalaceae

Santalum album L.

Local Name

Chandan

Ethnobotanical Notes

Medicine

1. Half spoon powder of leaves is mixed in sugar and consumed to cure sore produced due to unpurified blood.
2. Juice of leaves, about one spoonful, is given at morning and evening to cure dysuria.
3. Spoonful of stem powder with sugar is consumed for two times to cure blood-pressure till cure.
4. Spoonful of stem powder alongwith cold water is drunk to reduce body heat.
5. Half cup of decoction of wood is drunk with honey and boiled rice water once a day regularly to cure flatulence.
6. Paste of wood is applied onto swellings and itching body parts.

Notes from Classics

Wood is bitter, cooling, astringent and useful in biliousness, vomiting, fever, thirst and heat of body. An emulsion of wood is cooling. Wood powder is rubbed over skin to cool it and allay the irritation of prickly heat. Wood is valuable remedy on gonorrhoea. Wood is cold and dry, cardiac, tonic astringent, alexipharmic, antiaphrodisiac, resolutions of inflammation and swellings etc.

(Anonymous, 1946-1976)

Wood and oil is employed medicinally. They are cooling, diaphoretic, diuretic and expectorant properties. Paste of wood is applied on burns, fever and head-ache. Oil is prescribed for vonorrhoea and skin trouble. Timber is useful.

(Anonymous, 1946-1976)

It is a finest wood for carving. Oil is extracted for heart wood and used in perfumery.

(Ambasta, 1886)

Seeds are abortifacient. Wood is employed for boils, fever, head-ache and skin disease. Incense is obtained from wood. Plant is used for dysentery and spleen complaints.

(Jain, 1991)

Critical Notes

Leaves are useful to treat sores and dysurea. Wood is employed in the treatment of blood, flatulence, swellings and itchings. Usefulness of this sandal wood tree on this line is not on record yet. Wood is purchased from markets, although live plants are found wild.

Family: Euphorbiaceae

Acalypha indica L.

Local Name

Khokli

Ethnobotanical Notes

Medicine

1. A cup of decoction of the entire plant is administered daily against cough and bronchitis till cure.

Notes from Classics

Leaves alongwith garlic (*Allium sativum*) are useful as anthelmintic. Leaf powder together with common salt is applied on scabies. Leaf juice is emetic. Leaf juice is mixed with mustard oil is useful in rheumatism. Plant is reputed expectorant.

(Watt, 1889-1899)

Leaves are used for constipation in children. Leaf juice is emetic. Leaves mixed with common salt are used as parasiticide and applied externally. Leaves mixed with limejuice is used against ring-worm. Fresh and dry plant is gastro-intestinal irritant. A decoction of entire plant is useful as laxative.

(Anonymous, 1946-1976)

Herb is used against cough, associated with bleeding from nose and incipient phthisis. Leaf juice is employed for cutaneous troubles.

(Ambasta, 1886)

Leaves are used for ear-ache and skin diseases like scabies. Whole plant is useful for asthma bronchitis, cough, pneumonia, rheumatism and as a laxative.

(Jain, 1991)

Critical Notes

The plant species is reputed source of medicine for cough in our country. It holds true as folk medicine in this district and it is similarly warranted by its local name 'Khokli' which itself means cough.

Acalypha malabarica Muell.-Arg.

Local Name

Fulki

Ethnobotanical Notes

Miscellaneous

1. ***Fodder***: The entire plants constitute good fodder for domestic animals.

Critical Notes

The plant is a common weed in crop fields. While weeding out manually, it is not thrown by the farmers. They collect and fed to animals. It is thus a useful weed.

Baliospermum raziana Keshav. et Yog.

Local Name

Datti, Danti

Ethnobotanical Notes

Medicine

1. Young twig is used as tooth-brush to cure tooth-ache.

Notes from Classics

Roots are cathartic and used in dropsy, anasarca and jaundice. Leaves are applied to heal wounds and also to treat asthma. Seeds are drastic purgative, stimulant and rubefacient. Seed oil is hydrogogue, cathartic externally applied to rheumatic joints.

(Watt, 1889-1899)

Roots are cathartic. Seeds are drastic purgative. Seed oil is hydrogogue, cathartic, useful in rheumatism.

(Anonymous, 1946-1976)

Ambasta (1986) reports similar uses as those in Anonymous (*loc.cit.*).

Roots are used for dropsy and jaundice. Root is veterinary medicine and uses as laxative. Stem is used to treat tooth-ache. Leaves are used to cure asthma. Seeds are laxative, used for gastric complaints, rheumatism, gout and snake-bite. Plant is applied on wound for healing.

(Jain, 1991)

Bridelia airy-shawii P.T.Li.

Local Name

Asan

Ethnobotanical Notes

Medicine

1. Stem bark, about spoonful, is mixed in half cup of water and drunk every morning to decrease level of sugar in case of diabetic patient.

2. A pinch of stem bark powder mixed in spoonful of honey and few drops of lime juice [*Citrus aurantifolia* (Christm,) SW] is taken at early morning to reduce extra weight.
3. A small quantity of stem bark paste is mixed in a cup of milk and given to kids for three to five days twice a day. It expels tape warms and stops abdominal pain.
4. Seed oil is applied on head to increase length of hair and also to dye hair by tribal women.
5. ***Food***: Ripe fruits are eaten by aborigines.

Miscellaneous

1. Green leaves are used as fodder for domestic animals.
2. Wood is useful for making cart and for poles of huts.

Notes from Classics

Stem bark is used as anthelmintic.

(Watt, 1889-1899)

Wood is heavy suitable for turnery, carving and furniture. It is also used for making house post, rafter and cart etc.

(Anonymous, 1946-1976)

Fruits are edible. Leaves are used as fodder.

(Ambasta, 1986)

Stem bark is used against anti-fertility, arthritis debilia, diarrhoea, earache, pain in waist and rheumatism. Stem bark is also useful for cuts, tuberculosis, wounds and as antidote on snake bite. It is a good tonic. Fruits are laxative, edible and used against stomach-ache. Leaves are used as fodder for domestic animals. Dye obtained from bark is used to dye coarse wool.

(Jain, 1991)

Critical Notes

Stem bark is useful for diabetes, to reduce body weight, to kill intestinal worms and are being reported here for the first time. It is interesting to note use for hair. It should be critically assessed on experimental grounds. It is interesting to note use of seed oil for hair.

Chrozophora prostrata Dalz.

Local Name

Unnali, Betna

Ethnobotanical Notes

Medicine

1. Decoction of roots, about half cup, is given twice a day for cough problem in children till cure.
2. Infusion of roots is applied on cuts and wounds for quick-healing.

Notes from Classics

Ash of roots is given to children for cough. Leaves are considered depurative. Seeds are used as purgative. Whole plant is used for leprous affection.

(Watt, 1889-1899)

Ash of root is given to children for cough. Leaves are depurtive. Seeds are purgative and cathartic.

(Anonymous, 1946-1976)

This herb is acrid and poisonous. Seeds are used as cathartic.

(Ambasta, 1986)

Roots are used for asthma, burns, cuts, cold and cough. Wounds are cured by roots. Leaves are used for dermatitis.

(Jain, 1991)

Emblica officinalis Gaertn.

Local Name

Awala

Ethnobotanical Notes

Medicine

1. A spoonful of fruit powder with spoonful wood powder of sandal wood (*Santalum album* L.) is homogenized in honey. It is taken orally to stop vomiting.
2. Fruit ash and seasum oil (*Sesamum orientale* L.) is mixed and applied on scabies.

3. Equal amount of fruits and fruits of *Withania sommnifera* (L.) Dunal are powdered and about 10 gm of it taken with milk to cure weakness and improve complexion.
4. About 500 gm of leaves of *Centella asiatica* (L.) Lirb. mixed in bout 250 gm of fruit powder of *Emblica officinalis* Gaertn. are boiled in one litre of coconut oil. This oil is applied onto scalp for dandruff, hair growth on bald, and to prevent hair fall.
5. Twigs are used as tooth brush.

Notes from Classics

Roots are astringent. Bark is cooling, astringent and useful for gonorrhoea. Leaves are bitter tonic and used in chronic dysentery. Flowers are cooling, used as electuary with other articles. Fresh fruit is used for bilious affection, also used in cerebral disturbance, blood circulation, burning, head-ache, confusion of thought and loss of hair. Mustard oil is mixed with fruit powder, is allowed to remain sometime. Oil is applied on head gives good result in burning sensation of head. Dried fruits are astringent, used in haemorrhagia together with iron. They are valuable remedy on anemia, jaundice and dyspepsia. Fruit is refrigerant, cardial, blood purifier, used in diarrhoea and dysentery. Juice of fruit is cooling refrigerant, diuretic, laxative and used in dyspepsia. Infusion of seeds is used for eye disease.

(Watt, 1889-1899)

Root bark is astringent. Leaves and fruits are used as fodder. Flowers are cooling, refrigerant and aperiant. Fruits are acrid, cooling, refrigerant, diuretic and laxative. Dried fruits are used in haemorrhage, diarrhoea, dysentery, acute bacillary dysentery. Fruits are used in eye inflammation. Fruits are used in eye inflammation. Fruits are used for ink and hair dye. Wood is used for well-work, agricultural implements and fuel.

(Anonymous, 1946-1976)

Fruits are cooked and pickled. Fruits are used in hair dye, dried onse are used as detergent, used in shampoo. Wood is used in agricultural implements, poles and inferior quality furniture.

(Ambasta, 1986)

Leaves are applied on wounds and in case of animals leaves are employed for stomach disorders. Fruits are antiemetic, used in bronchitis, constipation diabetes, eye complaints, head-ache and indigestion. They are diuretic, laxative, purgative and tonic. Fruits are employed for liver complaints, stomach and vaginal complaints. It revives taste and quenches thirst. Fruits and seeds cure asthma. Fruits are pickled and used as local drink.

(Jain, 1991)

Critical Notes

Fruits are beneficial to cure scabies and improve complexion. Stem are useful as tooth-brushes. These uses still await documentation. Fruits are collected on large scale by the tribals from wild trees of forest areas. They are sold locally. The natural dispersal is truly reduced. Therefore, number of wild trees of this species have reduced. This warrants to check up the depleting species.

Euphorbia neriifolia L.

Local Name

Sabar, Sabarkand

Ethnobotanical Notes

Medicine

1. Stem axis is warmed, few drops of warmed extract are dropped in ears for ear-ache.
2. Latex is homogenized in mothers, milk. Half spoon at night for three days is given to baby orally for stomach-ache.
3. Five drops of latex with gram flour (*Cicer arietinum* L.) is consumed at night for habitual constipation.
4. Two drops of latex in spoonful of cow-ghee only at night for four days is advised for asthma.
5. Two drops of mixed latex with jaggary is taken at morning only for few days to cure jaundice.

Notes from Classics

Milky juice is purgative and rubefacient. Roots and black pepper are employed in snake-bite internally as well as externally. Latex mixed in ginger is given for mad dog-bite. Latex is used for ear-ache.

(Watt, 1889-1899)

Latex is acrid, rubefacient, purgative and expectorant. Cutaneous eruptions and warts are cured by latex if applied locally. Leaf juice is dropped in ears for ear-ache. Leaf juice mixed in soot is applied to opthalmia.

(Anonymous, 1946-1976)

Ambasta (1986) reported similar properties as those are given in Anonymous (*loc.cit.*).

Stem bark cures excessive thirst after child-birth. Stem is helpful for expelling intestinal worms. It is also used as purgative and vermifuge. Stem is employed for whooping cough and as fish poison. Leaves are used as expectorant. Latex is anthelmintic and used as antidote for snake-bite. Latex is a remedy against asthma, jaundice and skin diseases.

(Jain, 1991)

Critical Notes

Latex is being noted for the first time as beneficial in stomach-ache. Use of latex against ear-ache, constipation, asthma and jaundice have been also noted from other places in our country. The wide cross-country utilities of latex should be evaluated on scientific ground.

Euphorbia nivulia Buch.-Ham.

Local Name

Sabar

Ethnobotanical Notes

Miscellaneous

1. It is planted as live hedge along border of fields and farmyards.

Notes from Classics

The bark is suitable for making composite clock boards.

(Anonymous, 1946-1976)

Similar uses as in Anonymous (*loc.cit.*) are reported by Ambasta (1986).

Latex is used as fish poison.

(Jain, 1991)

Euphorbia tirucalli L.

Local Name

Thorya, Thor, Sher

Ethnobotanical Notes

Medicine

1. Few drops of latex are dropped in ears for deafness.
2. Latex is warmed and then applied topically on knees for rheumatic swelling regularly.

Miscellaneous

1. It is planted around farmyards and fields, along roads, etc.
2. When the dead child or unmarried boys are buried, the twigs of this plant are placed around the buried body.

Notes from Classics

Juice of plant is used as remedy in rheumatism, tooth-ache and debility. Juice cures affections of the spleen. It is purgative, cathartic, emetic and antisyphillitic. Juice is applied externally on ulcer. Goats and camel eat both leaves and bark.

(Watt, 1889-1899)

Latex is vasicant, rubefacient, used in application for warts, rheumatism, neuralgia and tooth-ache. In small dose, latex acts as purgative but in large doses it is an irritant and emetic. Roots are administered for pains in colic and gastralgia. Latex is used to cure cough, asthma and ear-ache. Plant ash is applied on abscesses. Wood is used for rafter and toys.

(Anonymous, 1946-1976)

Ambasta (1986) reported similar properties as those in Anonymous (*loc.cit.*).

Latex is used for body pain, eczema ear-ache, scabies and rheumatism. It is also applied locally on teeth for tooth-ache and on wounds.

(Jain, 1991)

Decoction of leaves promotes secretion. Leaves warmed and smeared with cotton oil is useful for suppurative wounds. Decoction of leaves is used to wash eczema and ulcer. Seeds are purgative, milky juice is styptic.

(Watt, 1889-1899)

Seeds are poisonous, purgative, antihelminthic and used as antidote for snake poisoning. Seeds are used for catharasis. Seeds are fried, powdered and taken for stomach-ache. Seeds are used as illuminant lubricant, used in soaps and candles.

(Anonymous, 1946-1976)

Seed oil is purgative, also used for making candles, soap, varnishes, illumination, lubrication and in wool industries. Seed oil used in sciatica, dropsy, paralysis and externally for skin troubles and rheumatism, abortifacient. Plant juice is useful in scabies, eczema and ring-worm. Leaves rubefacient and lactogogue

(Ambasta, 1986)

Root is abortifacient. Stem is applied on sores for tooth troubles and on wounds for healing. Twigs are used as tooth-brush. Leaves are used for congestion of chest and inflammation. Latex is efficacious in cancer. Seeds are used to control rheumatism and are purgative. Whole plant is used in the treatment of anasarca, carbuncle, cramps, dropsy. Pleurisy, pheumonia, rheumatism and syphilis. Bark and stem yield black dye. Seed oil is used for lighting.

(Jain, 1991)

Critical Notes

Usefulness of leaves against head-ache, young stem for dental carries and seeds for washing hair and is not recorded earlier. It is usually planted as live hedge. It is a exotic, native of Tropical America.

Jatropha gossypifolia L.

Local Name

Chandra Jyot

Ethnobotanical Notes

Medicine

1. Half cup of decoction of leaves is used as purgative. Only single dose is prescribed.

Veterinary

1. Stem pieces are given to animals to eat alongwith fodder. They improve digestion and appetite.

Notes from Classics

Roots are used as an antidote for snake-bite. Decoction of bark is used as an emmenagogue. Leaves are useful for stomach-ache, veneral diseases and blood purifier. Leaves are also applied for carbuncle, eczema and itches. Juice is applied for sores on tongue of babies and cataspasm. Fresh leaves is useful on swollen mammae. They are used as febrifuge in intermittent fever in anthills. Latex is applied to ulcers. Seeds are eaten by doves and fowls. Seed oil is used in lamps and also for leprosy.

(Anonymous, 1946-1976)

Roots are employed against leprosy. Decoction of bark is used as a emmenagogue. Decoction of leaves is purgative and stomachic, used as febrifuge in intermittent fever, and as an applicant on swollen mammae. Latex is useful on ulcer. Seeds are purgative and emetic.

(Ambasta, 1986)

Roots are garlanded around the neck of cattle for healing wound.

(Jain, 1991)

Critical Notes

Veterinary use of stem to increase appetite and to improve digestion is hitherto not reported. It is native of Brazil, naturalized on waste places.

Kirganella reticulata (Poir.) Baill.

Local Name

Kali-kamhuni, Kali-takii

Ethnobotanical Notes

Medicine

1. Spoonful of root powder mixed in warm water is given once a day to cure cough till cure.

Miscellaneous

1. Ripe fruits are eaten by rural boys.
2. Stems are used for thatching roofs of huts.

Notes from Classics

Bark is alternative and attenuant. Leaves are diuretic and cooling. Juice of leaves is used for bleeding gums. Leaves are alterative for heat in blood.

(Watt, 1889-1899)

Decoction of root is used in cough, catarrah and for asthma in kids. Bark is alterative, astringent, diuretic and attenuant. Juice of stem is used for sore eyes. Leaves are cooling and diuretic. Juice of leaves is useful in bleeding gums. Leaves are also prescribed for diarrhoea in infants. Powdered leaf is applied on sores, burns, suppuration and chafing of skin. Fruits are astringent. Whole plant is used for small-pox and syphillis.

(Anonymous, 1946-1976)

Fruits are eaten in times of scarcity. Stems are used in making basketery.

(Ambasta, 1986)

Fruits are reputed medicine for catarrah and cough. Stem twigs are used for tooth-brush. Leaves are used for fish killing. They are cooling, used for diarrhoea and gum bleeding.

(Jain, 1991)

Ricinus communis L.

Local Name

Arandi, Arand, Diwaji

Ethnobotanical Notes

Medicine

1. Half spoonful of root extract is used once a day for three days to treat urethritis.
2. Equal amount of root and ginger (*Zinziber officinale* Rosc.) and bark of *Moringa concanesis* Nimmo. are pounded together and homogenized. The paste is applied over head for migraine.
3. Half spoon of leaf extract is taken with honey to reduce acidity.
4. Seed oil is applied on single entire leaf and applied on mother's chest for chest-swelling.

5. Extract of one leaf is added in half cup of fresh unboiled cow-milk and given for three days to cure jaundice.
6. Seed oil is used to massage for back-ache and joints pains.
7. Seed oil mixed with milk is used for massage of kids. This gives fairness.
8. Half spoon of powder of *Terminalia chebula* Retz. mixed with two or three drops of seed oil and consumed twice a day for two weeks to cure appendix.
9. Seed oil is applied locally on scabies till cure.
10. Consumption of few drops of seeds oil for a month at night help to regulate blood circulation.
11. Poultice of seed is applied for facial paralysis.

Notes from Classics

Dry fruit is regarded as febrifuge. Leaves are employed to arrest secretion of milk in women. Externally leaves are applied on wound and bruises which act as detergent. Leaves if heated and applied to the abdomen will promotes menstrual discharge. Leaves are useful in constiveness, tympanitis fever and inflammation. Oil and root paste are useful in for lumbago, pleurodynia and sciatica. Seed oil is powerful resolvent and purgative, useful in cold, palsy, asthma, colic, flatulence, rheumatism, dropsy and amenorrhoea. Oil is dropped in eyes in case of abrasion of conjunctiva. It is also applied on eruption of skin. The fresh juice leaves is used as an emetic in opium poisoning and other narcotics. Poultice of leaves with barley meal is applied to inflammatory affections of eye.

(Watt, 1889-1899)

Small quantity of castor seeds is used as mild purgative for children. Seed paste or poultice is applied to sores boil, gout and rheumatic swellings. Roasted seeds are used in veterinary practice. Seed oil is cathartic in the treatment of hook-worm, also given to induce labour pains in pregnant women. It is also used as abortifacient. Seed oil is applied externally on cutaneous affection. Seed oil is used in hair oils, hair lotions, tonic, aromatic perfumes and lubricants for heavy machinery.

(Anonymous, 1946-1976)

Poultice of leaves is applied on boils and sores. Seeds yield oil used in cathartic, also used as lubricants and illumination. Seed oil is purgative. Seed oil is used for dying cotton fabrics, textile, synthetic nylon and transparent soap. Drying oil is used in paints, varnishes, industries also used in polishes, candles, insulation material, adhesive, heat sealing, preservation of fruits and vegetable. Dried stems are used for thatching huts. Dried stems and seed hulls are fried and eaten by man.

(Ambasta, 1986)

Roots are prescribed for lumbago, rheumatism, sciatica, hydrocoel, injuries, joints pains, night blindness, pneumonia, wound, skin diseases and swellings. Leaves are also used as purgative. Seeds are antidote for food poisoning and act against antifertility. They are employed for constipation and contraceptive, also used in head-ache, boils and wounds. They act as laxative, purgative and vermicide. Seeds are used as spice. Seed oil is used for burns, cholera, delirium and dropsy.

(Jain, 1991)

Critical Notes

Uses of roots for urethritis, migraine and of leaves for reducing acidity are not documented in past. Likewise uses of seed oil for massage, to cure appendix, scabies, paralysis and for blood circulation are not hitherto reported.

Securingea leucopyrus (Willd.) Muell.- Arg.

Local Name

Pandhar-phali

Ethnobotanical Notes

Miscellaneous

1. *Food*: Ripe fruits are eaten by tribal people.
2. *Fodder*: Leaves are used as fodder for goats.
3. Stem bark is employed in fish poison.

Notes from Classics

Juice of leaves made into paste with tobacco and used to destroy worms in sores. Fruit is eaten. Wood is used as fuel.

(Watt, 1889-1899)

Leaves are eaten. Juice of paste of leaves is used with tobacco to destroy worms in sore. Berries are sweet and edible. Slender branches are used for making baskets and thatching huts.

(Anonymous, 1946-1976)

Ambasta (1986) reported similar uses as those in Anonymous (*loc.cit.*).

Securingea virosa (Roxb. ex Willd.)

Local Name

Sina

Ethnobotanical Notes

Miscellaneous

1. ***Food***: Ripe fruits are edible.
2. ***Fodder***: Leaves are used as fodder for goats.
3. Shoots are used for making partition wall of huts.

Notes from Classics

Bark is astringent. Fruits are eaten. Wood is used for making agricultural implements.

(Watt, 1889-1899)

Ripe fruits are eaten. Wood is used for agricultural implements, walking stick, tent pegs and chair legs. Branches are used for thatching huts.

(Anonymous, 1946-1976)

Ambasta (1986) gives similar uses as in Anonymous (*loc.cit.*).

Leaves are useful for bone fractures, constipation and also applied on sores to expel worms. Fruits are used to get to get relief from stomach-ache.

(Jain, 1991)

Family: Ulmaceae

Holoptelea integrifolia (Roxb.) Planch

Local Name

Papda

Ethnobotanical Notes

Medicine

1. Bark powder is applied on chronic wound for quick healing.
2. Leaves in powder form are applied onto boils.

Miscellaneous

1. Bark is used as fish poison by the tribals.
2. ***Food***: Seeds are eaten raw.
3. ***Fodder***: Leaves are used as green fodder for goats.
4. ***Wood***: Wood is employed for raising huts. Wood is also used as fuel.

Notes from Classics

Leaves are used for fodder. Oil is extracted from seeds. Wood is employed for building purposes for carving and cart making. Wood is also used for fuel and for making charcoal.

(Watt, 1889-1899)

Bark may be pulped and made into hard board and insulation boards.

(Anonymous, 1946-1976)

Leaves and young fruits are eaten though not relished by live-stock. Seeds yield oil. Timber is used for brush backs, handles of brooms, indoor building purposes, cheap furniture, cabinet work, carving, ploughs, yokes, carts and carriages, combs, shoe heals, match boxes, plywood, packing cases, bobbins, paper pulp. Wood is used for making charcoal.

(Anonymous, 1946-1976)

Ambasta (1986) reported similar uses as in Anonymous(*loc.cit.*).

Stem bark is tied on arm in hydrocele, also used for rheumatism, scabies, ulcer, for mending bone-fracture and to expels ring-worm. Leaves are employed for boils, inflammation of body and skin diseases. They are used as antidote on scorpion sting. Flowers are used as vegetable. Bark and leaves are used as fish-poison.

(Jain, 1991)

Critical Notes

Bark is medicinally important for healing wounds. Leaves are useful on boils. These uses are yet unreported. Leaves form good source of fodder. This tree species is good source of small timber and fuel wood. Seeds yields edible oil. This species is used multipurposely and hence it should be given priority in afforestation programmes of state Government and NGOs.

Family: Moraceae

Ficus arnottiana (Miq.) Miq.

Local Name

Amsa

Ethnobotanical Notes

Medicine

1. Paste of stem bark is applied locally on skin for ring-worms.

Miscellaneous

1. ***Fodder***: Leaves are used as fodder for cattle.
2. ***Food***: Ripe fruits are edible.

Notes from Classics

Leaves and bark are used in cutaneous affections. Leaves are lopped for fodder.

(Ambasta, 1986)

Leaves are used as steriliser. They are also used for preparing tea.

(Jain, 1991)

Critical Notes

Stem bark is used to treat ring-worm and ripe fruits are edible. These uses are being noticed for the first time.

Ficus benghalensis L.

Local Name

Wad, Wat.

Ethnobotanical Notes

Medicine

1. About spoonful of aerial root powder is taken thrice a day with milk for few months to increase fertility in women.
2. Paste of aerial roots is applied on scalp regularly at night to promote hair growth.
3. Decoction of bark, about half cup, is drunk once a day for about a month in the treatment of leucorrhea.
4. Powder of leaves is applied on gum for swelling regularly till cure.
5. Half spoon of latex is drunk after meal twice a day to bring down sugar level in diabetes.
6. Latex is applied locally on joints to reduce rheumatic pains.

Miscellaneous

1. Aerial roots are splited and then used to tie bundles of grasses or fodder.
2. *Fodder*: Leaves are used as fodder.

Notes from Classics

Milk juice is externally applied for pains and bruises and as an anodyne application in rheumatism and lumbago. It is a valuable application to the soles of seat for cricking and inflammation. Milk juice is also applied to teeth and gums for tooth-ache. Small quantity is useful for dysentery. It is good astringent. Root fibres are given in gonorrhoea. Tender ends of hanging roots are given for obstinate vomiting. Infusion of bark is tonic and used to treat diabetes. Leaves are applied as poultice to abscesses. Young buds are astringent and useful in diarrhoea. Milky juice is applied for pains and bruises and as an anodyne in rheumatism and lumbago.

(Anonymous, 1946-1976)

Infusion of bark is tonic astringent, used in diarrhoea, dysentery and diabetes. Leaves are cooling and tonic, and lopped for fodder. Latex is applied in rheumatism and lumbago. Fruits are eaten in times of scarcity. Wood is suitable for making paper pulp, in furniture, well carts wheels, yokes, carts etc. Fibres are obtained from bark and aerial roots.

(Ambasta, 1986)

Roots promote hair growth. Aerial roots are prescribed for diarrhoea, menorrhoea, leucorrhoea and used as tooth brush. Bark is used in the treatment of diabetes, urine complaints and snake-bite. Leaves are used for cough and cold. Fruits are edible and aphrodisiac. Latex cures blisters, boils, eye complaints, gum trouble, skin disease and spermatorrhoea. It is also used as tonic. Plant is used for dropsy, cholera, diarrhoea, dysentery, eye complaints, insanity, scabies, sore in mouth, as antidote for snake-bite. From bark fibres are obtained.

(Jain, 1991)

Critical Notes

Aerial roots are useful to increase fertility in women and also to promote hair growth. They are used to tie bundles of fodder by farmers. Leaf powder is beneficial for swelling. Latex is important to lower down sugar level for diabetic patients and to reduce rheumatic pains. All these uses have not been documented in past literature. Use of aerial roots to promote hair growth has been already picked up by druggists and industrialists.

Ficus glomerata Roxb.

Local Name

Umbar

Ethnobotanical Notes

Medicine

1. Roots sap comes out from cut end of a root. It is collected in earthen pot. Half cup of root sap is drunk once a day to cure piles.
2. Spoonful of bark powder is consumed with honey at night for a week to cure polyurea.
3. Spoonful of bark powder with honey is consumed twice a day for a month in case of bleeding lochia.
4. Latex is applied regularly for sprained body part till cure.

Miscellaneous

1. ***Fodder***: Leaves are collected for cattle fodder.

Notes from Classics

Roots cure dysentery and diabetes. Fluid obtained by its incison is tonic. Bark is astringent, used as wash for wounds. Infusion of bark is useful for manorrhagia. Powder of leaves mixed in honey is used for bilious affection. Figs are astringent, stomachic, carminative and given in manorrhagia and haemoptysis. Fresh juice of fruits is given for diabetes and urinary diseases. Milky juice is applied for piles, diarrhoea and with seasamum oil for cancer. Sap extracted from trunk is applied to mumps inflammation and enlargement.

(Watt, 1889-1899)

Bark is astringent, its decoction is used to wash wounds. Root are useful in digestion. Powder of leaves with honey is given in bilious affection. Fruits are stomachic and carminative. Milky juice is used in piles and diarrhoea. Leaves are used as fodder for cattle. Wood is used for rice mortals, plank, shutters, toys, sides of carts, frames, ploughs, oars, yokes, match boxes, cotton reels, fuse boxes for fitting etc.

(Anonymous, 1946-1976)

Roots are used for diabetes and diarrhoea. Bark is given to cattle in rinder pest diseases. Decoction of bark is vulnerary. Leaves are used in bilious affection and as fodder. Powder of roasted fruits is used as breakfast food. Latex is efficacious in piles and diarrhoea. Wood is used for well-curb, cheap furniture, fuse boxes fitting, match boxes, spokes of wheels, mortal, shutters etc.

(Ambasta, 1986)

Root is used for antifertility, veneral complaints and dysentery. Bark is used as veterinary medicine for plague. Stem is used as tooth-brush. Fruit is carminative and galactogogue, used for diabetes, leprosy, stomach and urine complaints. Leaves are used for bronchitis and pneumonia. They are used as galactogogue. Latex is used in the treatment of blisters, boils, dermatitis, diarrhoea and dysentery. Latex is febrifuge, used in pain of throat, piles, pulmonary discharge in children, skin crack on heels, lips and dislocated joints. Plant is used for blisters, boils, hydroceal, menstrual complaints, muscular pains, scabies, small-pox and spermatorrhoea. Fruits are edible, used as vegetable also.

(Jain, 1991)

Critical Notes

Stem bark is beneficial for polyurea and bleeding lochia. Young shoots constitute a good feed for cattle. They are sold in local markets also.

Ficus heterophylla L.*f.*

Local Name

Dedumber, Gathi umber

Ethnobotanical Notes

Medicine

1. About spoonful of root extract is consumed orally twice a day for three to four days to cure dysentery.
2. To treat tumours on any part of body, the extract of leaves and stem bark mixed in cow urine and is applied locally at morning and night till cure.

Miscellaneous

1. ***Food***: Fruits are consumed by tribals.

Notes from Classics

Juice of root is internally given in colic pains. Bark of the root are pulverized and mixed in coriander seeds. This mixture is considered good remedy in cough and asthma and chest affections. Juice of leaves mixed with milk is given in dysentery.

(Watt, 1889-1899)

Roots are given in colic problems. Fruits are edible.

(Ambasta, 1986)

Root bark is administered in asthma and cough.

(Jain, 1991)

Critical Notes

Roots are found useful to treat dysentery. Leaves and stem bark are used to cure body tumours. Both uses are not reported earlier.

Ficus religiosa L.

Local Name

Pipal, Pimpal Pippay

Ethnobotanical Notes

Medicine

1. A spoonful of stem bark powder with water is advised twice a day for two weeks in the treatment of gonorrhoea.
2. Decoction of stem bark, about half cup, is given orally for three consecutive weeks daily to treat scabies.
3. Equal amount of bark powder and mucilage of *Aloe vera* mixed properly and applied locally on fistula till cure.
4. A poultice of stem bark powder is applied on bone fractures for quick mending.
5. Bark extract is used as mouth wash for a week to treat bad smell of breath and freshness of mouth.
6. Young twigs are chewed regularly for tooth-ache. It also strengthens gums.

Miscellaneous

1. ***Fodder***: Leaves are young twigs for lopped for fodder for cattle.
2. ***Food***: The fruits are eaten at the time of scarcity.

Notes from Classics

Bark is astringent. Dried bark is used to cure fistula and asthma. Leaf and shoots are purgative and reputed for curing skin diseases. Ashes of growing shoots are applied on ulcers. Fruit is laxative and helps in digestion. Seeds are cooling and alternative. Leaves and branches are good fodder for elephant and buffaloes. Leaves and buds are eaten during famine period.

(Watt, 1889-1899)

Infusion of bark is used for skin diseases and ulcer. Bark is astringent. Leaves and shoots are used as purgative and also lopped for fodder for cattle. Fruits are laxative and eaten at the time of scarcity. Seeds are cooling, alternative and laxative. The tree is host for lack insect. Latex is used for sealing wax.

(Anonymous, 1946-1976)

Wood is used for making spoons, bowls and match boxes. Bark yields fibres and also used for paper-pulp.

(Ambasta, 1986)

Bark is prescribed for blisters, boils and ulcers. It is also given in gonorrhoea, haematuria, scabies and in whooping cough. Bark is given as an antidote for snake bite. Leaves are abortifacient and employed for skin diseases and tooth-ache. Leaves are edible. Latex is tonic. Fruits are given as laxative and edible. Seeds are useful in urine complaints. Plant is given for cholera, dysentery, fever, bronchitis, gonorrhoea, rheumatism, small-pox, sores in mouth and spermatorrhoea.

(Jain, 1991)

Critical Notes

Stem bark is used in different human diseases or ailments such as gonorrhoea, scabies and bone fractures. These are not yet reported.

Ficus virens Ait.

Local Name

Pimpri

Ethnobotanical Notes

Medicine

1. Half cup of decoction of stem bark is given twice a day to regularize menstrual cycle.

Miscellaneous

1. ***Fodder***: Leaves are used as fodder for goats, cows and buffaloes.
2. Ripe fruits are edible.

Notes from Classics

Leaves are used as fodder for cattle. Fruits are edible. Decoction of bark is used as gargle and to wash ulcer. Bark yields fibre.

(Anonymous, 1946-1976)

Bark is tonic and diuretic. Latex is applied in rheumatism and lumbago.

(Ambasta, 1986)

Fruits and shoot buds are edible. Shoots are pickled. Young leaves are used as vegetable.

(Jain, 1991)

Critical Notes

Stem bark is being reported for the first time to regularize menstrual cycle. This tree species is important as leaves are commonly used as fodder for pet animals especially when other live sources are not many in the months of summer.

Family: Musaceae

Ensete superbum (Roxb.) Cheesm.

Local Name

Rankela, Dev Kel

Ethnobotanical Notes

Medicine

1. Rhizomes and seeds are used as antidote on snake and dog bite. Ten gm of rhizome or 5-6 seeds are powdered and given with sugar or betal leaf (*Piper betel* L.) after every six hours till the poison is reduced.
2. Rhizome extract about half cup is mixed in two spoons of cow urine. The mixture is given twice a day for strangury till cure.
3. A cup of juice extract from the pseudostem is given after every three hours in a day for food poisoning.
4. A mixture of ten gm of powder of pseudostem is mixed in a cup of water and small amount of sugar is given twice a day for twenty one consecutive days to cure gonorrhoea.
5. Flour obtained from dried rhizome is used to prepare bread.

Miscellaneous

1. Leaves are used as dinner plate in ceremony in hilly tribal areas.
2. ***Food***: Flower buds and inflorescence axis are used as vegetable.
3. Fibres are obtained from the plant.

Notes from Classics

The scape and convolute leaf sheath and roots are also eaten. Plant yields fibres.

(Watt, 1889-1899)

Inflorescence are eaten as vegetable. Young fruits are pickled.

(Ambasta, 1986)

Fruits are edible. Flower buds, inflorescence and young shoots are used as vegetable.

(Jain, 1991)

Critical Notes

Use of rhizome for strangury and poisoning are being noted for the first time. The pseudostem is also beneficial for gonorrhea and poisoning. These uses are not reported in past. Use of seeds as an antidote for dog-bite is not also noted. The food value of different parts of this species is doubtless. It is to be noted that nearly all parts for some or other purposes are widely collected. This has resulted in marginalization of its wild populations in forested areas. The fruits for their seeds are sold near temples or in local village markets.

Family: Hypoxidaceae

Curculigo orchioides Gaertn.

Local Name

Kali-musli

Ethnobotanical Notes

Medicine

1. Spoonful of rhizome powder is consumed at early morning with honey at tonic.
2. About 10-15 ml root extract is taken twice a day for three to four days to cure intermittent fever.

Notes from Classics

Rhizome is prescribed for asthma, piles, jaundice, diarrhoea, colic and gonorrhoea. Rhizome is demulcent, diuretic, tonic and aphrodisiac.

(Watt, 1889-1899)

Rhizomes are tonic, alterative, demulcent, diuretic and restorable. Rhizomes are used in the treatment of piles, diarrhoea, jaundice and asthma. Rhizomes are also useful on skin diseases and itch.

(Anonymous, 1946-1976)

Ambasta (1986) records similar properties as in Anonymous (*loc.cit.*).

Roots are used as abortifacient. They are useful in curing cough, dysentery, ear complaints, epilepsy, filarial, fever, impotency, leucorrhoea, nosal bleeding and veneral complaints. It is antidote for snake bite and scorpion sting. Rhizomes are used to cure asthma, blindness, eye complaints, skin diseases, and bleeding from piles. It is reputed tonic for unconsiousness of children. Rhizomes also cure eye diseases of animals. Rhizomes are edible, used as vegetable.

(Jain, 1991)

Critical Notes

Roots are not reported earlier as useful in intermittent fever.

Family: Amaryllidaceae

Crinum defixum Ker.-Gawl.

Local Name

Pankanda

Ethnobotanical Notes

Medicine

1. Four to five drops of juice of bulb are dropped in ears in case of pus formation. It is followed daily till cure.
2. Roasted bulb is taken daily and early morning for four days to cure strangury.
3. Paste of roasted bulb about on spoonful per day is prescribed for flatulence till cure.

Notes from Classics

Bulbs are eaten after boiling. Bulbs are emetic, emollient and used in the treatment of burns witlow and carbuncles.

(Anonymous, 1946-1976)

Leaves and roots are diaphoretic and emetic. Bulbs are used in biliousness, strangury and other urinary troubles.

(Ambasta, 1986)

Juice of leaf is dropped in ears for ear-ache. Tubers are used for abdominal diseases and body-ache.

(Jain, 1991)

Critical Notes

Utility of bulb to check pus formation in ears and for the treatment of strangury is being reported additionally.

Family: Agavaceae

Agave vera-cruz Mill.

Local Name

Ketki, Kektad

Ethnobotanical Notes

Medicine

1. About 5 ml leaf juice is mixed in equal quantity of common salt. It is boiled and then after cooling about 5 gm jaggery is added. It is consumed twice a day to cure constipation.

Veterinary Medicine

1. Paste of rhizome is applied on swelling or tumour on the neck of pet animals.

Miscellaneous

1. Fibres obtained from the leaves are used for cordage.
2. The plant is transplanted as live hedge along the borders of farms.

Notes from Classics

Roots are diuretic and used for gonorrhoea. Root and leaf juice is alterative. It is useful in syphilis. Leaf juice is laxative, diuretic and emmenogogue. It is useful in scurvy. Pulp of leaves is applied for conjunctivitis and internally for gonorrhoea. The plant is cultivated as hedge and also for fibres. Leaves and stem axis are used for roofing. Flowering stems are used as cork.

(Watt, 1889-1899)

Leaves yield fibres which are useful for ropes, cordage and mat making.

(Anonymous, 1946-1976)

Soft portion of stem is cooked with tamarind and jaggery and consumed by poors. Leaf fibres are useful for ropes, cordage and mat making.

(Ambasta, 1986)

Fibres obtained from the leaves are used for cordage.

(Jain, 1991)

Critical Notes

The medicinal uses of leaves and rhizome for human beings and pet animals respectively are new to the science of Indian Ethnobotany. The plant is usually planted as live hedge around fields and cultivated as a source of fibre. It is exotic and native of America.

Family: Taccaceae

Tacca leontopetaloides (L.) O.Ktze.

Local Name

Bramhan Zarya, Sasana-bhala, Diva

Ethnobotanical Notes

Medicine

1. Paste of leaves is applied locally on eczema regularly till cure.

Notes from Classics

The spilt stem, petiole and scape good braiding material after curing and used in hat making. Tubers are used for the treatment of piles. Bitter extract made from washing and the grated tubers in running water is a rubefacient.

(Anonymous, 1946-1976)

Ambasta (1986) reported similar properties and uses as those reported in Anynomous (*loc.cit.*).

Tubers are used for body-ache and head-ache. Tubers are edible.

(Jain, 1991)

Critical Notes

Medicinal importance of leaves for eczema is being reported for the first time.

Family: Dioscoreaceae

Dioscorea oppositifolia L.

Local Name

Kand

Ethnobotanical Notes

Medicine

1. Spoonful of powder of tuber is taken with water to cure dysentery.

Miscellaneous

1. ***Food***: Tubers are sliced and boiled in water then they are consumed.

Notes from Classics

Tubers are dried and applied to reduce swellings. Leaves are useful as a febriguge.

(Watt, 1889-1899)

Dried and pounded tubers are used in application of ulcer. They are used in piles, dysentery and syphilis and applied locally on sores.

(Anonymous, 1948-76)

Tubers are powdered and applied to swellings.

(Ambasta, 1986)

Tubers are consumed to cure abdominal pains. They are applied locally on boils, piles and bone fractures. Tubers are useful against dysentery and jaundice.

(Jain, 1991)

Critical Notes

For tubers entire plant are removed and thus destroyed. Very few plants are found in forest areas now-a-days. Collection should be avoided or stringent action to save it is necessary.

Family: Liliaceae

Allium cepa L.

Local Name

Pandhara Kanda

Ethnobotanical Notes

Medicine

1. Bulb juice is dropped in ears to treat the scorpion sting.
2. Juice of bulb is applied on wounds caused due to thorns.
3. Paste of bulb is applied on forehead and whole body for sunstroke.
4. Few drops of bulb juice are dropped in nose for epileptic fits.
5. Bulb juice is dropped in ears for ear-ache.
6. One drop of bulb juice is dropped in eyes for better health of eyes.

Veterinary Medicine

1. Bulbs are given to domestic animals when they are reluctant to eat fodder.

2. Bulbs are given to domestic animals for madness.

Notes from Classics

Bulbs contain volatile oil, which is stimulant, diuretic and expectorant. They are used in fever, dropsy, catarrh and bronchitis. They are reported snake repellant and similarly useful to venomous reptile. Bulb juice s used for ear-ache. It is aphrodisiac, carminative and emmenogogue.

(Watt, 1889-1899)

Fresh bulb juice has bactericidal properties. Root is useful in flatulence and dysentery.

(Anonymous, 1948-76)

Bulbs are considered stimulant, diuretic and expectorant. Bulbs are used against flatulence and dysentery. Roasted onions are applied as poultice. Bulbs yield essential oil. Bulbs are used as vegetable.

(Ambasta, 1986)

Bulbs are used as anthelmintic and expectorant. It stops nose bleeding. They are applied on blisters and boils. They are used against eye and ear complaints. Bulbs are used for insect bites, itch, piles and ring-worms. It is sedative, stops vomiting and is remedial on skin disease and stomach-ache. Bulbs have healing properties and is rubefacient. It is widely used as food and condiments.

(Jain, 1991)

Critical Notes

Usefulness of onion bulbs as human medicine for scorpion bite, sunstroke and epileptic fits is not on record. Likewise, their use as veterinary medicine is also noted for the first time. It is cultivated for bulbs. It is exotic, native of South Asia and Mediterranean Region.

Allium sativum L.

Local Name

Lasun, Losna

Ethnobotanical Notes

Medicine

1. Juice of bulblets is dropped in nose of a person who has loss sensation of smell.
2. To prevent pus formation sesamum oil (*Sesamum orientale* L.) and bulbets of garlic are boiled and then this oil is dropped in ears.
3. Juice of bulblets is mixed in turmeric powder (*Curcuma longa* L.) is dropped in ears for deafness.
4. Bulblets fried in cow ghee are eaten with bread to improve the lost taste.
5. A piece of cotton cloth is socked in fresh juice from bulblets. It is loosely tied on nose to cure tuberculosis.
6. Salted bulbets are eaten with meals to cure head-ache.

7. Bulbets with honey or sugar and cow-ghee are consumed regularly for quicker healing of fractured bone.
8. About ten bulbets are boiled in about 100 ml of water. This concentrate and bulblets are consumed to cure sciatica.
9. Bulblets are fried in groundnut oil (*Arachis hypogea* L.) and eaten with bajara [*Pennisetum americanum* (L.) K.Schum.] bread. This cures rheumatic pains.
10. Bulblet with rock salt is applied and massaged on swollen muscle to reduce pains.
11. About 50 to 100 ml sesamum oil is taken orally along with the bulbets immediately after paralytic attack. It lessens the intensity of paralysis attack and damage due to it.
12. From the first day of paralytic attack one bulblet is eaten with water. The number of bulbet is increased upto 21 for 21 days. Again reduce the number of bulblets in descending order. It cures paralysis.
13. Fresh paste of bulblets is kept under the sole of foot of a child. A pair of woollen socks and shoes are worn. It helps to cure asthma slowly.

Veterinary Medicine

1. Bulblets are crushed in wine and given to hens and cocks against dysentery.

Notes from Classics

Fruits are stimulant, diaphoretic, expectorant, diuretic and tonic. They are rubefacient when applied externally. They are anthelmintic, emmenogogue and given for fever, cough and piles. Externally juice is applied to ear pinnae for deafness and pain. It is largely used as liniment in convulsions and other nervous and spasmodic affection. They are employed as poultice in retention of urine. With vinegar garlic is used as an astringent in sore throat and relaxation of vocal cord. They are reputed for asthma, paralysis, gout, sciatica and colic. They also prevent hair turning gray. Juice of bulblet is used for sprains, ear-ache, bronchial and asthmatic complaints.

(Watt, 1889-1899)

Inhalation of fresh juice is useful in pulmonary tuberculosis, dyspepsia, flatulence and colic. External application of juice is rubefacient in skin diseases and ear-ache. Juice is used as lotion for washing wounds and foul ulcer. Juice of bulblet is given in pulmonary phthisis, gangrene of lung, whooping cough, laryngeal tuberculosis and duodenal ulcers.

(Anonymous, 1948-76)

Juice is applied in skin troubles and used as eardrops. It is diluted with water as a vulnerary. Bulblets are used for bronchitis, whooping cough, fever, ear complaints, eczema and flatulence. Juice of bulblets cure nervous disorders, respiratory disease, phthisis and rheumatism. They are also useful against skin eruptions, wounds, tumours, ring-worm and useful as expectorant.

(Jain, 1991)

Critical Notes

Bulblets are noted for the first time useful to recover loss of sensation and taste to prevent pus formation in ears and to treat head-ache. It is cultivated for its bulblets, especially used as condiments.

Aloe vera L.

Local Name

Korphad

Ethnobotanical Notes

Medicine

1. Turmeric powder (*Curcuma longa* L.), about 5 gm, is mixed in latex/juice of this plant. It is consumed for twenty one days to cure piles.
2. Latex is mixed in water. About fifteen to twenty ml of this mixture is taken orally to improve appetite.
3. Latex and honey are taken in equal ratio and homogenized. About 10 ml is given orally once to children to expel intestinal worms.
4. Half spoon of sugar is dissolved in latex and spoonful of this mixture is taken orally to improve appetite.
5. Latex is applied on burns to reduce burning sensation.

6. Latex is diluted in water and applied regularly to remove facial wrinkles and blackness, thereby improving complexion.
7. Latex is applied at night to stop burning sensation of palm foot sole in summer.
8. Latex is mixed in neem leaf powder (*Azadirachta indica* A.Juss) and turmeric powder (*Curcuma longa* L.). This mixture is warmed and tied as a poultice on boils and abscess.
9. Turmeric powder, stem bark powder of *Terminalia bellerica* (Gaertn.) Roxb. are homogenized in latex of *Aloe vera*. Paste is applied around eyes to treat conjunctivitis.
10. Sugar is dissolved in latex. A spoonful of this mixture is taken orally to cure habitual constipation and dropsy.

Notes from Classics

Fresh latex is used as cathartic, cooling and useful in fever. It is used for spleen and liver complaints. It acts powerfully on enlarged lymphatic gland. Externally latex is applied for eye disease, also applied on boils and useful as emmenogogue.

(Watt, 1989-1899)

Latex is stomachic and purgative, applied on piles and rectal fissures. Latex is cooling and used as poultice for inflammation. It acts as emmenoggogue.

(Anonymous, 1948-76)

Fresh juice of leaves is cathartic and refrigerant, used in liver and spleen complaints and eye trouble. Leaf juice is found useful in x-ray burns, dermatitis and skin disorders. Leaves yield fibre. Dye is obtained from the plant. Leaves, flowers and stalks are pickled.

(Ambasta, 1986)

Leaves are applied on boils and breast tissue which hardens wounds. Leaves are useful for constipation, jaundice, fever, gonorrhoea, liver complaints, piles, rheumatism and sexual vitality. Leaves are useful in spleen and menstrual complaints. They are vermicidal and also used in veterinary medicine. Leaf juice is used to wash hair

(Jain, 1991)

Critical Notes

Usefulness of latex in cough, cold, to improve appetite, to repel intestinal worms, to improve complexion, in conjunctivitis and dropsy are not recorded in the classical literature. It is exotic, native of North America.

Asparagus racemosus Willd.

Local Name

Shatavari, Sasadatali

Ethnobotanical Notes

Medicine

1. About 50 gm root powder is mixed in 2 cups of milk with a pinch of sugar. This is advised orally for a month at morning to control blood pressure.
2. Ripe dried fruits of *Ziziphus mauritiana* Lamk. are crushed with the roots of this plant. This paste is applied on body for flatulence.
3. Roots are pounded in milk and given to a mother to increase lactation.
4. Root extract is mixed with equal quantity of cow milk. A cup of this mixture is given as a febrifuge.
5. Roots of *Tribulus terrestres* L. and of *Asparagus* in equal amount are powdered. Three gm of this powder with goat milk is boiled and taken regularly to cure bleeding leucorrhoea.
6. Root extract is mixed with equal amount of milk. Some grapes (*Vitis vinifera* L.) and seeds of *Buchnania lanzan* Spreng. are added in it. After boiling it, one spoonful of this is consumed with honey per day to cure asthma, cough, jaundice and hysteria.

Veterinary Medicine

1. Roots are mixed in fodder and given to milching animals to increase lactation.

Food

1. Young sprouting stem apices are used as vegetable.

Notes from Classics

Roots are refrigerant and demulcent. It is antidiarrhoeal, antidysentric and antispasmodic. Roots are aphrodisiac, also useful in veterinary medicine.

(Watt, 1889-1899)

Roots are tonic, diuretic and galactogogue. A mixture of honey and root juice is given as demulcent in dyspepsia. Roots are largely used for nervous and rheumatic complaints.

(Anonymous, 1948-76)

Ambasta (1986) documented similar medicinal uses to those mentioned by Anonymous (*loc.cit.*).

Roots are anthelmintic and aphrodisiac. It stops bleeding from nose and blood in urine. Roots are popularly used as galactogogue. They are used against cough, cold, diarrhoea and dysentery. They are used against cough, cold, leucorrhoea, impotency, gonorrhea. Roots act as a febrifuge and minimize gastric complaints. Pleurisy, urine complaints, and somach-ache are cured by roots. Fruits are used as antidote on snake bite. The plant is useful for head-ache and ulcer of tongue.

(Jain, 1991)

Critical Notes

Roots to control blood pressure, asthma, jaundice and hysteria are noteworthy additions to the medicinal uses of this taxon. Utility of roots to increase lactation in case of human being is already well-known. However, for animals recorded by the present investigator can be also easily understood.

Chlorophytum borivilianum Sant. and Fernad.

Local Name

Safed Musali

Ethnobotanical Notes

Medicine

1. Root tubers are consumed to overcome impotency in man.
2. They are also consumed as general body tonic.

Notes from Classics

Roots are edible and used as tonic.

(Anonymous, 1948-76)

Leaves are used as vegetable.

(Jain, 1991)

Critical Notes

Root as tonic is already reported. Present observation reiterates it. To overcome impotency in men root tubers are consumed. This is however not recorded earlier. Recently, it is being cultivated by some farmers in this district because of good economic returns.

Gloriosa superba L.

Local Name

Kal-lawi, Khadyanag

Ethnobotanical Notes

Medicine

1. A paste of roots is applied onto foot soles and hand palms which helps in removal of delayed placenta after child-birth.
2. Decoction of roots, about half cup, is given once at early morning to cure gonorrhoea till cure.
3. Stem piece is tied along the waist to facilitate delivery.

Notes from Classics

Root is considered as minor poison. It is anthelmintic, also effective as antidote for cobra poisoning. Root is externally applied to navel at suprapubic region to promote labour and removal of placenta. They are also applied on skin disorders. Juice of leaf is used to kill lices in hair.

(Watt, 1889-1899)

Tuber is tonic, anthelmintic, stomachic, but in large doses it is poisonous. Dry tubers are used in gastro-intestinal disorders to stop vomiting, as purgative and to promote labour pain. It is abortifacient, useful in chronic ulcers and piles. Tubers are used externally for parasitic skin diseases.

(Anonymous, 1948-76)

Tubers are externally used for neuralgic pains and skin troubles.

(Ambasta, 1986)

Rhizomes are abortifacient and antidote to snake bite and poisonous insects. It is controls cholera, intermittent fever, gonorrhoea, gout, leprosy and piles. Rhizome is also used to treat ulcer, skin diseases, stomach-ache and veterinary diseases like small-pox. Leaves are used for asthma, gout, rheumatism, scabies, sore and skin complaints. Plant is useful to cure tumour.

(Jain, 1991)

Critical Notes

Usefulness for treating gonorrhoea is not documented earlier. Rhizomes are, however, used for similar purpose.

Family: Commelinaceae

Commelina benghalensis L.

Local Name

Keni, Kena, Kenpat

Ethnobotanical Notes

Miscellaneous

1. *Food*: Leaves are used as vegetable.
2. *Fodder*: This plant species is also useful as fodder for domestic animals.

Notes from Classics

Leaves are eaten by poor people as pot herb.

(Watt, 1889-1899)

Leaves are used vegetable and fodder.

(Anonymous, 1948-76)

Leaves are useful as vegetable and fodder.

(Ambasta, 1986)

Leaves are used as vegetable.

(Jain, 1991)

Critical Notes

The food and fodder value is proved by its cross-country observations. It is a common weed of waste places and cropped fields.

Family: Typhaceae

Typha domingensis Pers.

Local Name

Deodanda

Ethnobotanical Notes

Medicine

1. Root extract, about half cup, is taken internally for a week to cure leucorrhoea.
2. Flowers are crushed and applied on injury.

Notes from Classics

Roots are eaten, also used as fodder for elephant. Soft wooly inflorescence of male are spadix applied like cotton on wounds and ulcer. Young shoots are eaten. The ripe fruits are also eaten.

(Watt, 1889-1899)

Rhizomes are eaten. They are astringent, diuretic, employed in dysentery, gonorrhoea and measles. The leaves are used for thatching, soft mats, ropes and stuffing. Dried stalks are used for pens. Shoots are edible, favourite food for elephants. Soft and wooly floss of male spadix is used as absorbant in wounds and ulcer. Ambasta (1986) gave the same properties as those in Anonymous (*loc.cit.*).

(Anonymous, 1948-1976)

Shoots are edible.

(Jain, 1991)

Critical Notes

Root extract for curing leucorrhea is unnoted yet.

Family: Arecaceae

Cocos nucifera L.

Local Name

Naral, Naray, Shriphal

Ethnobotanical Notes

Medicine

1. Paste of fresh kernel mixed with turmeric powder (*Curcuma longa* L.) is applied on joints to reduce swelling and pains.
2. A mixture of ash of husk of coconut fruit, rock salt and camphor is used as tooth powder to strengthen gums.

Notes from Classics

Roots are diuretic, used in uterine diseases. Young roots boiled with ginger and salt are used in fever. Flowers are astringent. Flowers mixed with sugar and 'chandan' (Sandal) is good for bilious fever, to check vomiting and produce cooling sensation. Flowers are useful in dysentery, diarrhoea, menorrhoea and stomatitis. Fruits are used to expel worms. Juice of immature fruits reduces acidity and gastric irritation. Ash of coconut is antacid and digestive, used to expel ring-worm. Milk is used to stop vomiting, in debility and incipient phthisis. Fruit pulp is nourishing, cooling and diuretic. Oil from fruits are used for phthisis, tuberculosis, skin diseases and to promote hair growth. Flower buds are nourishing, also used as vegetable. Coconut water is cooling, refrigerant, useful in thirst, fever, urinary diseases and blood purifier.

(Watt, 1889-1899)

Decoction of root is astringent and used as mouth wash and gargle. Roots are roasted, ground and used as dentifrice. Shells are used for bowls, scoops and fuel. Coconut water is refreshing, aperiant, diuretic and destroy intestinal worms.

(Anonymous, 1948-76)

Meat of seed is eaten raw or used in sweet-meats, kitchen preparations, pastries and confectionary. Coconut oil is used in soups, cosmetics, shampoos, creams and hair oil preparations. Coconut fibres are used for mats, ropes, baskets and brushes. Coconut milk is a refreshing drink. Leaves are woven into mats and baskets. Wood is used for cabinets, building and as fuel.

(Ambasta, 1986)

Roots cure dysentery, bronchitis and uterine diseases. Leaves are useful for thatching huts. Fruit fibres are used for menstrual disorders. Fruits are useful to increase sex-potancy. Coconut flesh is edible. From fibres cordage and fish traps are made.

(Jain, 1991)

Critical Notes

Kernels with turmeric are useful to reduce pains and swellings of joints. Husk of fruit is useful to strenghen gums. Both uses are yet unnoticed else-where. It is cultivated mainly for its fruits but has various special importance to cottage industries.

Phoenix sylvestris (L.) Roxb.

Local Name

Tadi, Shindi

Ethnobotanical Notes

Medicine

1. Spoonful of root powder is consumed twice a day for a month to cure menostaxis.

Miscellaneous

1. ***Food***: Tree is tapped for sap which is called 'Toddy'. It is a nutritious drink, if drunk before sunrise. Afterwards it is drunk as 'Neera' a fermented drink.
2. Ripe fruits are edible.
3. Leaves are used for thatching huts and brooms.

Notes from Classics

Fruits are pounded and mixed with almonds, quince seeds, pistachio nuts, spices and sugar form 'paushthik'. Roots are used in tooth-ache and nervous debility. Paste of kernels and roots of *Achyranthes aspera* are consumed with betel leaves as a remedy for ague. Central tender part of root is used in gonorrhoea and gleet. Reddish fruits are eaten.

(Watt, 1889-1899)

Ropes are made from petiole. Leaf buds are eaten. Leaves are lopped for fodder. Trunk freed from pith is made excellent water conduits. Nira is a good source of vitamin-B. It is restorative juice. Fruits are edible.

(Anonymous, 1948-76)

Tree is tapped for sap which is used for making jaggary and sugar sap called 'Nira'. Leaves are used for thatching huts and making fans, baskets, bags and brooms. Leaves are lopped for fodder. Fruits are edible and restorative. Wood is used for temporary construction, bridges and tent pegs. Roots are used for tooth-ache.

(Ambasta, 1986)

Roots are used for gum trouble, tooth-ache menstrual complaints, veneral diseases, for feeling hot inside. Stems are used for gonorrhoea. Leaves are used for diarrhoea, dysentery and also used for basketary. Stem and fruits are edible. Beverage is prepared from fruits.

(Jain, 1991)

Critical Notes

Roots are remedial for menostaxis, a use hithreto not recorded. A word about caution about the use of leaves on large scale for making brooms appears pertinent. The leaves are collected on very large scale, even before the tree species have attained proper stage. The leaves are cut and collected even before flowering, this obviously has affected natural dispersal of seeds, regeneration and propagation in this area.

Family: Cyperaceae

Cyperus pangorei Rottb.

Local Name

Lavali

Ethnobotanical Notes

Medicine

1. Tubers are powdered. Five gm of this powder is taken with warm water daily for eight days to overcome body swellings.

Miscellaneous

1. Entire plants are used to thatch roofs of huts.

Notes from Classics

Tubers are used as tonics and stimulating medicine.

(Watt, 1889-1899)

Culms are used for woven mats.

(Anonymous, 1948-76)

Uses noted by Ambasta (1986) are similar to those of Anonymous (*loc.cit.*).

(Ambasta, 1986)

Plant is used for controlling a atrophy. It is also reported as antidote in snake poisoning.

(Jain, 1991)

Critical Notes

Efficacy of tubers as a remedy against body swellings is interesting and noted for the first time. In the light of use of culms for making mats (*cf.* Anonymous, *loc.cit.*), their utility also for thatching in this district can be easily conceived.

Cyperus rotundus Linn. subsp. *rotundus*

Local Name

Nagarmotha, Lovala

Ethnobotanical Notes

Medicine

1. Ten gm of rhizome powder mixed with some rock salt is drunk with warm water only at night for four to six days to purify mother's milk.
2. Rhizome powder is employed to wash hair and for better growth of hair.
3. Rhizome is powdered and mixed with sorghum flour and turmeric powder (*Curcuma long* L.); this mixture is called "tiksa", wash with this "tiksa" cures skin diseases, wound, scars and improve complexion.

4. A cup of decoction of rhizome in cold water is advised for a week to reduce body heat.

Notes from Classics

Roots are diaphoretic, astringent, stimulant and diuretic. Roots are also used as vermifuge for disorder of stomach and irritation of bowels. Roots are used in fever. Bulbous roots are used for dysentery. Fresh tubers are applied to the breast as a galactagogue and diaphoretic, also used for febrile and dyspeptic affections. In large doses, tubers are anthelmintic and externally applied to ulcer.

(Watt, 1889-1899)

Roots are diuretic, diphoretic and astringent, used in stomach and bowel complaints. Roots yield essential oil. Tuberous roots are used in perfumes and agarbatties.

(Anonymous, 1948-76)

Ambasta (1986) mentioned similar properties as those given in Anonymous (*loc.cit.)*.

Roots are diuretic, astringent, used in bowel and stomach complaints, jaundice, sore, wound and as antidote for snake-bite. Roots are also used as masticatory. Bulbs are used for delirium and dysentery. Whole plant is used for heat stroke and urine complaints. Plants are used for making mats. Roots, bulbs and tubers are edible.

(Jain, 1991)

Critical Notes

Rhizome is commonly used for hair wash which is generally mixed with Shikakai (*Acacia concinna* DC.), another fruits used for hair washes. Its use for better hair growth, fair complexion to treat skin diseases and scars are all unnoticed. It also helps to reduce body heat, a use hitherto unreported.

Family: Poaceae

Apluda mutica L.

Local Name

Fularya

Ethnobotanical Notes

Miscellaneous

1. ***Fodder***: Entire plants are used as fodder for domestic animals when green. It is also useful when dry.

Notes from Classics

Grass is used as fodder in Band district.

(Watt, 1889-1899)

Grass may be used as raw material for rayon industries. Young grass is eaten by buffaloes. Stalk is used for making hut.

(Ambasta, 1986)

Whole plant is used to cure mouth sores of animals.

(Jain, 1991)

Aristida funiculata Trin. et Rupr.

Local Name

Kusal

Ethnobotanical Notes

Miscellaneous

1. ***Fodder***: Entire plant is useful fodder grass before flowering for domestic animals.

Notes from Classics

Plants when tender are grazed by cattle.

(Anonymous, 1948-76)

Plant is grazed by cattle when tender.

(Ambasta, 1986)

Critical Notes

Fodder value of this species is only before flowering and fruiting, since the sharp awns render this grass unsuitable as such.

Bambusa aurandinacea (Retz.) Willd.

Local Name

Bans, Bamboo

Ethnobotanical Notes

Medicine

1. The outer part of culm is crushed and applied on injuries.
2. From bark of culms white discharge is produced. This discharge in powder form is called "Wansha-lochan". One gm of it in spoonful of honey is given at morning daily to cure cough and cold till cure.
3. A pinch of "Wansha-lochan" in spoonful of honey is taken after meal to improve digestion.
4. A cup of water with about two gm of "Wansh lochan" powder is drunk twice a day to stop leucorrhoea.
5. One gm of "Wansh lochan" powder mixed in spoonful of honey is consumed regularly twice a day for six months to cure tuberculosis.

Miscellaneous

1. ***Food***: Seeds are collected called 'venuyaw' and ground. This flour is used for preparing bread.
2. Culms are employed in making bows, fish-containers, food-grain containers, pots, bowls, basketary, musical instrument, roof of house and huts etc.
3. ***Fodder***: Leaves are used as fodder.

Notes from Classics

White discharge from inner bark of culms is also called "Tabashir", is cooling, tonic and aphrodisiac, used against lung diseases. Tabashir is also efficacious in paralytic complaints, flatulence and in poisoning cases. It is cooling, removes thirst. Tabashir is used in fever, jaundice and as a febrifuge. Water accumulated in bamboo joints is used against bowel complaints.

(Watt, 1889-1899)

Young shoots are pickled. Leaves and twigs are also used as fodder.

(Anonymous, 1948-76)

Bamboo is mainly used for floating heavy timber and structural purposes. Bamboo yields good quality paper pulp. Leaves and twigs are used as fodder. Leaves are fed to horses for cough and cold.

(Ambasta, 1986)

Stems are used for antifertility. Black soot of culms is applied on wounds and cuts. Leaves and shoots are used for vegetable curry. Culms are used for preparing basketery and house buildings.

(Jain, 1991)

Critical Notes

The powder mentioned above is beneficiary for improving digestion, to treat leucorrhoea and tuberculosis. These medicinal uses have not appeared in earlier literature.

Chrysopogon fulvus (Spreng.) Chiov.

Local Name

Shahadi

Ethnobotanical Notes

Miscellaneous

1. *Fodder*: Plant is used as fodder for buffaloes, bullocks and cows.

Notes from Classics

This grass is a valuable fodder and it is cut just before flowering.

(Anonymous, 1948-76)

It is a good fodder grass and sand binder. For fodder, it is cut before flowering.

(Ambasta, 1986)

Critical Notes

It's utility as a fodder is only before flowering.

Chloris tenella Koenig ex Roxb.

Local Name

Gawat

Ethnobotanical Notes

Miscellaneous

1. *Fodder*: The whole plant is useful as fodder for goats, cows and buffaloes.

Notes from Classics

Plant is considered as good fodder.

(Watt, 1889-1899)

It is a good fodder grass for domestic animals. It is considered to be a nutritious fodder.

(Anonymous, 1948-76)

It is considered to be nutritious fodder.

(Ambasta, 1986)

Critical Notes

The fodder value throughout our country is notable.

Cymobopogon martinii (Roxb.) Wats.

Local Name

Rosha-gawat.

Ethnobotanical Notes

Medicine

1. Small pieces of green leaves are added in tea to control cough and cold.
2. Oil extracted from leaves is applied on knees for rheumatic pains.
3. Rosha oil is applied regularly on bald to promote hair growth.
4. Cough is controlled if the children inhaled the fumes of burnt leaves.

Veterinary Medicine

1. Fresh plant of 'Rosha' and leaves of Neem (*Azadirachta indica* A.Juss) are burnt. The domestic animals are fumigated in case they are reluctant to eat fodder.

Notes from Classics

Infusion of leaves is stomachic. Leaf oil is used in rheumatism and neuralgia and cures baldness. Oil is powerful stimulant when applied externally. Plant is cool, astringent, useful in skin diseases, bilious skin affections and fever. Decoction of plant used in case of fever.

(Watt, 1889-1899)

Leaves are used as remedy for lumbago, stiff joints and skin diseases. Leaves are taken internally, surve as a remedy for bilious complaints.

(Anonymous, 1948-76)

Plant is source of 'Rosh oil', used in soups and cosmetics. Rosha oil is used as adulterant in atter of Rose.

(Ambasta, 1986)

Leaves and leaf oil is applied externally for rheumatic pains. It is also used locally for skin disease.

(Jain, 1991)

Critical Notes

Medicinal utility for pet animals reported presently is not documented in past. This grass is quite reputed for its other medicinal uses which are also observed in this district of Maharashtra. It is demanded for its oil. Local people may be advised to grow them on waste or barren lands in their custody.

Cynodon dactylon (L.) Pers.

Local Name

Durva, Daudi, Harali

Ethnobotanical Notes

Medicine

1. Spoonful of leaf juice is mixed with few drops of buttermilk and pinch of salt. Few drops of this mixture is dropped in nose for bleeding.
2. Juice of leaves is applied on skin-rash produced by herpes.
3. Half cup of leaf juice is drunk twice a day for two days to cure acidity and cough.
4. Few drops of leaf juice are dropped in eyes to cure conjunctivitis.

Notes from Classics

Decoction of roots is used in syphilis. It is diuretic. Roots with curds are used in cases of chronic gleet. Leaves are astringent. Juice of plant is astringent and applied to fresh cuts and wounds. It is diuretic, used in dropsy anasarca, chronic diarrhoea and dysentery. Juice of grass in useful in catarrhal opthalmia and haemataresis. Fresh juice is valuable styptic in epitaxis. Juice is antiperiodic and applied on scabies. Cold infusion of grass stops bleeding from piles, also used for irrigation of urinary organ.

(Watt, 1889-1899)

Rhizomes are used for genito-urinary disorders. Decoction of plant is diuretic, useful in dropsy and anasarca. Juice of plant is astringent and applied to bleeding cuts and wounds.

(Anonymous, 1948-76)

It is important as a pasture and lawn-grass. Decoction is diuretic, used in anasarca. Rhizomes used in genito-urinary troubles. It is a good soil blinder.

(Ambasta, 1986)

Roots are cooling, used in dysentery, hysteria and opthalmia. Stem is tonic. Leaves are used for dropsy, piles, bleeding, urine complaints, vomiting. Plant is used for carbuncle, cramps, headache, snake bite, sore to srop bleeding, to treat ulcer in stomach, wound etc.

(Jain, 1991)

Critical Notes

Leaf juice stops nose bleeding, and applied to remove rash caused by herpes. It is also useful to treat acidity, cough and conjunctivitis. These are all additional medicine uses hitherto unreported.

Dendrocalamus strictus Nees

Local Name

Velu

Ethnobotanical Notes

Medicine

1. One cup of decoction of leaves if drunk daily for a month, it cures menorrhagia.

2. Extract of tender shoots alongwith the roots of *Adhatoda zeylanica* Medic, leaves of *Azadirachta indica* A.Juss and fruits of *Syzygium cumini* (L.) Skeels is prepared in water and half glass of it is drunk once a day for about a month to cure leucorrhoea.
3. Paste of outer parts of culms is applied on injuries.

Miscellaneous

1. Culms are employed in making pots, bows, musical instruments, roofs of houses and huts. They are also used in making handles of axes and other weapons.
2. ***Fodder***: Leaves are used as fodder for cattle.

Notes from Classics

Silicious matter found near the nodes and used as cooling tonic and astringent. Decoction of leaves is given to women after delivery to put uterus in order. Juice of leaves is used for abortion. Pulp is utilized for manufacture of brown paper.

(Anonymous, 1948-76)

Stem is employed for rafters, mats, baskets, sticks, furniture, tent poles, fishing and musical instruments. Leaves are used as fodder.

(Ambasta, 1986)

Roots are used against antifertility in fever. Shoots and seeds are edible. Shoots are tonic. Leaves are useful for cough and veterinary medicine. Bark paste is applied locally on injuries.

(Jain, 1991)

Critical Notes

Decoction of leaves, extract of tender shoots and paste of outer part of culms are beneficial medicinally for human ailments *viz.*, menorrhagia, leucorrhoea and injuries respectively. These utilities did not find place in the classical literature.

Dinebra retroflexa Panz.

Local Name

Khar-gawat

Ethnobotanical Notes

Miscellaneous

1. The plants are used as fodder for cattle.

Notes from Classics

Plant is regarded as valuable wild fodder.

(Watt, 1889-1899)

Grass is excellent fodder for cattle, particularly for buffaloes. It is reported to increase flow of milk.

(Anonymous, 1948-76)

It is an excellent cattle fodder, particularly for buffaloes. It is suitable for making hay or ensilage.

(Ambasta, 1986)

Critical Notes

Its fodder value is doubtless. It is a common weed of cultivated fields.

Eremopogon foveolatus (Del.) Stapf.

Local Name

Kadu-marvel

Ethnobotanical Notes

Miscellaneous

1. ***Fodder***: It is a good fodder grass relished by all domestic animals.

Notes from Classics

It is favourite fodder for cattle.

(Watt, 1889-1899)

It is a fodder grass relished by cattle. This grass is considered good desert fodder for camels.

(Anonymous, 1948-76)

It is fodder grass relished by cattle.

(Ambasta, 1986)

Critical Notes

Its fodder value is reiterated by the present observation.

Family: Cyperaceae

Eriophorum comosum (Wall.) Wall. ex Nees

Local Name

Gawat

Ethnobotanical Notes

Miscellaneous

1. The fresh plants are made into ropes and used daily for various purposes.

Notes from Classics

Ropes are made from plant which is strong when fresh.

(Watt, 1889-1899)

Herb is used for making ropes.

(Anonymous, 1948-76)

It is used for making ropes.

(Ambasta, 1986)

Whole plant is used for cordage.

(Jain, 1991)

Critical Notes

The plant is useful as rope or as a substitute for fibres locally although in crude form. Cross-country observations amply clarify its use.

Paspalum flavidum (Retz.) A.Camus

Local Name

Gawat

Ethnobotanical Notes

Miscellaneous

1. *Fodder*: Plant is used as fodder for domestic animals.

Notes from Classics

The genus is highly valued for grazing and stacking.

(Watt, 1889-1899)

The grass is considered to be excellent fodder. Grains are eaten in time of scarcity.

(Anonymous, 1948-76)

Ambasta (1986) reported similar uses those are given in Anonymous (*loc.cit.*).

Sorghum controversum (Steud.) Snowden

Local Name

Boru

Ethnobotanical Notes

Miscellaneous

1. Culms are used for thatching huts and for making partition of bathrooms.
2. *Fodder*: Culms are used as fodder for domestic animals.

Notes from Classics

It is good fodder grass for grazing. In Rajasthan, seeds are mixed in bajara flour and eaten by poorer classes.

(Watt, 1889-1899)

It is a fodder plant. It is used for feeding when cut at early full bloom stage. Seeds are used as food in times of scarcity.

(Anonymous, 1948-76)

Plant is grown for fodder. Grains are diuretic, demulcent, used in food in times of scarcity.

(Ambasta, 1986)

Grains are utilized for preparing local drinks.

(Jain, 1991)

Critical Notes

Culms are useful for raising huts and bathrooms. This use is not documented yet. The senior people of this district told about the use of culms for writing. The stems were cut to sharp points. By dipping in ink it was used for writing by school-boys. In recent decades this method has been totally neglected and stopped. Now-a-days, it is not in practice.

Sorghum bicolor (L.) Moench.

Local Name

Jowar

Ethnobotanical Notes

Medicine

1. Spoonful of ash of stalks is mixed with honey and given at early morning for 3 to 4 months to cure asthma.

Miscellaneous

1. Culms are used as fodder for domestic animals. Apart from this, culms are also used to thatching huts and temporary sheds in farms and farmyards.

Notes from Classics

Grains are reckoned heating.

(Watt, 1889-1899)

Seeds are diuretic and demulcent.

(Anonymous, 1948-76)

Grains are used in baldness.

(Jain, 1991)

Critical Notes

Medicinal utility of ear-stalks for asthma is new information. The plant is commonly cultivated. Apart from its classic uses as food and fodder, it has gone down in the ambient society as a medicine as well.

Triticum aestivum L.

Local Name

Gahu

Ethnobotanical Notes

Medicine

1. Wheat flour is mixed in water and applied as poultice on tumour, after two or three days, tumour burst out and gives relief.

Miscellaneous

1. Wheat straw cut is to small pieces and mixed in mud while preparing bricks.
2. Vanjara people use it in 'Tij' festival.

Notes from Classics

Flour is esteemed as external application in erysipelas, burns, scald and itching. Flour is mixed in water and used as antidote in case of poisoning by salts of mercury, copper, zinc, silver and Tin. Bran bread is laxative and used in dyspepsia and diabetes. Bread crumb is useful for giving bulk to pills and making poultice.

(Watt, 1889-1899)

Wheat flour is nutritious, wheat brans has latative properties. Wheat straw is fed to live stock as roughage. Straw is used as bedding for cattle, also used as packing for fragile goods.

(Anonymous, 1948-76)

Straw is used for thatching huts.

(Ambasta, 1986)

Straw is used for making soles of grass shoes.

(Jain, 1991)

Critical Notes

Wheat flour is useful to treat body tumours. Straw is also useful in making bricks. It is cultivated for food grains.

Family: Ustilaginaceae

Sphacelotheca sorghii (Link) Clin.

Local Name

Kani, Kahani

Ethnobotanical Notes

Medicine

1. Ears of *Sorghum bicolor* are infected with *Spacelotheca sorghii*. The smut spores removed from the ears mixed in coconut oil (*Cocus nucifera* L.) and is applied in case of piles.

Notes from Classics

The host plant species is a common crop in this region but infective fungal species is being reported for the first time as useful for piles. It is the only fungal taxon noted useful as medicine during the present investigation.

Family: Marsileaeae

Marsilea minuta L.

Local Name

Zabzaba

Ethnobotanical Notes

Miscellaneous

1. *Food*: Leaves when young are used as vegetable.

Notes from Classics

Plants are regularly eaten as pot herb. Stalks and leaves of the plant are eaten as pot herb in times of scarcity.

(Watt, 1889-1899)

Stalks are leaves of the plant are eaten as pot herb in times of scarcity.

(Anonymous, 1948-76)

Stalks and leaves are eaten as pot herb.

(Ambasta, 1986)

Leaves are used as vegetable.

(Ambasta, 1991)

Critical Notes

Food value is reiterated by the observations from this district of Maharashtra. This is the only pteriodophyte found useful in this district. However, the pteridophytic flora is not rich in this district.

Part II
ABSTRACT RELATIONSHIPS

Section I
Tribals

(A) Pawara Tribe

FESTIVALS OF PAWARA

1.a) ***Aegle marmelos*** **(L.) Corr. (Rutaceae) Bel**

Cocos nucifera **L. (Arecaceae) Naral**

b) ***Madhuca longifolia*** **(Koen.) Macbride var.** ***latifolia*** **(Roxb.) Chev. (Sapotaceae) Mahuwa, Mahu**

Ola Pujanu/Ol Puttar: On the festival day of 'Ola Pujana' in the month of June before rainy season the agricultural implements are worshipped by offering 'Bel' leaves (*Aegle marmelos* (L.) Corr., coconut fruits (*Cocos nucifera* L.), Kunku (Red Powder), fragrant sticks etc. Each agricultural implement is sprinkled by wine of Mahua flowers (*Madhuca longifolia* (Koen.) Macbride var. *latifolia* (Roxb.) Chevalier.

2. ***Cocos nucifera*** **L. (Arecaceae) Naral**

Shivardev: In the festival of 'Neelapi Poojnoo' Shivardev is worshipped by offering shendur (*Lead powder*), milk, egg and coconut fruit (*Cocos nucifera* L.). On the border of a village some spherical stones are smeared with shendur, established under tree. It is an idol of Shivardev.

3. *Zea mays* L. (Poaceae) Maka

Dudamakay: It is a festival celebrated after harvesting of maize (*Zea mays* L.). Maize ears are worshipped with kumkum (Red powder) and fragnant sticks. From the days of 'Dudamakay' pawara start eating of maize grains.

4. *Tectona grandis* L.*f.* (Verbanaceae) Sag
Madhuca longifolia (Koen.) Macbride var. *latifolia* (Roxb.) Chev. (Sapotaceae) Mahu, Mahuwa

Bapdev Poojnoo: This festival is celebrated in September specially on Friday in honour of Bapdev *i.e.* tiger-god. Priest or Poojara prays all plants, animals and tiger to protect them. He worship them with Kumum (red powder) and sprinkle wine of Mahu flowers (*Madhuca longifolia* (Koen.) Macbride var. *latifolia* (Roxb.) Chevalier) on 'Bapdev'. Pawaras offer goat or hen as a sacrifice and their flesh is distributed among people on teak leaves (*Tectona grandis* L.*f.*). On the same day they start using new teak leaves and teak wood.

5. *Cucumis melo* L. var.*utilssimus* Duthie and Fuller (Cucurbitaceae) Kakadi
Madhuca longifolia (Koen.) Macbride var. *latifolia* (Roxb.) Chev. (Sapotaceae) Mahu, Mahuwa

Naway: Naway means new. All newly harvested grains, vegetables and fruits are worshipped on this day especially cucumber and rice. Naway are of two types 'Kakadi naway' and 'Khatri naway'. In case of 'Kakadi naway' leaves of cucumbers, flowers and pieces of cucumber (*Cucumis melo* L. var.*utilissmus*, Dutthie and Fuller) are placed on them. These are worshipped with Kumkum, shendur and fragrant sticks. Wine of Mahua flowers [*Madhuca longifolia* (Koen.) Macbride var. *latifolia* (Roxb.) Chevalier] is sprinkled all over the house and pieces of cucumber are distributed as 'prasad'.

6. *Butea monosperma* (Lam.) Taub. (Papilionaceae) Palas
Oryza sativa L.(Poaceae) Bhat, Tandul
Vigna ungiculata (L.) Walp.(Papilionaceae) Chawli

Naway: This festival is celebrated on Monday. New rice grains (*Oryza sativa* L.) are boiled, pulse of Chawali [*Vigna ungiculata* (L.) Walp.] is cooked. Pulse and rice grains are kept on leaf of Palas.

[*Butea monosperma* (Lamk.) Taub.]. Poojara worship this leaf with Kumkum, one leaf with aforesaid 'Prasad' and kept on roof as offering for forefathers. From this days Pavara start eating new rice grains.

7. ***Setaria italica* (L.) P.Beauv. (Poaceae) Bhagar**
   ***Mangifera indica* L. (Anacardiaceae) Amba**
   ***Mitragyna parvifolia* (Roxb.) Korth. (Rubiaceae) Kalam**
   ***Sorghum bicolor* (L.) Moench(Poaceae) Jowar**
   ***Madhuca longifolia* (Koen.) Macbride var.*latifolia* (Roxb.) Chev. (Sapotaceae) Mahu, Mahuwa**

Indal: Festival 'Indal' is celebrated by all villages. People from other villages are invited by sending 'Bhagar' coloured with turmeric powder [*Setaria italica* (L.) P.Beauv.]. Village is decorated with wreaths of mango (*Mangifera indica* L.) and Mahu leaves [*Madhuca longifolia* (Koen.) Macrbride var. *latifolia* (Roxb.) Chev.].

All villagers visit to the forest. Small three boys cut off the three branches of Kalam trees [*Mitragyna parvifolia* (Roxb.) Korth.] One branch is cut off in the name of God Indradev (God of Rain), second branch is in the name of Kanhari mata (Goddess of grains) and third branch in the name of Hudovdev. These three branches are fixed in three pits which are filled with grains of Jawar [*Sorghum bicolor* (L.) Moench] Poojara (Priest) worship these branches. On the next day at evening these three branches are worshipped and offered to river ceremonially.

8. ***Zea mays* L. (Poaceae) Maka**
   ***Madhuca longifolia* (Koen.) Macbride var. *latifolia* (Roxb.) Chev. (Sapotaceae) Mahuwa, Mahu**

Diwali-Pola: On this day the bullocks are washed and decorated. They are compelled to drink wine of Mahua flowers [*Madhuca longifolia* (Koen.) Macbride var. *latifolia* (Roxb.) Chev.] and to eat boiled grains of maize (*Zea mays* L.)

9. ***Bambusa arundinacea* (Retz.) Willd. (Poaceae) Bamboo**
   ***Cucurbita maxima* Date. ex Lam.(Cucurbitaceae) Bhopla**

Holi Melda: Holi is important festival of Pawaras. Long Bamboo [*Bambusa arundinacea* (Retz.) Willd.] is kept in a pit and supported with branches of trees. All villagers dance around it at night when 'Holi' is lighted with fire. Holi is worshipped by villagers. 'Buwa' is a person who tie 'Bhopla' (*Cucurbita maxima* Dute.ex Lam.) at wiast, with the help of wooden sword and bamboo stick he dances around Holi. The 'Buwa' expresses his gratitude towards Holi by dancing.

10. *Sorghum bicolor* (L.) Moench (Poaceae) Jowar
***Aegle marmelos* (L.) Corr.(Rutaceae) Bel**

Karhan muwne: Karhan is a goddess of grains. Newly harvested ears of Jowar are collected and the heap of ears of Jowar is worshipped with Kumkum (red powder), fragrant sticks and lamp is lit up. Wine of Mahua flowers is sprinkled all over the ground. 'Shendur' is kept on leaf of Bel [(Aegle marmelos (L.) Corr.] Shendur is applied on forehead of bullocks as a 'tika'. All invitees consume the remaining wine.

GODS AND GODDESSES OF PAWARA

Pawara tribe is mostly settled in the forest of Satpuda mountain. They have to face many difficulties to satisfy their basic needs. So they can not develops temples. Small rocks, round stones, plant and animals are the Gods and Goddesses for them.

To protect village or hamlet, Pawara worship the Shivardev. To increase milk and milk product they worship Gawlidev. For prosperity and good production of food grains they worship Kansari mata. To protect villagers from tiger, tiger god *i.e.* Bapdev or Vaghdev is worshipped. Some of important ones are explained below about which plants are found associated:

Girhoon

Girhoon is a clan diety of each Pawara family. In the kitchen 2 ft to 3 ft long wood of teak *i.e. Tectona grandis* L.*f.* as idol of Girhoon is established. On every religious ceremony they first worship the 'Girhoon'.

Dongrya Dev

Dongryadev is settled on big rock and small temple of this god is called 'Gadhi'. The festival of Dongryadev is celebrated during 'Dipawali'. When the production of Nagli [*Eleusine coracana* (Retz.) P.Beauv.] is increased, it is considered as symbol that Dongryadev is pleased with the people. A person who wants to celebrate festival, he sends an areca-nut (*Areca catechu* L.) to Poojara, a priest of Pavara, and invites him for worship. People of all villages go to Gadhi to worship the god with Kumkum (red powder) and fragrant sticks. Two goats or hens are sacrificed in front of god. Coconut (*Cocos nucifera* L.) is splitted and distributed as 'Prasad'.

Yahi Mogi Mata

This is a clan diety of Pawara kings. She is goddess of grains and according to Pawara belief she stays in grains. A big temple is present in village Tikoni (taluka Sagbara, District Bharuch, Gujarat) During month of February at full moon day Poojara and followers carry the idol of Mata near river for holy bath. Yellow soil is mixed with leaves of Bel (*Aegle marmelos* L.) and is applied to 'Mata'. After bath, Mata is brought to temple uttering some 'Mantra' in Pawara dialect. She is worshipped with Kumkum, turmeric powder (*Curcuma longa* L.) and various flowers. Rice (*Oryza sativa* L.), Oos (*Saccharum officinarum* L.), Dalya (*Cicer arientinum* L.), Naral (*Cocos nucifera* L.), Churmure (*Oryza sativa* L.), wine of Mahuwa flowers [(*Madhuca longifolia* (Koen.) Macbride var. *latifolia* (Roxb.) Chev.] are offered as 'Naivaidya'. People brought small basket of Bamboo (*Bambusa arundinacea* (Retz.) Willd.) wrapped with white cotton cloth which contains newly harvested grains, materials of worship and offer to Mata. This is called 'Hijri'.

Waghdev

Festival of Waghdev is celebrated during August. Waghdev is usually situated near border of a village on the bank of river. At morning, Poojara and villagers gather near God Waghdev. They wash and worship the god with Shendur, rice grain (*Oryza sativa* L.). They lit up the lamps and fragrant sticks. The 'naivaidya' is offered which consist of boiled rice (*Oryza sativa* L.) vagetable of Tandulya (*Amaratanthus roxburghianus* Havskii), 'Bhakari' of Dadar [*Sorghum bicolor* (L.) Moench.] and other newly harvested vegetables. The coconut (*Cocos nucifera* L.) is splitted and distributed as 'prasad'. Branches of teak (*Tectona grandis* L.*f.*) are kept on roof of a house or hut or in fields. After the festival of Waghdev, Pawara start eating of 'Maka' (*Zea mays* L.).

MARRIAGES IN PAWARA

Marriage in Pawara tribe is celebrated in a peculiar way. Bride and bridegroom seat on shoulders of some elder persons. A long procession is with many invitees and musical band is followed. The musical band consists of special musical instruments such as Pawari, Mandal, Piparya, Thari, Pawa etc. This procession ends at bride's house 'Poojara' (Priest) worship sun, moon, soil, tiger-god, Indal-dev and their own forefathers. During worship, wine of Mahuwa flowers [*Madhuca longifolia* (Koen.) Macbride var. *latifolia* (Roxb.) Chevalier (Sapotaceae)] is used. On the various occasions of their marriage ceremony, the plants used are as follows:

1. At the time of engagement ceremony, relatives of bridegroom party give botanicals to the bride such as:
   (*a*) Coconut fruit or kernels (*Cocos nucifera* L., Arecaceae)
   (*b*) Two areca nuts (*Areca catechu* L., Arecaceae)
2. Cordage is tied around the trunk of Wad tree (*Ficus benghalensis* L., Moraceae). Then paste of turmeric powder is applied on the body of bride and bridegroom under Apta or 'Hengala' tree [*Bauhinia racemosa* Lam., Caesalpiniaceae].

   'Otti' is a symbol of commitment to the protection of a bridge. The following things are offered to her in Otti:
   (*a*) Rice grains (*Oryza sativa* L., Poaceae)
   (*b*) Fruits of date palm (*Phoenix dactylefera* L., Arecaceae).
   (*c*) Leaves of Hengala (*Bauhinia racemosa* Lamk., Caesalpiniaceae)
   (*d*) Some money is also given along with these.
3. Just before completion of marriage ceremony, the family deity 'Girhoon' is worshipped. The idol of deity is made up of wood of sag (*Tectona grandis* L.*f*., Verbenaceae). The leaves of Hengala [*Bauhinia racemosa* Lam., Caesalpiniaceae] are also worshipped.
4. Pavilion or pandal of marriage ceremony is decorated by the wreaths of leaves of (a) Amba [*Mangifera indica* L., Anacardiaceae (b) Ruchkin (*Calotropis procera* (Ait.) R.Br. or *Calotropis gigantea* (L.) R.Br., Asclepiadaceae) and (c) Leaves of Hengala (*Bauhinia racemosa* Lam., Caesalpiniaceae).
5. Peepal tree [Ficus religiosa L., Moraceae] on the day of marriage is worshipped. Its leaves also spread over the ground of marriage ceremony.

6. Dining plates made up of leaves of 'Wad' (Ficus benghalensis L., Moraceae) are used.
7. Leaves are Apta or Hengala (Bauhinia racemosa Lam., Caesalpiniaceae) are put by bridegroom near the dining plate of bride. Then they are allowed for enjoying the dinner.
8. The following things are always possessed by bride and bridegroom with them during marriage ceremony:
   (*a*) 'Jupna' (Cordage made up of coir of coconut, *Cocos nucifera* L., Arecaceae).
   (*b*) 'Katak' (A walking stick of *Dendrocalamus strictus* Nees, Poaceae.)
   (*c*) 'Supda' (A winnowing pan for shifting grains made up of culms of bamboo. [*Bambusa arundinacea* (Retz.) Willd., Poaceae].
   (*d*) 'Akhas pas' (An agricultural implement of iron).
9. On one occasion during marriage ceremony a bow made up of *Bambusa arundincea* (Retz.) Willd., Poceae) and arrow made up of culms of *Dendrocalamus strictus* Nees, Poaceae) are hold in the hands bridegroom.
10. Paste of turmeric powder (Curcuma longa L., Zingiberaceae, is applied to cotton threads. These threads are wrapped around 'Y' shaped or forked branch of samadi or Radkya plant [*Prosopis cineracria* (L.) Druce, Mimosaceae]. It is then placed on the right hand side of new earthen water container. This is completed to prior application of paste of turmeric to bride and bridegroom at their home.
11. After marriage ceremony, central pillar of a house or hut which is made up of 'Sag' (*Tectona grandis* L.*f*., Verbenaceae), called Girhoon, is worshipped. Then a ornamental head-gear 'Mundal' or 'Bashing' is tied on forehead of the bride and bridegroom.

FUNERAL CEREMONY IN PAWARA

1. After death of men or women, various rituals are performed. Before reaching to cremation ground, meal of

pulse [*Vigna unguiculata* (L.)Walp.] and rice (*Oryza sativa* L.) is offered to dead person. On the bed of dead person an bow arrow [Bambusa arundinacea (Retz.) Willd.] and (*Dendrocalamus strictus* Nees) are kept.

2. At the time of funeral procession a near relatives carry earthern pot filled with maize grains (*Zea mays* L.) and Hal (*Oryza sativa* L.). This pot is called 'atthi'. Small bamboo [*Bambusa arundinacea* (Retz.) Willd.] basket is kept on the earthern pot. This basket is broken into pieces when funeral reach to the cremation ground.
3. After cremation the relatives return to home and sprinkle water in home and premises with the help of leaves of Pimpari (*Ficus virens* Ait.)
4. On the twelfth day of cremation a ritual is performed which is known as 'Undid Makano'. In this ritual relatives are invited and welcomed offering tea, tobacco and wine of Mahuwa flowers [*Madhuca longifolia* (Koen.)] Macbride var. *latifolia* (Roxb.) Chevilier.
5. Small platform of ash is made, called 'Undid Makano' A Bamboo (*Bambusa arundinacea* (Retz.) Willd.] basket is filled with fried rice (*Oryza sativa* L.) and glass of wine, pulse of Chawli [*Vigna ungiculata* (L.) Walp.] and is kept on it. This basket is covered with another basket. All the people go outside the host house and wait for the sole of dead person to come and eat these offerings. At night all people sing holy songs.
6. On the second day of 'Uttar Karya', two relatives and any widow go to river and bath in river water. After bath, widow (only widow) prepare rice (*Oryza sativa* L.) and dal [*Cajanus cajan* (L.) Milsp.] on cremation ground. The offerings of rice and dal are kept on several leaves of Palas [*Butea monosperma* (Lam.) Taub.]. One leaf is kept on 'Unddi' and remaining leaves are kept of roof of a house or hut for forefathers.
7. Near 'Unddi" small wooden seat, called 'Patli' is kept. Patli is made up of 'sag' wood. [*Tectona grandis* L.*f*.]. Invitees sprinkle rice grains (*Oryza sativa* L.) and wine of Mahua flowers [*Madhuca longifolia* (Koen.) Macbride. var.*latifolila* (Roxb.) Chev.] on it. This ritual is called 'Sak

Vandanoo'. Non-vegetarian lunch is served given to all invitees.

8. Leaves of Khakar or Palas [*Butea monosperma* (Lam.) Taub.] are carried with the corpse. They are worshipped in the cremetery and then only the corpse is cremated.
9. After death, wine prepared from Mahuwa flowers [*Madhuca longifolia* (Koen) Macbride var. *latifolila* (Roxb.) Chev.] Sapotaceae or/and Tobacco (*Nicotiana tabacum* L., Solanaceae) are placed near the corpse.

SPECIAL OCCASIONS IN PAWARA

1. On the new-moon day, the trees such as Amba (*Mangifera indica* L., Anacardiaceae); Kakad (*Garuga pinnata* Roxb., Burseraceae) and Pimpal (*Ficus religiosa* L., Moraceae) are worshipped.
2. A new hearth is worshipped by offering turmeric powder (*Curcuma longa* L., Zingiberaceae) and Sapta-Dhanya, seven different kinds of food grains like:
   (*a*) Jowar [*Sorghum bicolor* (L.) Moench, Poaceae]
   (*b*) Bajra [*Pennisetum americanum* Schum., Poaceae]
   (*c*) Maka (*Zea mays* L., Poaceae)
   (*d*) Dhan (*Oryza sativa* L., Poaceae)
   (*e*) [*Eleusine coracana* (Retz.) P. Beauv.]
   (*f*) Banti [Echinochola stagnina (Retz.) P. Beauv, Poaceae]
   (*g*) Rala [Setaria italica (L.) P.Beauv., Poaceae]
3. A threshold is made up of the wood of Hengala [Bauhinia racemosa Lam., Caesalpiniaceae]. It is the lower part of the door frame and is considered auspicious.
4. If the babies go on crying for longer period, the tree Samadi [*Prosopis cineraria* (L.) Druce, Mimosaceae] is worshipped and prayed for blessing to stop crying.
5. Banyan tree, called Wad (Ficus benghalensis L., Moraceae) is worshipped by newly married woman on the full-moon day of the month of Jeshtha.
6. On the day of 'Mahashivaratri', the 19th day of the 2nd fort-night of Magh month, the tribal pilgrims offer 108

inflorescences of Tulshi plant (*Ocimum tenuiflorum* L., Lamiaceae). to the idol of Mahadev at 'Shirvel' (Chopda taluka).

MAGIC AND BELIEFS IN PAWARA

This tribe believes in magic. Magic are said to be of two types *viz.*, white magic and black magic. White magic is good and useful to solve the problems of society. But black magic causes bad effect on person or society. It consists 'Muth marne' *i.e.* to kill person or loss of property etc.

1. *Muth Marne*: In this black magic, small doll of black cotton cloth is made. This doll is symbol of victim (The person is to be killed). Magicians take some materials like Kumkum (Red powder), Shendur (Orange lead powder), Rice (*Oryza sativa* L.), Udid (*Phaseolus mungo* L.), Jawar [*Sorghum bicolor* (L.) Moench] and an egg. They keep all these things on dish and utter some 'mantras' and stuck the pins to the body of a doll. As he starts stucking pins to the body of the doll, the victim dies due to blood-vomiting.
2. *Belief*: If small black sparrow is seen seating on Bor plant (*Ziziphus maurtiana* L., Rhamnaceae), the house where the man went to marry, is hunted by witches.

FOLK SONGS OF PAWARA

1. *Ficus religiosa* L. (Moraceae) Pimpal, Pipur *Bambusa arundinacea* Willd. (Poaceae) Bans, Bamboo

Telya Pipure.... patun bena

Tyove jarin ma pani voch

Mada chipawlan datun purya

Kutryan humnam pani boch

Mandalwalan bachashe kudale gayalo

Mandivajo padi gayalo.

['तेल्या पिपुरे ...... पातून बेना

त्योवे जारीन मा पाणी वोच

माडा चिपवलान दातुन पुऱ्या

कुत्र्यान हुमणाम पाणी बोच

मांदलवालान बाचशे कुदाले गयलो

मांदीवाजो पडी गयलो'.]

The marriage party while travelling for marriage ceremony sing this song. The bride brushes teeth using oily (shining) stem piece of 'Pipur' tree (*Ficus religiosa* L.) and washes mouth using water stored in golden water pot. But the bridegroom brushes his teeth by using dried stem of 'Chipawalan' (*Bambusa arundinacea* Willd.) and washes mouth using water from the pot in which water for dogs is stored. He dances madly and therefore the musicians and drummers run way. Thus the bridegroom is treated and teased by the party of a bride.

2. *Cannabis sativa* L. (Cannabaceae) Bhang
***Zea mays* L. (Poaceae) Maka**

Bayadivar tar ghar raha par mar

Aaweta jat bhang piye Nimbalya mar ghar

Chahparya ma kuw khad bhayade kamay kar

Chalwa puray kamay dekhane, kashi kamay kar

Mukayane wad velya vasan wad

Shiway ma awaji puray kashi maja awe.

['बयडीवर तार घर रहा परं मारं

आवता जात भांग पीये निंबळ्या मारं घर

छापऱ्या मा कुव खदं भयडे कामाय करं

चालवा पराय कामाय देखणे, कशी कामाय करं

मुकायने वाड वेल्या वसन वाड

शिवाय मां आवजी पुराय कशी मजा आवे'.]

Tribal ladies sing this song at the time of marriage ceremony. It is meant as follows: The bride returns home of her father in winter season. A tribal lady says that my home is atop the hill on your way.

Nimbalya, a person from another hamlet, visits my home time and again when he is narcotised by 'bhang' (*Cannabis sativa* L.). A well is dug out in stony field. Crops are in full bloom. Let us to go observe the crops. How are the crops? Stands of maize crop (*Zea mays* L.) are covered by green climbers. You please visit, you will enjoy greatly.

3. *Buchanania lanzan* Spreng. (Anacardiaceae) Charoli, Saroli

Badvi jodi Novsa Sarolya viseke
Sarolya dovo Novsa novri Sarolya viseke
Use sodi vabo dhosi sundari dodoke
Ingi nivdi Novsa novri sarolya viseke
Thisi thisi Novsa novri vate ledike re
Hiye palo Novsa novriko aaj lagoke re
Hite pato Novsa novri kardun ne livoke
Chopda Sahar Novsa novri vesan lageke
Misalya lovo Novsa novri samali lake.

['बाडवी जोडी नोवसा सारोल्या विसेके
सारोल्या डोबो नोवसा नोवरी सारोल्या विसेके
खोवसा खासाडो नोवसा नोवरी लोसो लागेके
उसे सोडी वाबो ठोशी सुंदरी दोडोके
इंगी निवडी नोवसा नोवरी सारोल्या विसेके
ठिसी ठिसी नोवसा नोवरी वाटे लेडीके रे
हिये पालो नावसा नोवरी को आज लागोके रे
हिते पातो नोवसा नोवरी कारदून ने लिवोके
चोपडा सहर नोवसा नोवरी वेसान लागेके
मुसल्या लोवो नोवसा नोवरी सामाली लेके'.]

Novsa's wife is collecting fruits of 'Charoli' (*Buchanania lanzan* Spreng.) in the cross-mountain. She climbs on the Charoli tree and twitches the fruits. She breaks open the fruits of Charoli. Now, she is going to Chopda city to sell the seeds of Charoli, Oh, Novsa's wife, there are many muslim people, be aware and take care of you. The entire folk song tells about the life of Pawara ladies and their work.

4. *Cocos nucifera* L. (Arecaceae) Naral
***Phoenix dactylifera* L. (Arecaceae) Khajur**
***Cicer arientinum* L. (Papilionaceae) Chana, Harbara**

Mavchadi patime dand ropyo madi bal

Mavchadi patime dand ropyo relo

Olil aay kay leni aau madi bal

Oli aay kay leni aau relo

Oli aay khobara leni aau madi bal

Olil Khobara lenu aau relo

Oli aay aala leni aau madi bal

Olil aay gulal, dalya leni aau madi bal.

['मावचाडी पाटेमे डांड रोप्यो माडी बाळ

मावचाडी पाटेमे डांड रोप्यो रेलो

ओलील आय काय लेनी आवू माडी बाळ

ओली आय काय लेनी आवू रेलो

ओली आय खोबरा लेनी आवू माडी बाळ

ओलील खोबरा लेनी आवू माडी बाळ

ओलील खोबरा लेनू आवू रेलो

ओली आय आला लेनी आवू गाडी बाळ

ओलील आय गुलाल, डाल्या लेनी आवू माडी बाळ'.]

A pawara girl purposefully says, "Holi mother, you arrived here in a common place. What should I bring for you to welcome? I will bring and offer you kernel (*Cocos nucifera* L.), dates (*Phoenix dactylifera* L.), parched gram (*Cicer arientinum* L.), garland, bracelet and red lead.

5. *Syzygium cumini* (L.) Skeels (Myrtaceae) Jambhul, Jamun

Vate vate janu chhe jamunyas vat

Jamunyas vat ga puray jayene hey thagayali.

['वाटे वाटे जाणू छे जामुण्यास वाट

जामुण्यास वाट न पुराय जायने हये ठगायली'.]

This song is sung while returning from the place of marriage. The friends of bride tease her. They say that there is grove of 'Jamhun; (*Syzygium cumini* (L.) Skeels) on the way you are passing. But mind well, a girl has been deceived on this way.

6. *Tectona grandis* L.*f.* (Verbenaceae) Sag

Pani phandale judo va bena

Jud todine chali va bena

Saganu lakud chare jud va bena

Jud todina chali va bena.

['पानी फंडाले, जुडो वा बेना

जुड तोडीने चाली वा बेना

सागनु लाकूड चारे जुड वा बेना

जुड तोडीने चाली वा बेना'.]

At the time of marriage, brother of a bride sings for his sister. He says that we used to fetch water together. We also used to gather fuel wood of Sag (*Tectona grandis* L.*f.*). But on account of your mariage, we are being separated. There will be no friendship henceforth. The brother thus feels sorry.

7. *Areca catechu* L.(Arecaceae) Supari
Citrus aurantifolia (Christm.) SW. (Rutaceae) Nimboo

Indi Raja chavrema bothy O Indi raja bandode boye O

Indi raja nimbune bhukyo O Indi raja bandode boye O

Indi raja lukdone bhukyo O Indi raja bandode boye O

indi raja kukedo ne bhukyo O Indi raja bandode boye O.

['इंदी राजा चवरेमे बेठयो इंदी राजा बांडोडे बोये ओ

इंदी राजा निंबुने भुकयो इंदी राजा बांडोडे बोये ओ

इंदी राजा लुकेडोने भुकयो इंदी रजा बांदोडे बोये ओ

इंदी राजा कुकेडोने भुकयो इंदी राजा बांदोडे बोये ओ'.]

This song is sung while worshipping God Indra. In the opinion of Pawara community, God Indra sits in foreground of one's house and observes everything critically. To please Indra, we must give as far as possible in worship. In turn, our crop yield will increase. He is hungry and fond of arecanut (*Areca catechu* L.), lemon [*Citrus aurantifolia* (Christm.) SW.], fowl and she-goat.

8. *Diospyros melanoxylon* Roxb. (Ebenaceae) Temburni *Ficus amplissima* J.E.Sm. (Moraceae) Pimpri

Tembarima bonome tembare khan geyali

Dadama maye chor mathe paderelo

Pimparima banome pimpari khawa geyali

Dadama maye chor mathe paderelo

Payome nodiyo mele dada dharado

Payrone sodwin latavun relo.

['टेंबरीमा बोनोमे टंबरे खान गेयली

दादामा माये चोर माथे पाडेरे लो

पिंपरीमा बोनोमे पिंपरी खावा गेयली

दादामा माये चोर माथे पाडेरे लो

पायोमे नोंडीयो मेले दादा धाराडो

पायरोने सोडविन लाटावून रेलो'.]

This song is sung at the time of 'Holi' Festival. A pawara boy says "I went to the forest of Tembare (*Diospyros melanoxylon* Roxb.) to enjoy the fruits (called 'Tembare'), but I was blamed for theft. I also went to the forest of Pimpari [*Ficus amplissima* J.E.Sm.] to eat the fruits, I was accused for theft.

9. *Sesamum orientale* L. (Pedaliaceae) Til

Tili otonho vadalo verya
Bado jug vetalyo verya
Panyambi nahan podaton verya
Dese ningi jahunra verya
Gati hidon bandine verya
Dese ningi jahunra verya.

['तिली ओतोंहो वादलो वेऱ्या
बादो जुग वेटाल्यों वेऱ्या
पायंबी नाहां पोडतों वेऱ्या
देसे निंगी जाहूंरा वेऱ्या
गाती हिदों बांदीने वेऱ्या
देसे निंगी जाहूंरा वेऱ्या'.]

Prior to rainy season, tribal farmer awaits for rains. he looks at clouds in the sky. The deity of water is called 'Verya' (Varundev). The tribal girls sing this song requesting 'Verya' and describe the facts. The girls say that the sky is covered by numerous Tili seed-sized (*Sesamum orientale* L.) clouds, still it is not raining. This will discontent and compel us to leave our country.

10. *Oryza sativa* L. (Poaceae) Bhat,Tandul

Kalya megu voradyen jahunva haru pandhore
Mahari mohari aavera, kalya megera
Halya hukay chalya ra, kalya megera
Mahari mohari aawesh, kalya megera
To je bhoruse hayara, kalya megera
Poyaren chavaren rodo ra, kalya megera
Topati bitume dobryo ra, kalya megera.

['काल्या मेगू वोराड्यें, जाहुंवा हारु पांढोरे
माहारी मोहारी आवेरा, काल्या मेगेरा

हाल्या हुकाय चाल्या रा, काल्या मेगेरा

माहारी मोहारी आवेरा, काल्या मेगेरा

तो जे भोरुसे हायरा, काल्या मेगेरा

पोयरें चावरें रोडो रा, काल्या मेगेरा

तोपती बिटुमे दोव्ये रा, काल्या मेगेरा'.]

The tribal people also refer the deity of water (Varnundev) as 'Kalyameg'. 'Haru Pandhor' is their family deity. They request their family deity to attend the marriage ceremony of 'Kalyameg' so that he will be happy and pleased to rain. The tribals say to Kalyameg that the crop of rice (*Oryza sativa* L.) is drying out as there are no rains. Children are also crying. You have hidden in the rocks of Tapi river. Please come out and rain for us.

11. *Bambusa arundincea* (Retz.) Willd. (Poaceae) Bamboo

Pode pany pode va poyari

Kenhdame dobi reheli

Pode pany va poyari

Huplyame dobi reheli

Huplyonbi goli podyonva

Poyari kenhdame dobi reheli

Pode pany pode va poyari

Chimbdame dobi reheli.

['पोडे पांय पोडे वा पोयरी

केंहडामे दोबी रेहेली

पोडे पांय वा पोयरी

हुपल्यामें दोबी रेहेली

हुपल्योंबी गोली पोड्योंबा

पोयरी केंहडामे दोबी रेहेली

पोडे पांय पोडे वा पोयरी

चिंबडामे दोबी रेहेली'.]

The tribals enquire their family deity 'Haru Pandhor'. There are heavy rains. Where you were hidden? Have you hidden under the winnowing pan. It was also leaking, then have you hidden in the jungle of bamboos [*Bambusa arundinacea* (Retz.) Willd.].

12. *Solanum melongena* L. (Solanaceae) Wange

Selbhaya maj viajlela dhadadhod

Selbhaya vengi molum rat rihunra

Selbhaya vengi ropto pajay giyonra

Selbhaya vengi valonta dond magera

Selbhaya chandal vechin dond dihura.

['सेलभाया गाज विजलेलो धाडाधोड

सेलभाया वेंगी मोलून रात रिहूंरा

सेलभाया वेंगी रोपतो पाजाय गियोंरा

सेलभाया वेंगी वालोंता डोंड मागेरा

सेलभाया चांदल वेचिन डोंड दिहूरा']

Wedding season continues in the month of June also. A tribal sister and her brother, who is a bridegroom, travel together by bullock carts. She sings this song. She says, brother, there is thundering and thunderclap. We must find shelter somewhere. it is better to pass the night in the brinjal field (*Solanum melongena* L.). We have passed the night but the growing plantlets are destroyed by us. The owner of brinjal field is demanding fine. Brother, let us sell our girl friend Chanda.

PUZZLES IN PAWARI DIALECT

1. *Ficus benghalensis* L. (Moraceae) Wad
***Bambusa arundincea* (Retz.) Willd. (Poaceae) Bamboo, Tokar**

Badan dukh ghatak rang vahani didane rang kon?

Answer: Parrot

['बडन डुख घटक रंग वाहणी दिडणे रंग कोण ?'

उत्तर : पोपट]

Who is that having colour like reddish apex of twig of Banyan tree (*Ficus benghalensis* L.). Who is that having also body like

2. *Ricinus communis* L. (Euphorbiaceae) Arand

Sisani inda melatali inda bulatala.

Answer: Seeds of castor

['सिसानी इंडा मेलतली इंडा बुलतला'

उत्तर : एरंडी बीया']

Sisani is said to be a bird which lays eggs. These eggs talk. Dried capsules containing seeds make noise. These are referred as if talking of eggs.

3. *Allium cepa* L. (Liliaceae) Kanda

Ayatali suri rajah radavatali?

Answer: Onion bulb

['आयतली सुरी राजाह राडावतली ?

उत्तर : कांदा']

Small girl who makes the King to weep. While cutting or slicing onion bulbs (*Allium cepa* L.) eyes become wet due to tears. This is referred as weeping.

4. *Allium cepa* L. (Liliaceae) Kanda

Ayatali suri kamar bharin katara.

Answer: Onion bulb

['आयातली सुरी कमर भरीन कातरा.

उत्तर : कांदा']

The onion bulb is first cut at middle. This is referred as cutting a small girl at lion.

5. *Cucurbita moschata* (Duch. ex Lam.) Poir. (Cucurbitaceae) Gangaphal

Ek dadne pet me date?

Answer: Red Pumkin

['एक दादने पेटम दाते?

उत्तर : तांबडा भोपळा']

Teeth are present in the stomach of an elder brother. The seeds are referred at teeth within cavity or hollow fruit of red pumkin (*Cucurbita moschata* Duch. ex Lam.)

6. *Cocos nucifera* L. (Arecaceae) Haral

Upar aad aay mah.

Answer: Fruit of coconut

['उपर आड आय माह'.

उत्तर : नारळ']

A strange situation is explained. The bone is outside and the flesh is inside. The stony endocarp is referred as bone, whereas the inside kernal is thought flesh.

PROVERBS IN PAWARI DIALECT

1. *Pennisetum americanum* L.(Poaceae) Bajra

Bulteri kankhadi bi vechay jay

Ni bule teri bajari bi ni vechay.

['बुलतेरी कानखडी बी वेचाय जाय

नी बुले तेरी बाजरी बी नी वेचाय'.]

An eloquent person can sell stones but a taciturn can not sell grains of bajara (*Pennisetum americanum* L.). A talkative person is more efficient in his work.

2. *Zea mays* L. (Poaceae) Maka

Khatomay sanen khat nakhali

Tinama mukay khub pikali.

['खतोमाय सानेन खत नाखली

तिनमा मुकाय खुब पिकली'.]

If you add compost in fields, you will get more yield of grains of maize (*Zea mays* L.). If you think in a proper way, proper results will be obtained.

(B) Bhil Tribe

MARRIAGES IN BHIL

1. At the time of betrothal a Bhil bridegroom offers to bride following things:
   (*a*) Coconut fruit (*Cocos nucifera* L., Arecaceae)
   (*b*) Leaves of Hengala (*Bauhinia racemosa* Lam., Caesalpiniaceae)
   (*c*) Jaggery (*Saccharum officinarum* L., Poaceae)
   (*d*) Parched gram (*Cicer arientinum* L., Papilionaceae)
   (*e*) Some quantity of rice grains (*Oryza sativa* L., Poaceae)
   (*f*) Areca-nut (*Areca catechu* L., Arecaceae)
   (*g*) Rhizome pieces of turmeric (*Curcuma longa* L., Zingiberaceae)
   (*h*) Wine of Mahua flowers [*Madhuca longifolia* (Koen.) Macbride var. *latifolia* (Roxb.) Chev., Sapotaceae]
2. Haladi ceremony in marriage, 'Turmeric' paste (Halad) (*Curcuma longa* L.) is applied on the body of bride and bridegroom at their house. 'Halad' is applied on five days

or occasion before marriage. Marriage ceremony is performed on Friday only. Before application of turmeric paste the Apta or Hengala tree (*Bauhinia racemosa* Lam. Caesalpiniaceae) and Bajranga (Maruti or Hanuman) is worshipped by the bridegroom.

3. On Thursday night, they dance whole night and before marriage ceremony bridegroom offers a coconut fruit (*Cocos nucifera* L., Arecaceae) to god Bajaranga alongwith the following:
   (*a*) Flowers of Ruchkin [*Calotropis gigantea* (L.) R.Br., Asclepiadaceae].
   (*b*) Udad dal [Pulse of *Vigna mungo* (L.) Hepper, Papilionaceae.
   (*c*) He also worships 'Samadi' or Radakya tree [*Prosopis cinerarcea* (L.) Druce, Mimosaceae].
4. Wine of Mahua flowers [*Madhuca longifolia* (Koen.) Macbride var.*latifolia* (Roxb.) Chev.] is offered to all guests and friends.
5. The pandal is decorated by the wreaths of 'Amba' leaves (*Mangifera indica* L., Anacardiaceae).
6. Dining plates made up of leaves of 'Wad' (*Ficus benghalensis* L., Moraceae) are used for dinner during marriages.

FOLK SONGS OF BHIL

1. *Acacia leucophloea* (Roxb.) Willd. (Mimosaceae) Hiwar

Tapi thadini hivar phul ba, hivar phul.

Kay dhol vaj tuna jivanbhar,

Hivar phul ba hivar phul.

Kay pawari vaj tuna jivanbhar,

Hivar phul ba hivar phul.

Kay pawa vaj tuna jivanbhar,

Hivar phul ba hivar phul.

Kay tal vaj tuna jivanbhar,

Hivar phul ba hivar phul.

Kay nach nacha tuna jivanbhar,

Hivar phul ba hivar phul.

Kay gana mhane tuna jivanbhar,

Hivar phul ba hivar phul.

['तापी थडीनी हिवर फुल बा, हिवर फुल,

काय ढोल बाज तुना जीवनभर

हिवर फुल बा हिवर फुल,

काय पावरी बाज तुना जीवनभर,

हिवर फुल बा हिवर फुल,

काय पावा वाज तुना जीवनभर,

हिवर फुल बा हिवर फुल,

काय टाळ वाज तुना जीवनभर,

हिवर फुल बा हिवर फुल,

काय नाच नाचा तुना जीवनभर,

हिवर फुल बा हिवर फुल,

काय गाना म्हणे तुना जीवनभर,

हिवर फुल बा हिवर फुल'.]

A Bhil wife asks questions time and again to her husband about what she has done throughout her entire life. She herself enlists her type of involvement in the life of Bhil husband. The word 'Hivar' [*Acacia leucophloea* (Roxb.) Willd.] recurs in the phrase of song. She says that I am beating drum, playing 'Pawari' (a kind of fife) and 'Pawa' (a flute), striking cymbals, dancing and singing throughout life.

2. *Madhuca longifolia* (Koen.) Macbride. var. *latifolia* (Roxb.) Chev. (Sapotaceae) Mahuwa

Dongar jangalma mana bhau

Mahua gaye

Mana mahuana phule, mana mahuana phule.

Koen melani chugali ki
Chopdana police una re
Mana mahuana phule, mana mahuana phule.
Chopdana police una
Mana bhau dharaya gaya
Mana mahuana phule, mana mahuana phule.
Aajubaju char police
Mana bhau madhi chalnare
Mana mahuana phule, mana mahuana phule.
Bhaula mi sod lagun
Dandkada thei dinnyat
Mana mahuana phule, mana mahuana phule.
Dandkada thei dinyat
Vyajana paisa linhat,
Mana mahuana phule, mana mahuana phule.

['डोंगर जंगलमा मना भाऊ
महुआ गाये
मना महुआना फुले, मना महुआना फुले.
कोन मेलानी चुगली की
चोपडाना पोलीस उना रे
मना महुआना फुले, मना महुआना फुले.
चोपडाना पोलीस उना
मना भाऊ धराया गया
मना महुआना फुले, मना महुआना फुले.
आजुबाजू चार पोलीस
मना भाऊ मधी चालनारे
मना महुआना फुले, मना महुआना फुले.
भाऊला मी सोड लागुन

दंडकडा ढेई दिन्यात

मना महुआना फुले, मना महुआना फुले.

दंडकडा ठेई दिन्यात

व्याजना पैसा लिन्हात

मना महुआना फुले, मना महुआना फुले'.]

This song of Bhils tells about the devotion and help extended by a sister to her brother. Her brother is arrested by the policemen as he is accused of preparing Mahua wine. To relieve him from custody of police, she mortagaged her bracelet. She also lent at interest. The word 'Mahua' [Madhuca longifolia (Koen.) Macbride. var.*latifolia* (Roxb.) Chev.] recurs in the phrase of song. This song also explains the life pattern of Bhil tribe.

(C) Vanjara Tribe

FOLK SONGS

1. ***Triticum aestivum*** **L.(Poaceae) Gahu**

Sawanme tu paida hui
Jeth liyo avatar
Gokul nagarim alakh jagai
Rani liyo avatar
Chhori ki devi bani
na gayi sasural
Kana kudal pather bhaye
Tij bhai gulal
Tij naveli na tute
Jen todo pachh Bhagwan
Tij todo, tij todo....
['सावनमे तू पैदा हुई
जेठ लियो अवतार
गोकूळ नगरीम आलख जगाई

रानी लियो अवतार

छोरी की देवी बनी

न गयी ससुराल

काना कुदळ पथर भये

तीज भयी गुलाल

तीज नवेली न टूट

जेन तोड पछ भगवान

तीज तोडो, तीज तोडे ....'.]

Vanjara people celebrate Tij festival in the month of 'Sharavana' each year. It is the young unmarried girls that participate actively in its celebrations. They grow seeds of wheat (*Triticum aestivum* L.). The young seedlings are suggestive of youthness. This song is sung by the unmarried young girls. After the festival, the seedlings are donated to the river.

2. *Citrus aurantifolia* (Christm.) SW. (Rutaceae) Nimboo, Limboo

***Cocos nucifera* L. (Arecaceae) Naral**

Yadi ye han hiyya

Ton kukar Chhodu ye yadi han hiyya

Chanda suriyari jodi

Ju ma betiri jodi han hiyya

Kidi mugi sapati

Pan tari beti koni sapati han hiyya

Limb naral vakage

Jyu tamari beti vakagi yadi ye han hiyya.

['याडी ये हं हिय्या

तोन कुकर छोडू ये याडी हं हिय्या

चांदा सुरीयारी जोडी

जु मा बेटीरी जोडी हं हिय्या

किडी मुगी सपाती

पन तारी बेडी कोनी सपाती हं हिय्या

लिंब नारळ वकगे

ज्यु तमारी बेटी वकगी, याडी ये हं हिरया'.]

This song is sung by a daughter for her mother. The daughter expreses a view that we are similar to the pair of sun and moon. Your daughter can not hide like the worms and insects. The lemons [*Citrus aurantifolia* (Christm.) SW.] and coconuts (*Cocos nucifera* L.) are sold. Similarly, your daughter is also married (sold.)

Section II
Non-Tribals

FESTIVALS IN THE DISTRICT

(1) Hindu People

1. *Azadirachta indica* A.Juss. (Meliaceae) Neem

The new year day is the festival called 'Gudhi Padva'. It is the first day of the first half of 'Chaitrya' month (according to Salivahan era). A long stick usually of Bamboo is erected in front or top of the house. A copper pot, a piece of new cloth and a flowering twig of neem plant (*Azadirachta indica* L.) are fastened to it. Few leaves of this plant are also consumed at morning with a belief of healthy action on the intestine.

2. *Zingiber officinale* Linn. (Zingiberaceae) Suntha *Coriandrum sativum* Linn. (Apiaceae) Dhane

'Ramanavami', the birth day anniversary of god Rama is celebrated on the ninth day of the first half of 'Chaitrya' month. A local article called 'Sunthavada' made up of powder of dry ginger (*Zingiber officinale* Linn. and seeds of coriander (*Coriandrum sativum* L.) and sugar is distributed as 'prasad'.

3. *Calotropis gigantea* (L.) R.Br. (Asclepiadiaceae) Ruchkin

A birth day of Lord Hanuman or Maruti (Hanuman Jayanti) is celebrated on fifteenth full moon day of 'Chaitrya' month. His birth is announced at day-break on this day. This idol is respectfully offered a garland of flowers of *Calotropis gigantea* L. especially of white flower variety. (Some people also worship especially on Saturday using garland of these flowers).

4. *Curcuma longa* L. (Zingiberaceae) Halad

In the month of 'Chaitrya' and 'Vaishakha' ladies celebrate 'Gauri Puja'. The idol of Parvati, wife of Lord Shankar, is decorated and illuminated in homes. Women, friends and relatives are invited for 'haladi-kunku' programme. Turmeric powder (*Curcuma longa* L.) and red powder is applied on the forehead of a married woman. Wet gram (*Cicer arientinum* L.) or groundnut (*Arachis hypogea* L.) is distributed among them. 'Panha', a beverage prepared from raw mangoes (*Mangifera indica* L.) mixed with jaggery or sugar (*Saccharum officinarum* L.) is also served.

5. *Mangifera indica* L. (Anacardiaceae) Amba (Mango)

Akhaji or Akshaya Trutiya is one of the auspicious festival day on the third of 'Vaishakh' month. The idol of Gavarai, a local deity, represents the goddess Parvati, is decked. A bunch of raw mangoes (*Mangifera indica* L.) is hung in front of it as a part of decoration. The married sisters who returned to their paternal homes sing song sitting on the swings and praise the goddess Gavarai.

6. *Ficus benghalensis* L. (Moraceae) Wad, Wat

This tree is worshipped by the Hindu married women on the full moon day of 'Jestha' month. This festival is called 'Vat Savitri Pournima' and observed as such in the memory of Savitri, a great mythological lady. It is believed that she regained her lost husband from the 'Yamas', God of Death, by her penance and wit.

7. *Santalum album* L. (Santalaceae) Chandan (Sandal)

Sandal paste, is used to draw images of cobra. It is worshipped on the bright fifth of Sravana month. This festival is known as 'Nag Panchami'.

8. *Cocos nucifera* L. (Arecaceae) Naral, Naray

People go to the river and offer coconut fruits with ceremonial worship to the god of the water, Varuna. This festival is celebrated on the full-moon day in the month of 'Shravana' and is called 'Narali Pournima'.

9. *Cynodon dactylon* (L.) Pers. (Poaceae) Durva

Ganpati festival is celebrated for ten days. Ganesh Chaturthi falls on the fourth day of month of 'Bhadrapada'. The Lord Ganesh, 'God of Learning', is worshipped offering especially a green called 'durva' [Cynodon dactylon (L.) Pers.] Rice flour balls stuffed with coconut kernel and jaggery or sugar, locally called 'Modaks', are distributed being favourite to Lord Ganesh.

10. *Oryza sativa* L. (Poaceae) Tandul (Rice)

A day on which one's father expired is selected and celebrated as 'Sraddha'. 'Pindas', rice flour balls, are offered to all dead elderly ancestors of one's family.

11. *Cocos nucifera* L. (Areaceae) Naral, Naray (Coconut)
***Oryza sativa* L.(Poaceae) Tandul (Rice)**

Navratra festival is celebrated for nine days honouring goddess Amba. The first day of 'Ashwin' month is the beginning and end on tenth day. The first day is known as 'Ghata-sthapana'. An earthen pot is filled with water. A coconut fruit (*Cocos nucifera* L.) is placed on the top of it. It is placed on a heap of rice grains (*Oryza sativa* L.). It is worshipped customarily.

12. *Bauhinia racemosa* Lamk. (Ceasalpiniaceae) Apta, Sone. (*i.e.*, Gold)

The tenth day after Navratra is an auspicious day of a year. It is called 'Dasara'. People collect leaves of sone (Apta) (*Bauhinia racemosa* Lamk.). Friends and relatives exchange leaves as a mark of respect for each other or younger receive blessings of the elders. The leaves on this eve are regarded as golden leaves and hence called 'Sone' instead of calling 'Apta', a local name for this species.

13. *Sorghum bicolor* (L.) Moench. (Poaceae) Jowar
***Pennisetum americanum* (L.) K.Schum. (Poaceae) Bajra**
***Gossypium herbaceum* L. (Malvaceae) Kapus**

Diwali or Dipawali is celebrated by illuminating every place. 'Dhan trayodashi' is one such day of this thirteenth day of 'Ashvin' month. The farmers on this day worship the agricultural produce like ears of Jowar [Sorghum bicolor (L.) Moench.] Bajra [*Pennisetum americanum* (L.) K.Schum. Cotton (*Gossypium herbaceum* L.) and few others. These are regarded as wealth by the farmers.

14. *Coriandrum sativum* L.(Apiaceae) Dhane (Coriander)
***Saccharum officinarum* L.(Poaceae) Oos (Sugarcane)**

Lakshmipoojan is also one such day of Diwali festival celebrated on the last day of 'Ashwin' month (*i.e.* Amavasya). Goddess Lakshmi is worshipped and a 'prasad' prepared from Coriander seeds (*Coriandrum sativum* L.) and Jaggery (*Saccharum officinarum* L.) are distributed among the friends, guests and neighbours.

15. *Cocos nucifera* L.(Arecaceae) Naral, Naray (Coconut)
***Musa paradisiaca* L. (Musaceae) Kela (Banana)**

In the festival of Bhaoobij the sisters wave a lamp before their brothers and brothers give some present to them. On this occasion, sister offers a cluster of banana (*Musa paradisiaca* L.) and coconut fruit (*Cocos nucifera* L.) along with silky thread, locally called Kardhoda.

16. *Ocimum tenuiflorum* L. (Lamiaceae) Tulshi
***Tamarindus indica* L.(Caesalpiniaceae) Chincha**
***Emblica officinalis* Gaertn. (Euphorbiaceae) Awala**
***Saccharum officinarum* L. (Poaceae) Oos**
***Ziziphus mauritiana* Lamk. (Rhamnaceae) Bor**

Tulashi plant (*Ocimum tenuiflorum*) is married with Lord Krishna ceremonially. This festival is called 'Tulashi Vivah' and celebrated on the twelth day in the bright half of 'Kartic' month. Only after this festival the Hindu marriage settlements are initiated. The people are allowed to consume fresh tamarind and jujub and emblic fruits. Sugarcane plants are used to decorate the pandal.

17. *Pennisetum americanum* L. (Poaceae) Bajara
***Cyamopsis tetragonoloba* (L.) Taub. (Papilionaceae) Gawar**
***Vigna radiata* (L.) Wileczek var. *sublobata* (Roxb.) Verdc. (Papilionaceae) Moog**
***Vigna vexillata* (L.) Rich. (Papilionaceae) Chawali**

A festival called 'Vasu Baras', 'Govasta Dwadashi' or 'Gay Gorhyachi Baras' is celebrated on the bright second day of 'Ashwin' month. Farmers observe fast on this day. They collect the ears of bajra (*Pennisetum americanum* L.) Pods of gavar (*Cyamopsis tetragonoloba* (L.) Taub.), *Vigna radiata* (L.) Wileze K. var. *sublobata* (Roxb.) Verdc. [Chawali [*Vigna vexillata* (L.) Rich.]. All are collected by hands. No instruments or utensils made up of iron are used while preparing bread and vegetable. The cow and calf is first worshipped and served with the bread and vegetable, then only the farmers eat food.

18. *Celosia argentea* L. (Amaranthaceae) Kombada

People of Rajput origin celebrate the festival of 'Gonvardhan Poojan': An idol lying flat in the foreground of a house is made using cow-dung. The weeds *viz. Celosia argentea* L. are placed erected on it. It is worshipped by these people. They circumambulate (*i.e.* pradakshina) with a pair of bullocks around this idol.

19. *Oryza sativa* L. (Poaceae) Sal, Tandul (Rice)
***Musa paradisiaca* L. (Musaceae) Kela (Banana)**
***Cocos nucifera* L. (Arecaceae) Naral, Naray (Coconut)**

The ladies belonging to the castes like Dhobi, Sutar, Nhavi, Mahar, Bhil, Lohar, Kumbhar of a village celebrate this festival of Diwali collectively. The first day of 'Kartic' month is called Pratipada. All ladies and some of their gents visit each family of a village early after midnight on this day. They wave lamp around the gents of a family, apply rice grains (*Oryza sativa* L.) and turmeric powder (*Curcuma longa* L.) and red powder on forehead. They also offer a cluster of Banana (*Musa paradisiaca* L.) fruits and coconut (*Cocos nucifera* L.) to them. The elder chief of a family give them presents especially in the form of money.

20. *Saccharum officinarum* L. (Poaceae) Oos (Sugarcane)

On the 11th day of bright half of 'Kartik' month, a festival known as 'Khopadi' or 'Khopadi Ekadashi' is celebrated. A hut using culms of Sugarcane (*Saccharum officinarum* L.) is erected. Idols of family deities of each family are brought and kept inside this hut. A Brahmin is invited to worship them. An unmarried body sits in the hut. He is asked questions enquiring the rates or cost of food-grains, cotton, banana etc. Afterwards the pieces of culms of sugarcane are distributed among the members of family.

21. *Sorghum bicolor* (L.) Moench. (Poaceae) Jawar

On the eve of Dasara festival which is celebrated on the tenth day of Ashwin month, farmers offer entire plants along with ears of Jawar [*Sorghum bicolour* (L.) Moench.] to all gods and goddess of a village.

22. *Aegle marmelos* (L.) Corr. (Rutaceae) Bel

People worship of Lord Shiva and also are on fast on the thirteenth day of month of 'Magha'. This festival is known as Shivaratri. Lord Shiva's idol is worshipped especially offering the leaves of Bel plant (*Aegle marmelos* L.). The tree leaflets resemble the trishul or trident of Lord Shiva. Trident is thought as a symbol of the powers attributed to Lord Shiva.

23. *Sesamum orientale* L. (Pedaliaceae) Til

Makar Sankrant is celebrated on fourteenth January every year. This is a day, the sun enters 'Makar' (the zodiac sigh of Capri-corns). People exchange Til-gul or Halva on this occasion and say 'Til-gul ghya aani god god bola'. It is prepared from seeds of til (*Sesamum orientale* L.) and Jaggery or sugar (*Saccharum officinarum* L.). Ornament prepared from Seasamum seeds and Sugar are worn by ladies.

24. *Oryza sativa* L. (Poaceae) Tandul (Rice)
***Vigna radiata* (L.) Wilczek var.*sublobata* (Roxb.) Verdc. (Papilionaceae) Mug**
***Solanum melongena* L. (Solanaceae) Vange (Brinjal)**

The day previous to 'Makar-sankrant' is observed as 'Bhogi'. People eat 'Khichadi' as a custom which is prepared from the grains of rice (*Oryza sativa* L.) and Mug (*Vigna radiata* L.) Wilczek var.*sublobata*). Vegetable prepared from brinjal (*Solanum melongena* L.) is also eaten.

25. *Ricinus communis* L. (Euphorbiaceae) Arand (Castor)
***Cocos nucifera* L. (Areaceae) Naral, Naray (Coconut)**

'Holi' is the festival in the month of Phalgun especially on full moon day. The Holika goddess is worshipped ceremonially. People collect fire-wood and cow-dung cakes in a heap around a central pole. This central pole is nothing but the entire plant of Castor (*Ricinus communis* L.). It is burnt at night and people shout and dance around it. Coconut (*Cocos nucifera* L.) are offered to it.

26. *Pennisetum americanum* L. (Poaceae) Bajra

'Pola' is the festival when bullocks are worshipped by the farmers. The bullocks are not employed for any agricultural work. They are provided with good feed, washed in rivers, different colours are used to colour for their horns and body. At evening, they visit different temples of gods and goddesses. After returning home of their owner, they are worshipped and fed with sweet 'puran-poli'. On the same day bullocks made up of clay are also coloured. Bajara grains (*Pennisetum americanum* L.) are spread on a carpet or cot and coloured clay bullocks are placed on it in rows. They are also worshipped.

27. *Annona reticulata* L. (Annonaceae) Ramphal, Ramphay

The fruit is an auspicious offering in the worship of 'Laxmi-poojan" at the time Diwali festival.

28. *Citrullus lanatus* (Thunb.) Matsmura and Naki (Cucurbitaceae) Kalingad, Tarbuz

In the memory of forefathers a dinner is arranged on the festival of 'Akhaji' or 'Akshay Trutiya'. A new small earthern pot is filled with water for them. Fruit of this plant is placed atop earthen pot which is then worshipped. It is called 'Ghagar Bharane'.

29. *Daucus carota* L. (Apiaceae) Gajar (Carrot)
***Ziziphus mauritiana* Lamk.(Rhamnaceae) Bor**
***Triticum aestivum* L.(Poaceae) Gahu (Wheat)**
***Emblica officinalis* Gaertn. (Euphorbiaceae) Awala**

During the celebration of 'Makar-Sankrant' festival, the ears of wheat (*Triticum aestivum* L.), Carrot (*Daucas carota* L.) fruits of Bor (*Ziziphus mauritiana* Lamk.) and Awala (Emblica officinalis Gaertn.) are kept in a earthen pot. They are distributed to the married women in the name of 'Wan'.

30. *Benincasa hispida* (Thunb.) Cogn. (Cucurbitaceae) Kohla

On the ninth day of navratra a day before 'Dasara' festival in the month of 'Ashvin' fire is worshipped. It is called Nav-chandi Yadnya. The fruits of *Benincasa hispida* (Thunb.) Cogn. is cut with knife as a sacrifice.

31. *Benincasa hispida* (Thunb.) Cogn. (Cucurbitaceae) Kohla
***Phaseolus mungo* L. (Papilionaceae) Udid**
***Sorghum bicolor* (L.) Moench. (Poaceae) Jawar**

On the eve of 'Dasara' festival the fruit of *Benincasa hispida* (Thunb). Cogn. is cut into pieces by the people of a village. The fruit is regarded as 'Hela' (male buffaloe). Afterwards, they are mixed with the grains of udid (*Phaseolus mungo*) and Jawar [*Sorghum bicolor* (L.)] Moench. This mixture is called 'bakola'. First it is thrown around the village and then distributed to each family of a village to throw it in houses. It is believed that it helps to ward off evils from the village and houses.

32. *Oryza sativa* L. (Poacee) Tandul

The rural part of Jalgaon district 'Ghata' festival is celebrated by the farmers and farm labours to satisfy village deities like Marimata, Asara and Mhasoba. It is celebrated after good rain or sowing of all crops in the month of 'Ashadha' on the first Tuesday or Friday. All people of a village together decide for its celebration. Curd, milk and cooked rice (*Oryza sativa* L.) are offered to these deities.

33. *Cyperus rotundus* Linn. subsp. rotundus (Cyperaceae) Lawali
Calotropis gigantea (L.)R.Br. (Asclepiadaceae) Local Name: Ruchkin, Rui

Bhaldev festival is celebrated by the farmers for five to nine days. They collect cow-dung and placed it in foreground or varandah. It is sprinkled with ash. The culms of *Cyperus rotundus* Linn. and twigs and flowers of *Calotropis gigantea* (L.) R.Br. are placed on the heap of cow-dung. An idol is prepared from this dung. It is kept near the place in one's home. Whey and butter are prepared. The culms of *Cyperus rotundus* Linn. are also placed on this idol. Food articles prepared for feasts are placed near it. Lamps prepared from flour are lit in front of it. The curd, milk or butter are also not sold on the days of festival. Feast is served to all individuals after worship.

34. *Sesamum orientale* L. (Pedaliaceae) Til
Pennisetum americanum L.(Poaceae) Bajara

A day previous to the festival of 'Makar-Sankrant' is Bhogi. Ladies eat breads of bajara (*Pennisetum americanum* L.) to which seeds of *Sesamum orientale* L. *i.e.* Til and butter are applied.

35. *Triticum aestivum* L. (Poaceae) Gahu (Wheat)

Seven days after 'Pola' festival, Bhadimay or Bhadibhaddev festival is celebrated. However this festival is rarely celebrated now-a-days. The people of a village who make public announcement, take care of visiting officers and dispose dead bodies of animals. They kneed wheat flour (*Triticum aestivum* L.) and prepare large lumps of it. They are prepared from it the combs, kum-kum container, idols of cow and calf etc. They cooke all these and consume afterwards.

36. *Tagetus patula* L. (Asteraceae) Zendu

In the Navratra, a 'Kalash' is established as the idol 'Devi'. This kalash, an earthen pot filled with water and coconut fruit is placed on it. This earthen pot is kept on soil with Gahu (*Triticum aestivum* L.). Nine garlands of Zendu flowers (*Tagetus patula* L.) are offered for nine days on the Kalash and is worshipped.

37. *Phaseolus vulgaris* L. (Papilionaceae) Shravan Ghewda

During 'Datta Pournima' fast is observed and vegetable of *Phaseolus vulgaris* L. is essential as 'Naivaidya'.

38. *Manihot esculenta* Crantz. (Euphorbiaceae) Sabudana

From the tubers of *Manihot esculenta* Crantz. sago *i.e.* 'Sabudana' is prepared. These granules are used to make ornaments of ladies. Halwa *i.e.* sugar is coatead on sago granules. The ornament of Halva are worn by ladies especial who are pregnant during festival of 'Makar Sankrant' (Plate 22).

FASTS AND WORSHIPS IN THE DISTRICT

1. *Limonia acidissima* L. (Rutaceae) Kavath
***Jasminum auriculatum* Vahl (Oleaceae) Jai**
***Chrysanthemum indicum* (L.) Des. (Asteraceae) Shewathi**
***Crossandra infundibuliformis* (L.) Nees (Acanthaceae) Aboli**

Leaves or fruits of *Limonia acidissima* L. various seasonal flowers mentioned above are used during the worship of 'Hartalika' and 'Anant pooja'. These are the fast days especially of ladies in the month of 'Shravana'.

2. *Mangifera indica* L. (Anacardiaceae) Amba

Leaves find place in 'Anant pooja' in the month of 'Sharavana' and 'Navratra utsav' in the month of October. The wreaths of leaves are employed as decoration at the time of 'Satya-narayana pooja'.

This worship is performed after some auspicious occasion like marriage.

3. ***Musa paradisiaca* L. (Musaceae) Kela**

Suckers are placed erect around the place of worship like 'Satyanarayana'. There are also used to decorate Chariot during the procession of deities or pilgrims.

4. ***Prosopis cineraria* (L.) Druce (Mimosaceae) Saundad**

The fruits are used to prepare garland which is employed to decorate the idol of Gavarai or Gauri (Wife of Lord Shankar) during the festival days of 'Akshya Trutiya' or 'Akhaji' in the month of 'Vaishakha'

5. ***Ipomoea quamoclit* L. (Convolvulaceae) Ganesh Phool, Ganesh Pushp**

The flowers are used especially during worship of Lord Ganesha.

6. ***Sesbania grandiflora* (L.) Poir. (Papilionaceae) Hadga**

Flowers are used in the worship of 'Hartalika' and 'Anant pooja'.

7. ***Nerium indicum* Mill. (Apocynaceae) Kanher**
***Triticum aestivum* L. (Poaceae) Gahu (Wheat)**

The flowers are used to worship a local deity 'Kanbai' in the month of 'Shravana'. Thick breads, locally called 'rot' are prepared from wheet flour. (*Triticum aestivum* L.). These and other sweets are offered to the goddess 'Kanbai'.

8. ***Aegle marmelos* (L.) Corr. (Rutaceae) Bel**

Unmarried girls performs 'Shiv-pooja' by offering 108 leaves of this tree to get married.

9. ***Barleria prionitis* L. (Acanthaceae) Katekoranti**
***Impatiens balsamina* L. (Balsaminaceae) Terda**
***Mimusops elengi* L. (Sapotaceae) Bakul**

The flowers are used in the worship of Lord Shankar.

1. Pawara ladies with traditional dress and ornaments.

2. Another Pawara lady with traditional dress and ornaments. Note the tiled roof of house and a cattle shed. Also note pillars of cattle shed and a 'charpai' made up of Sag (*Tectona grandis* L.*f.*).

3. A group of Pawara dancers at the time of 'Holi Melda' festival at Vaijapur.

21. Goddess Kanbai worshiped by using leaves of (a) *Nerium indicum* Mill. (b) *Musa paradisiaca* L., (c) Piper betle L., (d) Fruit of *Cocos nucifera* L. at Janave.

22. Seeds of Til (*Sesamum orientale* L.) and grains of Sago (*Manihot esculenta* Crantz.) used while preparing ornaments of ladies at the time of 'Sankrant' festival.

8. *Triticum aestivum* L. (Poaceae) Gahu (Wheat)
***Bauhinia racemosa* Lamk.(Caesalpiniaceae) Shami**
***Oryza sativa* L. (Poaceae) Tandul (Rice)**

Ekod is brother of the goddess Kanbai. Festival in his name is celebrated especially before the period of the eldest son's marriage. Four day ceremony is celebrated after full moon in the month of 'Phalgun'. It is believed that this ceremony reminds the eldest son about his wordly affairs after marriage. Few articles necessary for this worship are brought from carpenter (Sutar) and potter (Kumbhar) of the village. Wheat (*Tritium aestivum* L.) is ground by all relatives using stone-mill. Five couples visit Shami tree (*Bauhinia racemosa* Lamk.) They dig out and lit a limp there. This lamp is brought at home and placed in the place of worship. It is maintained as such for five weeks. A copper container full of rice grains (*Oryza sativa* L.) is kept on earthen pot. On the fifth day, the 'Kalash' is offered to river. This ceremony is, however, rarely celebrate in modern period.

9. *Triticum aestivum* L. (Poaceae) Gahu (Wheat)
***Piper betel* L. (Piperacea) Nagin Pan (Betle leaf)**
***Oryza sativa* L. (Poaceae) Tandul (Rice)**
***Cocos nucifera* L. (Arecaceae) Naral (Coconut)**

A 'gondhal' a recreational programme, is performed before the marriage of the eldest son of a family. Gondhalis are hereditary devotees of the goddess 'Ambabai' in whose honour they perform 'gondhal'. A copper cup is placed on wheat (*Triticum aestivum* L.). Betle leaves (*Piper betle* L.) are placed in the cup. Half coconut (*Cocos nucifera* L.) filled with rice grains (*Oryza sativa* L.) is also placed on the cup. The gondhalis sing songs praying the goddess and play the double drum (Sambal) throughout the night.

DEATHS AND BIRTHS IN THE DISTRICT

Plants Related with Death

1. *Ocimum tenuiflorum* L.(Lamiaceae) Tulashi

Few Tulashi leaves (*Ocimum tenuiflorum* L.) are placed in the mouth of dead person.

2. *Piper betle* L. (Piperaceae) Nagin Pan

Leaves (*Piper betel* L.) are stringed and kept on the cloth covering the dead body.

3. *Santalum album* L. (Santalaceae) Chandan (Sandal)

Leaves of sandal (*Santalum album* L.) are kept below the head or neck of a dead body before cremation.

4. *Azadirachta indica* L. (Meliaceae) Neem

The women mourners go out of the village. After the return of men mourners, the women are given leaves of neem plant (*Azadirachta indica* L.) to eat. Then they return home.

5. *Oryza sativa* L. (Poaceae) Tandul (Rice)

On the tenth day, a stone is propitiated with a 'Sraddha' ceremony at the burning ghat. Rice balls [*Oryza sativa* (L.)] are offered to the dead. Mourners have to wait for crows to until they consume it.

6. *Ocimum tenuiflorum* L. (Lamiaceae) Tulashi

Tulashi plants are grown in the foreground on a special structure called 'Tulashi Vrindavan'. They are not watered after death for thirteen days.

Plants Related to Birth

1. *Mimosa hamata* Willd. (Mimosaceae) Arati
***Ziziphus mauritiana* Lamk. (Rhamnaceae) Borati**
***Pennisetum americanum* (L.) K.Schum. (Poaceae) Bajara**
***Hibiscus cannabinus* L.(Malvaceae) Ambadi**
***Santalum album* L. (Santalaceae) Chandan (Sandal)**

On the fifth day after childbirth, a deity is worshipped using leafy twig of Arati (*Mimosa hamata* Willd.) and Borati (*Ziziphus mauritiana* Lamk.) The mother is given bread of Bajara (*Pennisetum americanum* L.) and leafy vegetable of Ambadi (*Hibiscus cannabinus* L.). It is believed that on the fifth day deity presides over the fate of a new born baby and she is propitiated. An arecanut cutter and sword or sickle are kept near the deity. Sandle paste (*Santalum album* L.) and flowers are offered to these.

2. *Sorghum controversum* (Steud.) Snowden (Poaceae) Boru

On the fifth day after childbirth, the sixth day deity is thought to inscribe future of a baby on his forehead. Blank paper, a reed pen [*Sorghum controversum* (Stud.) Snowdan] and ink are kept on fifth day after delivery. Mother prays for blessings.

CUSTOMS IN THE DISTRICT

1. *Curcuma longa* L. (Zingiberaceae) Halad (Turmeric)
***Oryza sativa* L. (Poaceae) Tandul (Rice)**
***Cocos nucifera* L. (Arecaceae) Naral (Coconut)**
***Areca catechu* L. (Arecaceae) Supari (Arecanut)**

Good wishes are expressed for married women in the form of 'oti' custom. A married women sits facing east and head covered with a portion of saree. Another married woman applies on her forehead turmeric powder (*Curcuma longa* L.) and red powder. Then she gives rice grains (*Oryza sativa* L.), Coconut (*Cocos nucifera* L.), Areca-nut (*Areca catechu* L.), dry pieces of turmeric, fruits and flowers. The married woman accepts these in her portion of saree held near abdomen. These things are usually given five times on the same occasion. Instead of rice grains, wheat (*Triticum aestivum* L.) is also given. This custom is thought indication of conception and motherhood.

2. *Phoenix sylvestris* (L.) Roxb. (Arecaceae) Tad
***Brassica juncea* (L.) Czern. et Coss. (Brassicaceae) Mohri**

If all of sudden, small boys or girls feel feverish, then their mothers pass over their body a broom made from *Phoenix sylvestris* L., grains of *Brassica juncea* L., common salt and even leather foot wears. This process is repeated seven times from head to toes. This custom is known as 'Najar Utarane'. It is believed that it wards off evil.

3. *Citrus arunatifolia* (Cristm.) SW. (Rutaceae) Nimboo (Lemon)
***Capsicum annuum* L. (Solanaceae) Lal-mirchi (Chillies)**
***Semecarpus anacardium* L.*f.* (Anacardiaceae) Bibba**

A black doll alongwith a lemon fruit [*Citrus* arunatifolia (Christm.) Saw.], a Dhobis nut (*Semecarpus anacardicum* L.*f.*) and seven chillies (*Capsicum annum* L.) are stiched together. This doll is hanged at the entrance of a newly constructed house to ward off evil.

4. *Ficus benghalensis* L.(Moracee) Wad
***Ficus religiosa* L. (Moraceae) Pimpal**
***Azadirachta indica* L. (Meliaceae) Neem**

These plants are avoided as fire or fuel wood in this region.

5. ***Citrus arunantifolia*** **(Christm.) SW. (Rutaceae) Nimboo (Lemon)**
Curcuma longa **L. (Zingiberaceae) Halad (Turmeric)**
Capsicum annuum **L.(Solanaceae) Lal-mirchi (Chillies)**

Some people keep a lemon [*Citrus aurantifolia* (Christm.) SW.] turmeric powder (*Curcuma longa* L.), chillies (*Capsicum annum* L., red powder and black silky threads on the path or road. These are usually kept secretly at night. It is believed that if somebody crosses these on the next day morning, bad evils are passed to him. This custom is called 'Olanda takane'.

6. ***Ocimum tenuiflorum*** **L. (Lamiaceae) Tulashi**

A Hindu family usually plant it in foreground or on elevated structure called 'Tulashi Vrindavan'. After bath in the morning, it is worshipped by the Hindus and circumambulated.

7. ***Curcuma longa*** **L. (Zingiberaceae) Halad (Turmeric)**
Oryza sativa **L. (Poaceae) Tandul (Rice)**
Cocos nucifera **L. (Arecaceae) Naral (Coconut)**
Piper betle **L. (Piperacee) Nagin Pan (Betle Leaf)**

After construction of new home, a custom of 'Kalash pooja' is generally followed. A copper water container is decorated with leaves of betle (*Piper betle* L.). Then a coconut (*Cocos nucifera* L.) is placed on it. It is worshipped by applying turmeric powder (*Curcuma longa* L.) and rice grains (*Oryza sativa* L.). This coconut is then used as an oblation into fire.

8. ***Sorghum bicolor*** **(L.) Moench. (Poaceae) Jowar**
Oryza sativa **L. (Poaceae) Tandul (Rice)**
Phaseolus mungo **L. (Papilionaceae) Udid**
Capsicum annuum **L. (Solanaceae) Mirchi (Chillies)**
Arachis hypogaea **L. (Papilionaceae) Bhuimug (Groundnut)**
Allium cepa **L. (Liliaceae) Kanda (Onion)**

On the new moon day especially at morning, ladies belonging to 'Mang' caste, demand 'amawashya' which consists of jowar grains [*Sorghum bicolor* (L.) Moench.], rice grains (*Oryza sativa* L.), Chillies (*Capsicum annum* L.), groundnut oil (*Arachis hypogea* L.), Onion (*Allium cepa* L.) etc. It is believed that if their demands is satisfied, evils of a family are ward off.

9. *Lawsonia inermis* L. (Lythraceae) Mehandi

Hand palms or even foot soles are stained using leaf-paste of this plant. This custom is followed on some auspicious occasion like marriage, festival like Diwali, Raksha Bandhan etc.

10. *Azadirachta indica* L. (Meliaceae) Neem

During famine period, farmers of a village collectively visit each house requesting water. The leafy twigs of neem in the form of wig is placed on the head of a boy or senior man. He dances visiting every house. All others follow him. Water is poured on his head. Presents in the form of food grains are given to them. All followers sing song requesting 'Varun' god of rains. Following songs are commonly heard: (*i*) Baras re, baras re, Meghraja baras re. (*ii*) Dhundi Dhundi Pani De, Sayama Shiku de ....

GAMES IN THE DISTRICT

1. *Caesalpinia bonduc* (L.) Roxb. (Caesalpiniaceae) Sagar-goti

The big round glazy seeds of Sagargoti [*Caesalpinia bonduc* (L.) Roxb.] of any odd number (*e.g.* 5, 9 etc.) are thrown up into the air. The player catches them on the back of hand palms.

SONGS OF MARRIAGES IN AHIRANI DIALECT

1. *Plumeria acuminata* Ait. (Apocynaceae) Chapha
***Sesamum orientale* L.(Peadliaceae) Til**
***Madhuca longifolia* (Koen.) Macbr. var. *latifolia* (Roxb.) Chev. (Sapotaceae) Mahu**
***Albizia lebbeck* (L.) Willd. (Mimosaceae) Chinchola**

Nandurbarna Nandya teli

Kheti kayi re nayadi

Aage payari hawari tie

Mang kay payari Chafa kai

Yana ravanya pavanya bail

Yani chicholani dhal

Yana ambiyana ghana

Yani tambyani ghagar

Var rupana davana
    Tel ikas mohana
Ya telanya pachi dhara
    Navardev aabavari dhara
Hai kay sutaran pat
    Hai kay kasarani vati
Hai kay barnin pan
    Hai kay saravnyan kand
Yale kachchi sute gund
    Ya pachi suwasini
Hya kay asana sobhane
    Hya kay guravana nandi
Hya kay bamanana yede
    Nandurbarna Nandya teli....

[‘नंदुरबारना नंद्या तेली
    खेती की रे नयाडी
आगे पयरी हावरी तीइ
    मंग काय पयरी चाफा कई
याना रवन्या पवन्या बैल
    यानी चिचोलानी ढाल
याना आंबीयाना घाना
    यानी तांब्यानी घागर
वर रुपाना डवना
    तेल इकस मोहना
या तेलन्या पाची धारा
    नवरदेव आबावरी धरा
हाई काय सुतारन पाट
    हाई काय कासारनी वाटी

हाई काय बारनीन पान

हाई काय सरावन्यान कांड

याले कच्ची सुते गुंड

या पाची सुवासिनी

ह्या काय आसाना सोभने

ह्या काय गुरवाना नंदी

ह्या काय बामणना येदे (वेद)

नंदुरबाना नंद्या तेली ...']

This is another song for singing prior to application of turmeric paste to the body of a bridegroom. A men from teli caste of Nandurbar city cultivates fields in Sakri taluka of Khandesh region. He cultivates sesamum (*Sesamum orientale* L.) and plants like Chapha (*Plumeria acuminata* Ait.) for its flower buds. His healthy bullocks are fed by the leaves of 'Chichola' [*Albizia lebbeck* (L.) Willd.]. He sits under the silvery tree of fragrant flowers and sells oil of 'Moh' [*Madhuca longifolia* (Koen.) Macbr. var.*latifolia* (Roxb.) Chev.]. This oil is advised to shower on the bridegroom. Carpenter's wooden seat, brazier's bowl and other things are brought for this purpose. The five married women are rolling up threads. The bull of Gurav (worshipper of Lord Shiva) is also present. The 'Vedas' of Bramhins are also available for this ceremony.

2. *Gossypium herbaceum* L. (Malvaceae) Kapus

Ganga Jamana doni khete

Tathe kay dev payana rope

Tathe kay Sitabai kate

Isnu Kisnu Kandya bhare

Rai Rukhmani poyat kare

Tathla poyat konle aana

Tathla poyat Aabale aana

Tathla poyat Aabale aana

Aabana bap ghode una

Bashing moti jadi una

Vaja gayarat una

Moti payarat una.

['गंगा जमना दोनी खेते

तठे काय देव पयाना रोपे

तठे काय सीताबाई काते

इस्नू किस्नू कांड्या भरे

राई रुख्मनी पोयतं करे

तठल् पोयतं कोनले आनं

तठल् पोयत आबाले आनं

आबाना बाप घोडे उना

बाशिंग मोती जडी उना

वाजा गयरत उना

मोती पयरत उना'.]

Before application of turmeric paste (*Curcuma longa* L.) to the body of a bridegroom, this song is sung by the married women. It is narrated as follows: The sacred (river) Ganga and Jamuna appeared in cotton field (*Gossypium herbaceum* L.) Sitabai is drawing out into threads. These threads are being rolled up by God Vishnu and Lord Krishna. Goddess Rukhmini prepared fine threads with beads of gold or glass. This thread is brought for Aaba (here it is meant for the bridegroom). His father arrived on a horse. His ornamental head-gear is decorated by the pearls. He came a in the company of a musical party and sowing pearls on his way.

3. *Santalum album* L. (Santalaceae) Chandan (Sandal)

Pach paratanani mati, pach paratanani mati,

Chul chandan kapati, chul chandan kapati

Aan shiji savvakhandi, aan shiji savvakhandi

Ya aanani kirti mothi, aanani kirti mothi

Navardevana bap dunyadari, Navardevana bap dunyadari

Char pangati dinyat dari, char pangati dinyat dari

Aakhand tupanya layat dhari, aakhand
tupanya layat dhari

Chikhol yeshichya baheri, Chikhol yeshichya baheri.

['पाच परतननी माटी, पाच परतननी माटी

चुल चंदन कपाटी, चुल चंदन कपाटी

आन शिजे सव्वाखंडी, आनाशिजे सव्वाखंडी

या आनानी किर्ती मोटी, या आनानी किर्ती मोठी

नवरदेव ना बाप दुन्यादारी, नवरदेवना बाप दुन्यादारी

चार पंगती दिन्यात दारी, चार पंगती दिन्यात दारी

आखंड तुपन्या लायात धारी, आखंड तुपन्या लायात धारी

चिखोल येशीच्या बाहेरी, चिखोल येशीच्या बाहेरी'.]

This song describes how the father of a bridegroom served dinner for the invitee of marriage. Plenty of soil is used to construct hearth. Sandal wood (*Santalum album* L.) is also made available as fire wood. Large quantity of food is cooked as the father of the bridegroom is a popular man in his society. He arranged good dinner for which there was a provision of continuous flow of ghee. Ghee was served in such a large quantity that its flow crossed the boundary of village.

4. *Curcuma longa* L. (Zingiberaceae) Halad, Hayad (Turmeric)

Samadirsuktta panimuktta

Hayad kuku kulawjyat

Khanderavwar telan pade

Mhaysale balawjyat

Samadirsuktta panimuktta

Hayad kuku kalawjyat

Kanbaiwar telan pade

Ranbaile balawjyat

Samadir suktta panimuktta

Hayad kuku kalawjyat

Kuimatale balawjyat.

['समदीरसुक्त पानीमुक्त

हयद कुकु कालावज्यात

खंडेरावबर तेलन पडे

म्हायसाले बलावज्यात ॥१॥

समदीरसुक्त पानीमुक्त

हयद कुकु कालावज्यात

कानबाईवर तेलन पडे

रानबाईले बलावज्यात ॥२॥

समदीरसुक्त पानीमुक्त

हयद कुकु कालावज्यात

कुईमाताले बजावज्यात ॥३॥']

Gods and goddesses are invited for marriage. They should be welcomed by using turmeric powder (*Curcuma longa* L.) and red powder. Gods like Khanderao and goddesses like Mhayasa, Kanbai and Ranbai are suggested to invite. The family deity is also suggested for invitation of marriage ceremony.

5. *Tectona grandis* L.*f.* (Verbanaceae) Sag
Oryza sativa L. (Poaceae) Tandul
Piper betle L. (Piperaceae) Pan

Uth Uth re Kabadi

Khandi ghetalya kurhadi

Doi ghetalya bhakari

Gaya dongary parwadi

Sag todjya niwadi

Sag gadyavar ghatala

Sag nadyan guphila

Sag shelyan zakila

Sag shelyan zakila

Sag chaubaile jumpila

Pandav dhurkari basala
    Yana ravanya pavanya bail
Yan zun zun puran
    Taka sutarna dari
Aare tu sutar re bhau
    Umarat ghadi de lavakari
Managhari shobhanani tatadi
    Umarat sutar ghadiya
Umrat lavhar jadiya
    Umrat tuibandi patiya
Umrat vajat gajat lay
    Umrat majhare roya
Umrat kukuni punjal
    Umrat tanduye aachil
Umrat vaparev kon
    Umrat vapare Resam
Kapay kukuna bijora
    Hathi narayina chuda
Matha ina bhaur juda
    Payi begadina bula
Mukhe naginana ida
    Bai Kheyadat jai
Shela layadat jai
    Surya shipadat jai.

['उठ उठ रे कबाडी
    खांद्या घेतल्या कुऱ्हाडी
डोई घेतल्या भाकरी
    गया डोंगर पारवाडी
साग तोडज्या निवाडी
    साग गाड्यावर घातला

साग नाड्यान गुफिला
    साग शेल्याने झाकिला
साग चौबैले जुंपिला
    पांडव धुरकरी बसला
याना रवन्या पवन्या बैल
    यान झुन झुने पुरान
टाका सुतारना दारी
    अरे तु सताररे भाऊ
उमरट घडी दे लवकरी
    मनघर शोभननी तातडी
उमरट सुतार घडीया
    उमरट लव्हार जडीया
उमरट तुईबंदी पाटीया
    उमरट वाजत गाजत लय
उमरट माजहारे रोया
    उमरट कुकूनी पुंजल
उमरट तांदूये आचिल
    उमरट वापरेव कोन
उमरट वापरे रेसम
    कपाय कुकूना बिजोरा
हाथी नारयीना चुडा
    माथे ईना भऊर जुडा
पायी बेगडीना बृला
    मुखे नागिनना ईडा
बाय खेयाडत जाय
    शेला लोयाडत जाय
सुऱ्या शिपडत जाय'.]

This song also gives an account of preparations for a marriage ceremony. A wooden, ornamental threshold is generally fixed at ground level in the door frame. This song narrates the whole story of fashioning the threshold. A senior lady of the family orders for cutting of teak wood (*Tectona grandis* L.*f*.). The workers cut selected teak wood, carry in the bullock cart and give to the carpenter. He carves out a threshold. It is brought from carpenters house ceremonially. The blacksmith fixed it properly in the place. It is then worshipped by the married women using rice grains (*Oryza sativa* L.) and red powder. This woman has wore green bangles in hand, red powder is applied on her forehead and has a well plaited round on her head.

6. *Oryza sativa* L. (Poaceae) Tandul
***Santalum album* L. (Santalaceae) Chandan**

Kachcha sutana badala
    Laya devana dechare
Taka mando tyana dari
    Sadak tanduy dhoyaja
Sarge nichat deyajya
    Varya narya band karajya
Haushi varu palharajya
    Rangit vahejya jupajya
Bahin bhashile muy jayajya
    Tapate anghoyi karajya
Chandan ghasun layajya
    Gay bahule bandhajya
Dudh dudhane dhoyajya
    Dudh tape gahire gahire
Ya dudhani pivaji say
    Tyani vasana sarge jay
Sarge vadil mhanati
    Mirge gajar kasana

Aapala vashila nandati
    Kanya putra ujavati
Aamana lokhandi daraja
    Aamale yewani nahi raja.

['कच्चा सुतना बादला।
    लया देवना देव्हारे॥
टाका मांडो त्याना दारी।
    सडक तांदुय धोयजा॥
सर्गे निचतं देयज्या।
    वाऱ्या नाऱ्या बंद करज्या॥
हौशी वारु पल्हारज्या।
    रंगीत वहेज्या जुपज्या॥
बहिण भाशीले मुय जायज्या।
    तपते अंघोयी करज्या॥
चंदन घासून लायज्या।
    गाय बहुले बांधज्या॥
दुध दुधाने धोयज्या।
    दूध तापे गहिरे गहिरे॥
या दुधानी पिवयी साय।
    त्यानी वासना सर्गे जाय॥
सर्गे वडील म्हणती।
    मीर्गे गजर कसाना॥
आपला वशीला नांदती।
    कन्या पुत्र उजवती॥
आमना लोखंडी दरजा।
    आमले येवानी नही रजा॥']

This is a song informing preparations for a marriage ceremony. It also notes about feelings of one's forefathers on such occasion. The preparations such as garland of cotton threads for god, to raise pandal and wash the path as white as rice (*Oryza sativa* L.) a decorated horse, invitation for sisters and her daughters, a cow applied with sandal paste (*Santalum ablum* L.) and washed with milk etc. The smell of boiling milk will reach the heaven and the forefathers will enquire about their descendants. They will think that their daughters and sons are happy. The forefathers also say that their doors of heaven are made of iron. They are not allowed even on leave.

7. *Mangifera indica* L. (Anacardiaceae) Amba
***Jasminum sambac* (L.) Ait. (Oleaceae) Mogra**
***Jasminum officinale* L. var.*officinale* (Oleaceae) Chameli**

Ukhalla mhane mi ukhalla dev
Ukhalla punja kon take
Ukhalle punja Reshma take
Tini sonani dalaki
Tini motyani chomay
Hati rupana kharota
Zadi uni gay gotha
Ini ukhalla kaya motha
Kon hanas kudai
Hiraman hanas kudai
Kon ghalas duwai
Ashok ghalas duwai
Khat gadi bail jai
Khat Amba bane jai
Amba bane na mogara
Rane bane ni chameli
Phul tode suvasini
Dei navardevarla kani.

['उखल्ला म्हणे मी उखल्ला देव
उखल्ला पुंजा कोन टाके
उखल्ले पुंजा रेसमा टाके
तिनी सोनानी डालकी
तिनी मोत्यानी चोमय
हाती रुपाना खरोटा
झाडी उनी गाय गोठा
इनी उखल्ला कया मोठा
कोन हानस कुदाई
हिरामण हानस कुदाई
कोन घालस दुवाई
अशोक घालस दुवाई
खत गाडे बैल जाई
खत आंबा बने जाई
आंबा बने ना मोगरा
राने बने नी चमेली
फुल तोडे सुवासिनी
देई नवरदेवना कानी'.]

The heap of rubbish is regarded as god. A lady, Resma cleaned the cowpen and collected the dung with the help of silvery stick-broom and placed it in a golden pan. She put this pan on a ring of pearls above her head. She threw this dung on the heap of rubbish. She thus increased the size of a heap. This rubbish is carried by the bullock cart and spread in the grove of mangoes (*Mangifera indica* L.). Mogra [*Jasminum sambac* (L.)Ait.] and Chameli (*Jasminum officinale* L. var. *officinale*) also grow in this grove. Their flowers are plucked by the married women and offered to the bridegroom.

8. *Plumeria acuminata* Ait. (Apocynaceae) Chapha
***Mangifera indica* L. (Anacardiaceae) Amba**

Navara navari ga kasha nahati
Tej pani ga bane jati
Tej pani ga mor piti
Tathun more gu udiyela
Amba chafyachi hirvi kadi
Navari mangas hirvi sadi.

['नवरा नवरी ग कशा नहाती
तेज पानी ग बने जाती
तेज पानी ग मोर पिती
तथून मोरं गे उडियेला
आंबा चाफ्याची हिरवी काडी
नवरी मांगस हिरवी साडी'.]

Bride and bridegroom bath in the foreground of bridegroom's home. The water after their bath flows towards the forests. The peacock drinks this water. This peacock flys from this place and rests on the green twigs of mango (*Mangifera indica* L.) and (*Plumeria acutifolia* Ait.). The bride then demands green saree to wear.

9. *Santalum album* L. (Santalaceae) Chandan
***Piper betle* L. (Piperaceae) Nagin pan, Vida Pan**

Navara navirile balaito
Makhamalicha ida deto
Shidi lavali chandanachi
Bangayi tangali motiyachi.

['नवरा नवरीले बलाईतो,
मखमलीचा इडा देतो
शिडी लावली चंदनाची
बंगयी टांगल मोतीयाची'.]

This song informs about the welcome of a bride by the bridegroom. Bridegroom invites her bride and offers velvety roll of betal leaf (*Piper betle* L.). He decorates the swing with pearls and to reach it a ladder made up of sandal wood (*Santalum album* L.) is employed.

10. *Santalum album* L. (Santalaceae) Chandan *Cocos nucifera* L.(Arecaceae) Naral

Todila chandan, ghadile daran

Saga motha fandi, sapan jodile

Jayajya parole, layajya narale

Tatya navardevale, motina narale.

['तोडीला चंदन, घडीले दारन

सगा मोठा फंडी, सपन जोडीले

जायज्या पारोळे, लयज्या नारळे

तात्या नवरदेवले, मोतीना नारळे'.]

Ladies sing this song to praise relatives at the time of marriage. The ladies say that the father-in-law cut sandal tree (*Santalum album* L.) and fashioned the door of his home. He is the great desirous man, he fullfilled his dream. These ladies also expect to purchase coconuts (*Cocus nucifera* L.) studded with pearls from Parola city of Khandesh region.

11. *Triticum aestivum* L. (Poaceae) Gahu

Gahuj nahi ya, gahunya kadya

Parnale jati ya, kothena hadya

Parnale jatiya, Mhasadina hydya.

['गहुज नही या, गहुन्या काड्या

परनाले जाती या, कोठेना हाड्या

परनाले जाती या, म्हसदीना हाड्या'.]

During marriage ceremony, one party provokes other. This is reflected in this song. One party calls the another party at marriage in the name of wheat straw (*Triticum aestivum* L.) or 'crows'. They thus make amusement in such ceremony.

SONGS OF AKHAJI FESTIVAL IN AHIRANI DIALECT

1. *Jasminum officinale* L. (Oleaceae) Jai

Vatvar hirkani handi va may
Shankar Rajani khandi va may
Vatvar jai koni layi va may
Sankar Rajani layi va may
Jaile pani koni ghal va may
Sankar Rajani ghal va may
Jaile phul koni aan va may
Sankar Rajani aan va may
Gavaraini may koni vhayan va may
Sankar Rajani vhyan va may
Gavaraina gayama may koni ghali va may
Sankar Rajani ghali va may.

['वाटवर हिरकनी खंदी व माय
संकर राजानी खंदी व माय
वाटवर जाई कोनी लायी व माय
संकर राजानी लायी व माय
जाईले पाने कोनी घालं व माय
संकर राजानी घालं व माय
जाईले फुल कोनी आनं व माय
संकट राजानी आनं व माय
गवराईनी माय कोनी व्हयनं व माय
संकर राजानी व्हयनं व माय
गवराईना गयामा माय कोनी घाली व माय
संकर राजानी घाली व माय'.]

Gavrai is a form of goddess Parvati. This song is sung while worshiping the goddess Gavarai by the married girls. Unmarried girls also participate in the programme of singing. The song describes how the goddess Gavarai is respected by Lord Shankar. A diamond bit is dug out by Lord Shankar for her. He plants and waters the 'Jai' plants (*Jasminum officinale* L.). The garland of the fragrant flowers of Jai is placed around the neck of Gavarai by Lord Shankar.

2. *Nerium indicum* Mill. (Apocynaceae) Kanher

Chaitrya Vaishakhach unha va may
Vaishakhach unha
Khadak tapun zale lal va may
Tapun zale lal
Aaichya payi aale phod va may
Payi aale phod
Aaichi begadi vavhan va may
Begadi vavhar
Tathe kay kanherach zad va may
Kanherch zad.

['चैत्र वैशाखाचं उन्हं व माय
वैशाखचं उन्हं
खडक तपून झाले लाल व माय
तपून झाले लाल
आईच्या पायी आले फोड व माय
पायी आले फोड
आईची बेगडी वाव्हन व माय
बेगडी वाव्हन
तठे काय कन्हेरचं झाड व माय
कन्हेराचं झाड'.]

Newly married girls are eager to return their maternal home to enjoy the 'Akhaji' festival. Their experience while returning to home is described in this song. There is hot sun. The rocks have turned red due to hot sun. Boils have appeared on their foot sole. Their footwears are not good. They note a kanher plant (*Nerium indicum* Mill.). They rest for a while under the shade of this plant. This is how they travelled from their husband's home to maternal home.

3. *Mangifera indica* L. (Anacardiaceae) Amba, Kairi

Aathani kairi, tathani kairi
Kairi zoka khay va
Kairi tutani, khadak phutna
Tathe zuizui pani vhya va
Thathe Ratan dhobi dhoy va
Ratan dhobi dhoy
Tathe kasana bazar va
Tathe lachchyasna bazar va
May male lachchya li thewjo
Badhun khatma di de jo.

['आथानी कैरी, तथानी कैरी
कैरे झोका खाय व
कैरी तुटनी, खडक फुटना
तईं झुईझुई पानी व्हाय व
तठे रतन धोबी धोय व
रतन धोबी धोय, तठे कसाना बजार व
तडे पच्छारना बजार व
माय माले लच्छया ली ठीवजो
बंधून हातमा दि देजो'.]

During this period of year, the mango trees (*Mangifera indica* L.) are usually in full fruit-set. The bunches of unripe mangoes sway in the strong wind. These fall on the underlying rocks. The rocks break open and water flows from these rocks. A washerman, named Ratan, use to come there for washing clothes. This is a place of weekly bazar. The newly married girls remind their mothers to purchase 'Lachchya' (bracelet to wear at ankle). This song is sung before day of 'Aakhaji' festival while swinging.

4. *Phoenix dactylifera* L. (Arecaceae) Kharik (Dates) *Areca catechu* L.(Arecaceae) Supari (Arecanut)

Gadgad rathe chale Ramacha

Ni bahut lavannyacha

Sola sakhalya rathala

Ni bavanna khidkjya tyala

Baini lavali Kharik[a]

Bap se barik

Baini lavali Supari[b]

Bay se bepari.

['गडगड रथ चाले रामाचा

नि बहुत लावण्याचा

सोठा साख1या रथाला

नि बावन्न खिडक्या त्याला

बाईनी लावली खारीक

बाप से बारीक

बाईनी लावली सुपारी

बापसे बेपारी'.]

The days of festival of 'Akhaji' are passed. After this festival, husband of a married girl arrives at her maternal home. The chariot of her husband is described in the song. Her husband is referred as Ram (Lord Rama). His chariot is very graceful. It has sixteen chains and fifty two windows. Her parents are careful and have purchased for her dates (*Phoenix dactylifera* L.) and areca-nuts (*Areca catehu* L.) and few other things.

5. *Echinochoa frumentacea* Link. (Poaceae) Raya

Khadake khadake raya peru,

Ugana ka nai va

Bhau mana vanaspati,

Una ka nai va

Gaurai sarakha toda male,

Aanat ka nai va.

['खडके खडके राया पेरु

उगना कां नई वं

भाऊ मना वनस्पती,

उना कां नई वं

गौराई सारखा तोडा माले,

आनात कां नई वं'.]

The married sister is busy at work at her father-in-law's home, but dreams for invitation from her brother for 'Aakhaji' festival. She awaits for her brother. The sown plants of 'Raya' (*Echinochloa frumentacea* Link.) sprout out even from the rocks. Likewise, whether my brother is coming out of many difficulties of life. She also asks question to herself whether my brother has purchased bracelets, like Govarai, for me or not.

6. *Mangifera indica* L. (Anacardiaceae) Amba (Mango)

Mai re bhau tule mi gau

Vatvar Amba nako lau

Mani gaurai lekurvai

Tunha ambasna ronday hoi.

['माई रे भाऊ तुले मी गाऊ

वाटवर आंबा नको लावू

मनी गौराई लेकुरवाई

तुन्हा आंबास्न रोंदय होई'.]

A mother advises her son that your sister will come here for 'Aakhaji' festival. She will come with her kids. They will disturb your mango plantation (*Mangifera indica* L.). Therefore, you should not transplant mangoes on their way.

7. *Cocos nucifera* L. (Arecaceae) Naral, Naray
***Capsicum annuum* L. (Solanaceae) Mirchi**
***Nicotiana tabacum* L. (Solanaceae) Tambakhu**
***Piper betle* L. (Piperaceae) Widyache pan, Nagin Pan**

Evhadhishi Gaurai thumakani
Va may thumakani
Narayana zadkhale jai basani
va may jai basani
Kachch-pakka naray todi layani,
Va may todi layani
Ekale gayi tu ded paisa,
Va may ded paisa
Sasuni sangya mith mirachya,
Va may with mirachya
Sasarani sangi tambakhu
Va may tambhaku
Derani sanga zingi bhavara
Va may zingi bhavara
Nanani sange ain dora
Va mai ain dora
Patini sanga pan puda,
Va mai pan puda
Ya Sansarale hat joda
Va may hat joda.

['एवढीशी गौराई ठुमकनी
व माय ठुमकनी
नायरना झाडखाले जाई बसनी
व माय जाई बसनी
कच्चा पक्का नारय तोडी लयनी

व माय तोडी लयनी

इकाले गई त देड पैसा

व माय देड पैसा

सासूनी सांग्या मिठ मिरच्या

व माय मिठ मिरच्या

सासरानी सांगी तंबाखू

व माय तंबाखू

देरनी सांगा झिंगी भवरा

व माय झिंगी भवरा

ननिननी सांगा ऐन दोरा

व माय एन दोरा

पतिनी सांगा पान पुडा

व माय पान पुडा

या संसारले हात जोडा

व माय हात जोडा'.]

Gaurai collected unripe coconut fruits (*Cocos nucifera* L.) and sold for few paisa. She gained less but she has to expend more. She faces the difficulties of worldly affairs of her home. Mother-in-law ordered her for salt and red pepper (*Capsicum annum* L.), while father-in-law asked her to purchase tobacco (*Nicotina tabacum* L.). Husband's sister ordered for threads. Husband himself asked for roll of betal leaf (*Piper betle* L.). And therefore, she is wearisome.

8. *Cocos nucifera* L. (Arecaceae) Naral (Coconut)

Kaya ghodani kay mani Gaurai

Ena Pratap chalas thai thai

Pratap konya vadi gela

Pratap shimpi vadi gela

Shimpi uthala ghai ghai

Sadya kadhalya navalai

Sonar uthala ghai ghai

Navsar har lai yeyi

Vani uthala ghai ghai

Naray lai payi payi.

['काया घोडानी काय मनी गौराई

इना प्रताप चालस ठाई ठाई

प्रताप कोन्या वाडी गेला

प्रताप शिंपी वाडी गेला

शिंपी उठला घाई घाई

साड्या काढल्या नवलाई

सोनार उठला घाई घाई

नवसर हार लई येई

नारय लई पायी पायी'.]

Goddess Gaurai is graceful and brave. She went to tailor's shop and purchased sarees. She also purchased nine-stringed necklace from goldsmith's house. Likewise, she took coconut fruits (*Cocos nucifera* L.). Gaurai says goodbye to all and returns to her husband's home. Everybody responded her by offering one or other things.

9. *Mangifera indica* L. (Anacardiaceae) Amba

Aakhaji diwai saha mahinani laman

Bhause pavhan mazar dasara jamin

Dhakala murai nako dhadjo

May bai ambani amrai

Ragho mainana jiv bhyai.

['आखाजी दिवाई सहा महिनानी लामन

भाऊसे पाव्हना मझार दसरा जामिन

धाकला मुराई नको धाडजो

माय बाई आंबानी आमराई

राघो मैनाना जीव भ्याई'.]

There is a long period of six months between 'Aakhaji' and 'Diwali' festivals. A sister suggests not to send younger brother. He will be frightened in the grove of mangoes (*Mangifera indica* L.) on his way to my home. Therefore, you send elder brother to invite me for 'Diwali' festival.

SONGS OF GODDESS KANBAI IN AHIRANI DIALECT

1. *Nerium indicum* Mill. (Apocynaceae) Kanher
***Cocos nucifera* L.(Arecaceae) Naral**
***Rosa indica* L. (Rosaceae) Deshi**

Kanbai sange kanherale, may kanherale

Chala javu may mana maherale

Mani maherani hava

Bag bijorana lava

Bijora todu tabakale, tabakale

Chala javu manha maherale

Kanbai sange kanherale, may kanherale

Chala javu my manha maherale | | 1 | |

Mani maherani hava

Bag narayana lava

Naray todu tabakale, tabakale

Chala javu manha maerale

Kanbai sange kanherale, may kanherale

Chala javu my manha maherale | | 2 | |

Mani maherani hava

Bag Kanherna lava

Kanher todu tabakale, tabakale

Chala javu manha maerale

Kanbai sange kanherale, may kanherale

Chala javu my manha maherale | | 3 | |

Mani maherani hava
Bag gulabna lava
Gulab todu tabakale, tabakale
Chala javu manha maherale
Kanbai sange kanherale, may kanherale
Chala javu my manha maherale | |4| |

[‘कानबाई सांगे कन्हेरले, माय कन्हेरले
चला जाऊ माय मन्हा माहेरले
मनी माहेरनी हवा
बाग बीजोराना लावा
बीजोरा तोडू तबकले, तबकले
चला जावू मन्हा माहेरले
कानबाई सांगे कन्हेरले, माय कन्हेरले
चला जावू माय मन्हा माहेरले ॥१॥
मनी माहेरनी हवा
बाग नारयना लावा
नारय तोडू तबकले, तबकले
चला जावू मन्हा माहेरले
कानबाई सांगे कन्हेरले, माय कन्हेरले
चला जावू माय मन्हा माहेरले ॥२॥
मनी माहेरनी हवा
बाग कन्हेरना लावा
कन्हेर तोडू तबकले, तबकले
चला जावू मन्हा माहेरले
कानबाई सांगे कन्हेरले, माय कन्हेरले
चला जावू माय मन्हा माहेरले ॥३॥

मनी माहेरनी हवा

बाग गुलाबना लावा

गुलाब तोडू तबकले, तबकले

चला जावू मन्हा माहेरले

कानबाई सांगे कन्हेरले, माय कन्हेरले

चला जावू माय मन्हा माहेरले ॥४॥']

In this region, an idol of goddess Kanbai is worshipped. There is also procession of Kanbai. Khandeshi people sing songs to praise this goddess. In this folk song, Kanbai says that let us go to my maternal home. There is good weather to grow plants and raise gardens of Coconut (*Cocos nucifera* L.), Kanher (*Nerium indicum* Mill.) and Rose (*Rosa indica* L.).

2. *Piper betle* L. (Piperaceae) Widhyache pan, Nagin Pan
***Areca catechu* L. (Arecaceae) Supari**
***Nerium indicum* Mill. (Apocynaceae) Kanher**

Surya chamakana chamakana zadana mhajari

Yanha zaya re zaya re gavana patwari

Tyanhi dini re dini re pan supari

Kanher yevu dya yevu dya mandav mhajari

Kanbai punju dya punju dya rajaskumari

Surya chamakana chamakana zadana mhajari ||1||

Surya chamakana chamakana zadana mhajari

Yanha zaya re zaya re gavana patwari

Tyanhi dini re dini re pan supari

Kanher yevu dya yevu dya mandav mhajari

Kanbai punju dya punju dya rajaskumari

Surya chamakana chamakana zadana mhajari ||2||

['सुर्या चमकना चमकना झाडना म्हजारी

यान्हा झाया रे झाया रे गावना पटवारी

त्यानी दिनी रे दिनी रे पान सुपारी

कन्हेर येऊ द्या येऊ द्या मांडव म्हजारी

कानबाई पूंजू द्या पूंजू द्या राजसकुमारी

सूर्या चमकना चमकना झाडना म्हजारी ॥१॥

सूर्या चमकना चमकना झाडना म्हजारी

यान्हा झाया रे झाया रे गावना पटवारी

त्यानी दिनी रे दिनी रे पानसुपारी

कन्हेर येऊ द्या येऊ द्या मांडव म्हजारी

कानबाई पूंजू द्या पूंजू द्या राजसकुमारी

सूर्या चमकना चमकना झाडना म्हजारी ॥२॥']

The sun shun among the trees and this transformed into a village 'Patwari'. He offered roll of betal leaf (*Piper betle* L.) and betal nut (*Areca catechu* L.). The plant kanher (*Nerium indicum* Mill.) is expected to arrive in the pandal to worship the goddess Kanbai. Likewise, the sun shun among the trees and this transformed into a village merchant. He also offered betal leaf and betal nut. The kanher plant is similarly expected.

3. *Curcuma longa* L. (Zingiberaceae) Halad,Hayad

Dongarale padi gayi vat, vat mani maherale

Dongarale padi gayi vat, vat mani Kanhadale

Kasan bhari una tat, tat mana Kanhadale

Gulal bhari una tat, tat mana Kanhadale ||1||

Dongarale padi gayi vat, vat mani maherale

Dongarale padi gayi vat, vat mani Kanhadale

Kasan bhari una tat, tat mana Kanhadale

Kuku bhari una tat, tat mana Kanhadale ||2||

Dongarale padi gayi vat, vat mani maherale

Dongarale padi gayi vat, vat mani Kanhadale

Kasan bhari una tat, tat mana Kanhadale

Hayad bhari una tat, tat mana Kanhadale ||3||

Dongarale padi gayi vat, vat mani maherale
Dongarale padi gayi vat, vat mani Kanhadale
Kasan bhari una tat, tat mana Kanhadale
Phule bhari una tat, tat mana Kanhadale | |4| |

['डोंगरले पडी गयी वाट, वाट मनी माहेरले
डोगरले पडी गयी वाट, वाट मन्ही कान्हडले
कसानं भरी उनं ताट, ताट मनं कान्हडले
गुलाल भरी उनं ताट, ताट मनं कान्हडले ॥१॥

डोंगरले पडी गयी वाट, वाट मनी माहेरले
डोगरले पडी गयी वाट, वाट मन्ही कान्हडले
कसानं भरी उनं ताट, ताट मनं कान्हडले
कुकू भरी उनं ताट, ताट मनं कान्हडले ॥२॥

डोंगरले पडी गयी वाट, वाट मनी माहेरले
डोगरले पडी गयी वाट, वाट मन्ही कान्हडले
कसानं भरी उनं ताट, ताट मनं कान्हडले
हयद भरी उनं ताट, ताट मनं कान्हडले ॥३॥

डोंगरले पडी गयी वाट, वाट मनी माहेरले
डोगरले पडी गयी वाट, वाट मन्ही कान्हडले
कसानं भरी उनं ताट, ताट मनं कान्हडले
फुले भरी उनं ताट, ताट मनं कान्हडले ॥४॥']

Khandeshi women sing this song telling about the welcome preparations for the goddess Kanbai. A way or path has been cleared out in the mountain to reach to the place of Kanbai. Red powder ('Gulal' and 'Kunku'). turmeric powder (*Curcuma longa* L.) and flowers have been brought to worship the goddess.

4. *Nerium indicum* Mill. (Apocynaceae) Kanher *Rosa indica* L. (Rosaceae) Deshi Gulab

Akas varanya tin chandanya, chauthi chandani parmale
An janha ghar Kanbai pavani ni tathe gulab parmale
Kanbai mayani mandao dinha, var kasan zad va
Kanbai mayani mandap dinha, var bijoran zad va
Bijorale phul lagan, phul kay ghammagor va ||1||

Akas varanya tin chandanya, chauthi chandani parmale
An janha ghar Kanbai pavani ni tathe gulab parmale
Kanbai mayani mandao dinha, var kasan zad va
Kanbai mayani mandap dinha, var Kanherna zad va
Kanherle phul lagan, phul kay ghammagor va ||2||

Akas varanya tin chandanya, chauthi chandani parmale
An janha ghar Kanbai pavani ni tathe gulab parmale
Kanbai mayani mandao dinha, var kasan zad va
Kanbai mayani mandap dinha, var Gulabna zad va
Gulable phul lagan, phul kay ghammagor va ||3||

['आकास वरण्या तीन चांदण्या, चौथी चांदनी परमळे
अन् जान्ह घर कानबाई पावनी, नी तठे गुलाब परमळे
कानबाई मायनी मंडप दिन्हा, वर कसानं झाड व
कानबाई मायनी मंडप दिन्हा, वर बिजोरानं झाड व
बिजोराले फूल लागनं, फूल काय घम्मागोर व ॥१॥

आकास वरण्या तीन चांदण्या, चौथी चांदनी परमळे
अन् जान्ह घर कानबाई पावनी, नी तठे गुलाब परमळे
कानबाई मायनी मंडप दिन्हा, वर कसानं झाड व
कानबाई मायनी मंडप दिन्हा, वर कन्हेरनं झाड व
कन्हेरले फूल लागनं, फूल काय घम्मागोर व ॥२॥

आकास वरण्या तीन चांदण्या, चौथी चांदनी परमळे

अन् जान्ह घर कानबाई पावनी, नी तठे गुलाब परमळे

कानबाई मायनी मंडप दिन्हा, वर कसानं झाड व

कानबाई मायनी मंडप दिन्हा, वर गुलाबनं झाड व

गुलाबले फूल लागनं, फूल काय घम्मागोर व ॥३॥']

This is another song praising the goddess Kanbai. If this goddess steps in one's house and pleased with such family, it is perfumed by rosy smell (*Rosa indica* L.). Kanbai raises pavilion and different plant flouish on it. Especially, the plants like 'Kanher' (*Nerium indicum* Mill.) and rose (*Rosa indica* L.) are growing on the pavilion. Their flowers are very fragrant. Thus the delightful place of Kanbai is described in the song.

5. *Cinnamonum zeylanicum* Bl. (Lauraceae) Dalchini

Daychinini day range bhare va

Kanbaina mukhada koni ghada va

Hau sonar kalidar, yanhi ghada va

Mangaywar, mangaywar Dhuyana hau bhare bajar

Kasamise jau bajar

Bindimise jau bajar

Dhuyana hau hausi sonar, hausi sonar

Bindi ghadi yani nakshidar, yanhi nakshidar

Ghar yewale zahi dupar, may zahi dupar

Bhojanale zaya pahila par, may pahila par

Bhojan bhare Jamna nar

Kashinath phiraye tupani dhar, may tupani dhar

Gavarani basanyat hat var hat, my hat var hat

Bhojan laye hai konati nar

Lhisan bhojan dhari tini vat

Utari unyat awaghad ghat, may awaghad ghat

Kanbai mayana motha that, may motha that.

['दायचिनीनी दाय रंगे भरे व
कानबाईना मुखडा कोनी घडा व
हाउ सोनार कळीदार, यान्ही घड व
मंगयवार मंगयवार, धुयाना हाऊ भरे बाजार
कसामिसे जावू बजार
बिंदीमिसे जावू बजार
धुयाना हाऊ हौसी सोनार, हौसी सोनार
बिंदी घडी यान्ही नक्षीदार, यान्ही नक्षीदार
घर येवाले झाई दुपार, माय झाई दुपार
भोजनले झाया पहिला पार, माय पहिला पार
भोजन भरे जमना नार
काशिनाथ फिराये तुपनी धार, माय तुपनी धार
गवरानी बसन्यात घट वर घट, माय घट वर घट
भोजन लये हाई कोणती नार
ल्हिसन भोजन धरी तीनी वाट
उतरी उन्यात अवघट घाट, माय अवघड घाट
कानबाई मायना मोठा थाट, माय मोठा थाट'.]

This song conveys that the idol of Kanbai is fashioned by a skilful goldsmith of Dhule city. For this idol, 'Bindi' (an ornament for the forehead of a Hindu lady) is also fashioned by a desirous and expert goldsmith of Dhule city. While purchasing it, the lady returned home very late in the noon. The devotee women crossed the rugged mountain to visit the place of Kanbai. They are pleased on account of magnificience of the goddess Kanbai.

6. *Piper betle* L. (Piperaceae) Nagin Pan

Pach naginana pan va may naginana pan
Pan varyan udal va may varyan udal
Aaichya angani padal va many angani padal
Aaina tamhane ghatal va may tamhine ghatal

Aaine rumale zakil va may rumale zakil

Aaine chaurangi theval va may chaurangi theval

Jyot kapurani layi va may kapurani layi

Aai kanbai vavai va may Kanbai vavai.

['पाच नागीननां पान व माय नागीननां पान

पान वाऱ्यानं उडालं व माय वाऱ्यानं उडालं

आईच्या आंगनी पडलं व माय आंगनी पडलं

आईने तम्हाने घातलं व माय तम्हाने घातलं

आईने रुमाले झाकिलं व माय रुमाले झाकिलं

आईने चौरंगी ठेवलं व माय चौरंगी ठेवलं

ज्योत कापूरनी लायी व माय कापूरनी लायी

आई कानबाई ववाई व माय कानबाई ववाई'.]

Five betal leaves (*Piper betle* L.) are flown by the wind and carried to the foreground of goddess Kanbai's home. She collected them and placed in a plate of worship. This plate is covered by a towel and then kept on a square sheat. Camphor is alighted by goddess Kanbai in front of it.

7. *Cocos nucifera* L. (Arecaceae) Naral

Aada zaya darvaja te

Kathain ishi mani kanbai may

Darvajale bandhasu toran

Nichit mankhal isu va may

Aada zaya maroti te

Kathain ishi mani kanbai may

Marotile phodasu naray

Nichit mankhal isu va may

Aadi zayi Bhavani te

Kathain ishi mani Kanbai may

Bhavanile bharasu chuda

Nichit mankhal isu va may
Aada zaya ghardhani te
Kathain ishi mani Kanbai may
Ghardhinile jiwadisan
Nichit mankhal isu va may.

['आडा झाया दरवाजा ते
कथाईन ईशी मनी कानबाई माय ।।धृ।।
दरवाजाले बांधसू तोरन
निचित मनखाल इसू व माय ।।१।।
आडा झाया मारोती ते
कथाईन ईशी मनी कानबाई माय ।।
मारोतीले फोडसू नारय
निचित मनखाले इसू माय ।।२।।
आडी झाली भवानी ते
कथाईन ईशी मनी कानबाई माय ।।
भवानीले भरसू चुडा
निचित मनखाल इसू व माय ।।३।।
आडा झाया घरधनी ते
कथाईन इशी मनी कानबाई माय ।।
घरधनीले जेवाडीसन
निचित मनखाल इसू व माय ।।४।।']

A devotee woman asks the goddess Kanbai that if your entry is obstructed by the gate itself, then how you will enter. Answer for this question is given by the devotee woman herself. She says that I will hang a wreath of leaves on that gate. If the God Maruti obstructs your entry, I will pray him and offer coconut fruit. If the goddess Bhavani, obstructs your way, then I wil offer green bangles to her. If the head of house refuses your entry, then I will serve him a dinner. Thus, I will be free from anxiety and then will come to you for blessings.

SONGS ON DIFFERENT OCCASIONS IN AHIRANI DIALECT

1. *Ocimum tenuiflorum* L. (Labiatae) Tulashi

"Mana anganama shobhe 'Tuyasina brindavan'
Kare sarva gharale pavitar rup tinha".

['मना आंगनमा शोभे 'तूयसिनं' ब्रिंदावन
करे सर्व घराले पवितर रुप तिन्ह'.]

Tulshi plant is generally grown in specially made pots or pedestal pots in Hindu homes. It is usually decorated in the foreground of a house. The house or the surrounding of a house is purified by the existence of tulshi plant.

2. *Ocimum tenuiflorum* L. Tulashi

"Tuyasin rop dole varavar thayathaya,
Kanman gan mhana papi rog dur jay".

['तुळसिन रोप डोले वारावर थयाथया,
कानमान गानं म्हना पापी रोगं दूर जाय'.]

Tulashi plantlet (Holy Basil) is waggling joyfully. You sing song near ears, the sinful disease or evil spirit will run away.

3. *Ocimum tenuiflorum* L. (Lamiaceae) Tulashi

"Tuyasimay bayi vatoy tunha pan,
Tunha sawalima dev basas Bhagwan".

['तुयसिमाय बायी वाटोय तुन्ह पांन,
तुन्हा सावलीमा देव बसस भगवान'.]

Oh, Tulashi mother, your leaves are orbicular. Lord Vishnu (Bhagwan) sits in your shade. You are very dear to him.

4. *Anthocephalus cadamba* Miq. (Rubiaceae) Kadamb, Kayamb

"Radhika nari haka mare deva deva.
Kayamban zadavari kushna vaje pava".

['राधिका नारी हाका मारे देवादेवा,

कयंबानं झाडावरी कृष्न वाजे पावा'.]

According to Hindu mythology, there is a legend of Lord Krishna stealing the clothes of the gopies (mentioned in the Bhagavata and the Vishnu Puranas) who were bathing in the river Yamuna. Krishna picked up their clothes and climbed the tree 'Kadamba' (*Anthocephalus cadamba* Miq.). The same situation is represented by this folk song. Radhika, a gopi, is requesting him, to return the clothes however, Krishna is sitting on 'Kadamba' tree and playing flute.

5. *Tectona grandis* L.*f.* (Verbenaceae) Sag

"Sitale vanavas dharu kakare dongare,

Sagana pane sita andhare pinghare".

['सिताले वनवास धरु काकरे डोंगरे,

सागना पाने सिता आंधरे पांघरे'.]

In the great epic Ramayana, Lord Rama and his wife Sita were ordered to lead a forteen year's forest life (Vanvas). They faced many difficulties on account of rugged life. One such situation is depicted in this song. Sita used to wear the leaves of teak (*Tectona grandis* L.*f.*) as her clothes.

6. *Acacia nilotica* (L.) Willd. ex Del. subsp.*indica* (Bth.) Brenan. (Mimosaceae) Babhul

"Vakad tikad hai babhuyan khod,

Dekhama nahi dhad todale jas avaghad".

['वाकडं तिकडं हाई बाभूयनं खोड,

देखामा नही धडं तोडाले जास अवघड'.]

The Babul tree is crooked. It is not good or beautiful to look at and also very hard to cut. A crooked man is likewise recognisable by his external appearance.

7. *Solanum melongena* L. (Solanaceae) Wange
***Trigonella foenum-graecum* L. (Papilionaceae) Methi**

"Jathi dekhu tathi vangyabandhuni kheti,

Khetima pasari methi bigar paninya layarya deti".

['जथी देखू तथी वांग्याबंधुनी खेती,

खेतीमा पसरी मेथी बिगर पानीन्या लयऱ्या देती'.]

A lady observes crop field of egg plants or Brinjal (*Solanum melongena* L.) everywhere. She refers Brinjal plant as her brother. Methi or fenygreek (*Trigonella foenum-graecum* L.) is also sown in between. Therefore, the yield of brinjal is also increased. We know, legume plant bear root nodules, rich in nitrogen. If the legumes are sown as mixed crop with other crops, it increases its yield. Probably, these people know this 'Science'.

8. *Phaseolus mungo* L. (Papilionaceae) Udid

"Tanha kaya ni savaya jasa udidana dana,

Sunyabai tule sangu yale kusandev mhana".

['तान्हा काया नी सावया जसा उडिदना दाना,

सुन्याबाई तुले सांगू याले कुसनदेव म्हना'.]

A mother-in-law says to her daughter-in-law that your son or child is black to fairly black like the grains of black gram (*Phaseolus mungo* L.). Therefore, you call him Kusandev (Lord Krishna, who is also said to be black in colour).

9. *Aegle marmelos* L. (Rutaceae) Bel

"Sakali uthasani chidya karati chivchiv,

Belana zadkhale Deva tu Mahadev."

['सकळी उठसनी चिड्या करती चिवचिव

बेलना झाडखाले देवा तू महादेव'.]

At early morning, the sparrow's chirp is generally heard. A natural scene is picturized in the first line of song. Lord Shiva sits under the Bel tree (*Aegle marmelos* L.), hence the tree is held religious or the tree has received divinity.

10. *Cocos nucifera* L. (Arecaceae) Naral
Artocarpus integrifolia L.*f.* (Moraceae) Phanas

"Varin naray kathin tyanha potma jivan,

Toche phanasna kata madhma raj amrutan".

'वरीन नारय कठिन त्यान्या पोटमा जीवन,

टोचे फणसना काटा मधमा राज अमृतन'.]

This song describes the nature of person and warrants for not to consider only external appearance. Coconut fruit (*Cocos nucifera* L.) is very hard outside, but life-giviing nutritious water is present inside. It is also the case of jackfruit (*Artocarpus integrifolia* L.*f*.). It bears pinching spines all over, however, inside sweet pulp is present.

11. *Bauhinia racemosa* Lamk. (Caesalpiniaceae) Apta

"Aare, aapatan pan, panale aaj man,

Don jiv hota ek shilangan Ani tech son".

['आरे, आपटानं पानं, पानाले आज मानं,

दोन जीव होता एक शिलांगन आणि तेच सोनं'.]

On the day of Dasara festival, leaves of Apta (*Bauhinia racemosa* Lamk.) are particularly valued as gold. It is respected so on this eve. Two halves of a leaf meet each other. Likewise, the fellowmen or friends offer leaves of this plant to each other. They embrace and express good wishes for each other.

12. *Prosopis cineraria* L. (Mimosaceae) Saundad

"Aaj Dasara Dasara, saundadale aaj aana,

Aate esara talwari, kay badalana jana".

['आज दसरा दसरा, सौंदडले आज आना,

आते इसरा तलवारी काय बदलना जाना'.]

Today is the festival of Dasara. The Saundad tree (*Prosopis cineraria* L.) which is referred as 'Shami' tree in Hindu mythology, on this day Lord Rama and Laxamana placed their swords and other weapons on this tree). We must forget the swords and quarrels because the world has changed. We should welcome the 'new world' with love.

13. *Sesamum orientale* L. (Pedaliaceae) Til *Saccharum officinarum* L.(Poaceae) Oos

"Aaj utaran utaran, tihi gui wata bai,

Tihi lai ang dhava, pap tumhan pai jai".

['आज उतरानं उतरानं, तिही गुई वाटा बाई,

तिही लाई अंग धवा, पाप तुम्हनं पई जाई'.]

Today is the day of "Makar-Sankrant" [Passage of the sun from Sagittarius (Dhanu) to Capricorn (Makar)]. It is always falls on 14th day of January each year. The seeds of Sesamum (*Sesamum orientale* L.) alongwith jaggery (*Saccharum officinarum* L.) or their preparations are distributed amongst the friends, relatives and neighbours. One should bath using sesamum oil so that the diseases or sins will disappear.

14. *Sesamum orientale* L. (Pedaliaceae) Til

"Tihi tihi karu tihi may tun son,

iv dhakulsa tun vay jun jun son".

['तिही तिही करु तिही, माय तून सोनं.

जीव धाकुलसा तूनं वय जूनं जूनं सोनं'.]

Oh, Tihi mother, you are so small in size. But you are great by virtues like the old gold. Medicinal utility is emphasized of the sesamum in this song.

15. *Sorghum bicolor* (L.) Moench. var.*bicolor* (Poaceae) Jondhala

***Musa paradisiaca* L. (Musaceae) Kela**

"Kaya vavarma dhavaya nandi rambole,

Pikana jondhaya keyi mayana khamb dole".

['काया वावरमा धवया नंदी रामबोले,

पिकना जोंधया केयी मायना खांब डोले'.]

Bullocks are working hard in the black cotton soils. Because of their efforts, yield of Banana or plantain (*Musa paradisiaca* L.) and 'Jondhala' [*Sorghum milliforme* (Hack.) Snowdon] has increased.

16. *Madhuca longifolia* (Koen.) Macbr. var.latifolia. (Roxb.) Chev. (Sapotaceae) Mahu

"Vatvaran mahu tap tap gaye,

Tathi mani Kanbai phugadya kheye".

[‘वाटवरनं महू टप टप गये,

तठी भनी कानबाई फुगड्या खेये’.]

There is a Mahua tree [*Madhuca longifolia* (Koen.) Macbr. var.*latifolia* (Roxb.) Chevalier]. It is in full blossom and its flowers and fruits are dropping incressantly. Goddess Parvati is playing ‘Phugadya’, a joint dance of girls or woman facing each other and crossing hands with antiflockwise movement.

17. *Anogeisus latifolia* (Roxb.) ex DC. Wall. ex Guill. (Combretaceae) Dhawda
***Musa paradisiaca* L. (Musaceae) Kela**

"Lokasnya leki dhavadanya biyi,

Devani dini ek lek jasi Savadani Keyi".

[‘लोकसन्या लेकी धावड्यान्या बीई,

देवनी दिनी एक लेक जसी सासदानी केयी’.]

Jalgaon district is well-known for its banana farming. In the form of babana plants (*Musa paradisiaca* L.) this region has handsome and lovely daughter. Daughter of others are like the seeds of ‘Dhavda" [*Anogeissus latifolia* (Roxb.) ex DC. Wall. ex Guill] which are ugly to look at.

18. *Michelia champaca* L. (Magnoliaceae) Chapha

"Bap mana chafa, may mani champakayi,

Champakayina mandivar jap malc lagi gayi".

[‘बाप मना चाफा माय मनी चंपाकयी,

चंपाकयीना मांडीवर जप माले लागी गयी’.]

My father and mother both are very affectionate like Chafa tree (*Michelia champaca* L.) and its flower buds respectively. I slept putting my head on her thigh, forgetting all worries.

19. *Butea monosperma* (Lam.) Taub. (Papilionaceae) Palas

"Bap mhane lek tu nashiban hin,

Dongar dhundisan payasale pan tin".

['बाप म्हणे लेक तू नशीबानं हिनं,

डोंगर ढूंडीसन पयसाले पानं तिनं'.]

A father accuses the luck of his daughter. He tried to do his best but he always observed only three leaves for 'Payas' plant [Butea monosperma (Lam.) Taub]. He experienced similar circumstances everywhere.

20. *Microtyloma uniflorum* (Lamk.) Verdc. (Papilionaceae) Kulith

"Javaini jat jas kuithan khet

Jitaka ghalu mey ugari yani jat".

['जवाईनी जात जसं कुईथनं खेत,

जितका घालू मेय उगरी यानी जात'.]

It is always difficult to protect the field of 'Kuith' or Horse gram, (*Dolichos uniflorus* Lam.) Likewise son-in-law is ferocious although every care is taken of him.

21. *Cicer arientinum* L. (Papilionaceae) Harbara

"Bap kasa mhane lek harabarani day,

Janma devana adhikar tunha daivale karu kay".

['बाप कसा म्हणे लेक हरबरानी डाय,

जन्म देवाना अधिकार तुन्हा देवाले करु काय'.]

A father compares his daughter with the tender fruiting branch of Chick pea or Gram (*Cicer arientinum* L.). He is wearied and says that he has the (natural) right of giving birth to her daughter but he can not go ahead of her fortune.

22. *Oryza sativa* L. (Poaceae) Tandul

"Bap kasa mhane lek tanduyani phani,

Jai parghar nav kadhatil barajani".

['बाप कसा म्हणे लेक तांदुयनी फनी,

जाई परघर नाव काढतील बाराजणी'.]

A father finds her daughter as beautiful as rice-grains. In his opinion, if she goes her husband's home (in future), she will bring reputation or popularity to them.

23. *Pennisetum americanum* (L.) K. Schum. (Poaceae) Bajra

"Kaya wawarma hirvi bajarani gari,

Soyaran ghar mani Indurani pari".

['काया वावरमा हिरवी बाजारानी गरी,

सांयरानं घर मनी इंदूरनी परी'.]

A heap of green bajara Pearl millet (*Pennisetum americanum* (L.) Schum.) catches one's fancy. Likewise, my daughter, according to her father, will suit as if a nymph of Indore city.

24. *Santalum album* L. (Santalaceae) Chandan

"Kavaya kokaye chandan philakyavari,

Bandhu ga pahuna bahin Draupadichya dari".

['कावया कोकाये चंदन फिलक्यावरी,

बंधू ग पाहुना बहिन द्रौपदीच्या दारी'.]

Crow's caw is conceived in this region as a mark of arrival of some guests. Crow sat on the branch of Sandal tree (*Santalum album* L.) and its caw is perceived by a sister (Draupadi) as an indication or a warning of arrival of her brother.

25. *Cordia gharaf* (Forsk.) Ehrenb. and Aesch. (Boraginaceae) Gondani,Bhokar

"Bandhu she mhorai ghoda bandhajo gondhani,

Soyarale Ram Ram mag bhetjo chandanile".

['बंधु शे म्होराई घोडा बांधजो गोंधनीले,

सोयराले राम राम मग भेटजो चांदनीले'.]

Brother has arrival at sister's home to invite her (especially on the day of festival like Diwali or Akhaji). After his arrival, the sister guides him to fasten his horse to the trunk of 'Gondhani' tree [*Cordia gharaf* Forsk. (Ehrenb. and Asch.)]. Afterwards, she directs him to bow to his elderly relatives. Then, he can meet his sister.

26. *Allium cepa* L. (Liliaceae) Kanda

"Mayinya mayat bai Dada lawe kanda,
Nadan bal mhana bandhe rajeshahi bangala".

['मायीन्या मयात बाई दादा लावे कांदा,
नादानं बाळ म्हना बांधे राजेशाही बंगला'.]

This song tells about the dream of a mother about her son. In her dream, her son has cultivated onions in field. He will raise a princely bunglow after the good yield of onions.

27. *Triticum aestivum* L. (Poaceae) Gahu

"Pani bharata gahu yale varana bahu bhev,
Dhakala mana bhau yale kitik lavu jiv".

['पानी भरता गहू याले वाराना बहु भेव,
धाकला मना भाऊ याले कितीक लावू जीव'.]

When the plants of wheat (*Triticum aestivum* L.) are small, and if ones intends to water or irrigate them, there is a danger or fear of disturbing its own life. Likewise, a sister says, my brother is younger, I have to take every care of him.

28. *Cucumis melo* L. var.*agrestis* Naud. (Cucurbitaceae) Shendadi

"Kaya vavarma shendodi kare mat,
Nanind kamjat bhavajai mare lat".

['काया वावरमा शेंदोडी करे मात,
ननींद कमजात, भावजाई मारे लात'.]

'Shendodi' (*Cucumis melo* L. var. *agrestis* Naud.) flourishes well in black cotton soiled fields and dominates crop plants. Likewise, husband's sister always keeps on eye on sister-in-law and predominates.

29. *Citrus aurantifolia* (Christm.) SW. (Rutaceae) Limbu *Punica granatum* L. (Punicaceae) Dalimb *Carica papaya* L. (Caricaceae) Papayi, Popayi

"Tapila narayan jasa aginana puya,

Bandhuni bagshahi limb, dalimb, popaya".

['तपीला नारायण जसा आगीनना पूया,

बंधूनी बागशाही लिंब, डाळींब, पोपया'.]

The sun is sparkling in the sky. Likewise, a sister says, that my brother's orchard is glittering because of the fruit plants such as lemon [*Citrus aurantifolia* (Christm.) SW.], pomegranate (*Punica granatum* L.) and Papaya (*Carica papaya* L.).

30. *Azadirachta indica* A.Juss. (Meliaceae) Neem

"Mhana anganama gard nimbani savuli,

Kitale sangu bahin tule sasu nahi ti mauli".

['म्हना आंगनमा गर्द निंबानी सावली,

कितले सांगू बहिन तुले सासू नही ती माऊली'.]

Generally, mother-in-law is famliar for vexatious treatment to her daughter-in-law. However, a brother advises his sister, the neem tree (*Azadirachta indica* L.) gives us dark and cool shade. Likewise, mother-in-law is loving shade for yourself.

31. *Sesbania sesban* (L.) Merr. subsp. *sesban* (Papilionaceae) Shevara

"Khetana bandhala ubha dandina shevara,

Kamjat narile hat jodito navara".

['खेतना बांधले उभा दांडिना शेवरा,

कमजात नारीले हात जोडीतो नवरा'.]

There is a troublesome weed 'Shevara' [*Sesbania sasban* (L.) Merr. subsp.*sesban*.] along the bandhs of cultivated fields. It is just like a wearied husband whose wife is not obedient.

32. *Acacia leucophloea* (Roxb.) Willd. (Mimosaceae) Hiwar

"Hiwarana hete nako basu va narbai,
Hiwar beiman yale ghadina dhir nahi".

['हिवरना हेटे नको बसू व नारबाई,
हिवर बेईमान याले घडीना धीर नाही'.]

A woman advises another lady. Please, do not sit under the 'Hiwar' tree [*Acacia leuophloea* (Roxb.) Willd.] This tree is also dishonest like the gents. They don't have stamina, they can change their moods at any time.

33. *Citrullus colocynthis* (L.) Schr. (Cucurbitaceae) Kaduindravan

"Kadu indravan mhan khayan zay man,
Kiti kaku sakhar nahi janar kadupan".

['कडू इंद्रावन म्हनं खायानं झायं मन,
किती टाकू साखर नही जानारं कडूपन'.]

A lady wishes to eat the fruit of 'Kadu-indrayan' [*Citrullus colocynthis* (L.) Schr.]. However, she expresses a view that whatever quantity of sugar is added in it, it remains always bitter. In brief, the nature of a bitter person can not be changed.

34. *Zea mays* L.(Poaceae) Maka

"May may karu may mayamana maka,
Kitalya maru haka, thand paninya zuyakya".

['माय माय करु माय मयामाना मका,
कितल्या मारु हाका, थंड पानीन्या झुयक्या'.]

My mother is just like maize in fields. I am calling her again and again because she is like a gentle breeze in my life.

35. *Annona squamosa* L. (Annonaceae) Sitaphal

"Dekhale jau Badrinarayanana maya,
Pikana sitaphay eka phandile soya".

['देखाले जाऊ बद्रिनारायनना मया,

पिकना सिताफय एका फांदीले सोया'.]

A sister praises the prosperity of his brother. In his orchard, Custard apple (*Annona squamosa* L.) bore bunches of fruits on every branch.

36. *Cucurbita maxima* Duch. ex Lamk. (Cucurbitaceae) Bhopala

"Asturi purush sada iyani bhopaya,

Uni Lakshmi mari gayi kan taya.".

['अस्तुरी पुरुष सदा इयानी भोपया,

उनी लक्ष्मी मारी गयी कान टाया'.]

The relationship of sickle and red pumkin is always thought quarrelsome. This also finds parallel in the relationship of a husband and wife. Because of quarrels and dissatisfaction Lakshmi, the goddess of wealth according to Hindu mythology, leaves their home.

37. *Syzygium aromaticum* L. (Myrtaceae) Lawang *Areca catechu* L. (Arecaceae) Supari

"Sat talana madivar, Ram Sitani katkat,

Lavangdoda she tikhat, supari nighani tikhat".

['सात तालना माडीवर, राम सितानी कटकट,

लवंगदोडा शे तिखट, सुपारी निघनी तिखट'.]

Husband and wife has implied comparison with Lord Rama and goddess Sita. Although a couple dwells in a multistoried building and if nature of either of them is pungent like a clove (*Syzygium aromaticum* L.) and also of arecanut (*Areca catechu* L.), there are always controversies or quarrels.

38. *Nerium indicum* Mill. (Apocynaceae) Kanher

"Soyara Patil gav una manasange,

Kavaya kanher hau datanale mange".

['सोयरा पाटील गाव उना मनासंगे,

कवया कन्हेर हावू दातनले मांगे'.]

My son's father-in-law travelled and arrived with me. He is now, demanding young branch of 'Kanher' (*Nerium indicum* Mill.) to use it as tooth-brush.

39. *Hibiscus cannabinus* L. (Malvaceae) Ambadi

"Ambadina plulavar dev ratale ladana,

May mhana blaukarata doya asusana".

['आंबाडीना फुलावर देव रातले लडना,

माय म्हना भाऊकरता डोया आसूसना'.]

A sister observes dews on the flowers of Kenaf (*Hibiscus cannabinus* L.) and says that the god wept at night. I am also eager and waiting for my brother. My eyes are tearful.

40. *Ocimum basilicum* var.*thyrsiflora* Benth. (Lamiaceae) Sabja

"Navatini lek tunhi navati Banarasi,

Kanama sabedan phul was una chulipashi".

['नवतीनी लेक तून्ही नवती बनारसी,

कानमा सबेदानं फूल वास उना चुलीपाशी'.]

Your young daughter wore a flower of the sweet Basil (*Ocimum basilicum* L.) for the ears. It has filled with sweet smell everywhere.

41. *Coriandrum sativum* L. (Apiaceae) Kothambir

"Dhanai Kothambir yale dhane dhane vas,

Kabilana dhana dhana karu mi barmas".

['धनाई कोथंबीर थाले धने धने वास,

कबीलाना धंदा धंदा करु मी बारमास'.]

Fruits of coriander (*Coriandrum sativum* L.) are very fragrant. I (a house woman) can carry on my routine work easily due to these sweet smelling fruits.

42. *Cyamopsis tetragonoloba* (L.) Taub. (Papilionaceae) Gawar
Acaia leucophloea (Roxb.) Willd. (Mimosaceae) Hiwar

"Hirad phodi phodi kiti songu gavarale,

Chaitrya mahinama pala phutana hiwarale".

['हिरद फोडी फोडी किती सांगू गवारले,

चैत्र महिनामा पाला, फुटना हिवरले'.]

A sister (when she is staying at her home of father-in-law) waits for his brother who will invite her for the festival of Akshaytritiya (or Aakhaji in Khandesh region). This festival starts in the month of April (or Chaitrya month). The 'Hivar' tree [*Acacia leucophloea* (Roxb.) Willd.] generally bears leaves during this period. This suggests the probable arrival of one's brother. A sister is telling this situation to 'Gavar' plant [*Cyamopsis tetragonoloba* (L.) Taub.] from her base of heart.

43. *Cassia tora* L. (Caesalpiniaceae) Tarota

"Bhajima bhaji barik bhaji tarotani,

Andhya gaye gana khan kolatani".

['भाजीमा भाजी बारीक भाजी तरोटानी,

अंध्या गाये गानं खान कोलाटनी'.]

Tarota (*Cassia tora* L.) is a wild plant and sometimes its leaves are used as vegetable. Sometimes, wild or unpleasent vegetable is mixed with good vegetable varieties. Thus, the wild plants gain favour and popularity in association with good plants. Similarly, bad people also mix up with good people and thus gain benefits in their life.

44. *Echinochloa frumentacea* Link. (Poaceae) Raya

"Maranana veye khoy bharaja rayani,

Bhau kasa mhane bahin chalani dhanyani".

['मरनना वेये खोय भरजा रायानी,

भाऊ कसा म्हने बहिन चालली धन्यानी'.]

On the death of a sister, her brother mournfully says that she is from a well-to-do family. Therefore, please stuff and tie the grains of 'Raya' (*Echinochloa frumentacea* Link.) in front portion of her saree. This is a rite in the region of Khandesh.

45. *Triticum aestivum* L. (Poaceae) Gahu

"Gahu mandi gahu bansi gahuni bhesak,

Nasik jilhamandi manha bhauni thasak".

['गहू मंधी गहू बन्सी गहुनी भेसक,

नासिक जिल्हामंधी मन्हा भाऊनी ठसक'.]

Amongst the different varieties of wheat (*Triticum aestivum* L.), 'Bansi', a variety of wheat, is the best. My brother is the most prompt, energetic and smart in the entire Nasik district. The help of the best of qualities of a wheat variety is taken to show brother's prominance in the region.

46. *Syzygium aromaticum* L. (Myrtaceae) Lawang *Piper nigrum* L. (Piperaceae) Mire

"Khandesi bai tuna papadale vas,

Pikana dang des long mirile una ghos".

['खानदेसी बाई तुना पापडले वास,

पिकना डांग देस लोंग मिरीले उना घोस'.]

For the preparations like 'papad' (a thin crisp cake made of flour and spices), 'lonache' (spicy preserve of mango) and other preparations, the Khandeshi ladies are thought experts and always praised for them. One lady reminds Khandesh ladies that the plants of clove (*Syzygium aromaticum* L.) and black pepper (*Piper nigrum* L.) are in full bloom now in Dang country. Your 'papad' will emit fragrance.

47. *Areca catechu* L. (Arecaceae) Supari

"Chikan supari adkittat datani,

Bhola purusale nar basundi bhetani".

['चिकन सुपारी अडकित्तात दाटनी,

भोळा पुरुसले नार बासुंदी भेटनी'.]

Sometimes a husband is simple and gullible and his wife is clever and cunnig. A sister of such husband says that the nuteracker (the husband) can not accommodate and crack the spleepery arecanut (the wife). She again remarks that simple, gullible man got 'basundi' (a sweet dish of milk). The nature and relationship of husband and wife is explained in this folk song.

48. *Cuminum cyminum* L. (Apiaceae) Jire *Triticum aestivum* L. (Poaceae) Gahu

"Sheji ajibai usna de va jire,

Bandhava pavana vanyani gali dur.

Sheji ajibai usana de va gahu,

Usir nako lavu dur desana mana bhau".

['शेजी आजीबाई उसना दे व जीरे,

बंधवा पाव्हना वान्यानी गली दूर,

शेजी आजीबाई उसना दे व गहू,

उसिर नको लावू दूर देसना मना भाऊ'.]

A brother arrives at sister's home. The necessary food-grains and spices are not now available to her to welcome him. Therefore, she requests her senior neighbour lady for cumin seeds (*Cuminum cyminum* L.) and wheat (*Triticum aestivum* L.). She demands these on loan basis as the shops are far ahead from her home. Her hurry and bustle reflects affection for her brother.

49. *Sesamum orientale* L. (Pedaliaceae) Til

"Giranana masa randhu tihi lahi,

Bhause pahuna sadodit yet nahi".

['गिरनाना मासा रांधू तिही लाही,

भाऊसे पाहूना सदोदीत येत नाही'.]

This is another folk song depicting love of a sister in welcoming her brother. She cooks fishes brought from Girna river using sesamum seeds (*Sesamum orientale* L.). She feels that her brother visits her hourse only occasionally.

50. *Mangifera indica* L. (Anacardiaceae) Amba

"Kaya vavarma zopaya amba

Tathe kay basani Gavrai rambha

Gavarai may kay kay mange

Gavarai may lachchhya mange".

['काया वावरमा झोपाया आंबा

तठे काय बसनी गवराई रंभा

गवराई माय काय काय मांगे

गवराई माय लच्छा मांगे ...'.]

In Khandesh region, married sisters return to their father's home to celebrate the festival 'Aakhaji (or Akshaytritiya) in the month of April. These sisters sing this song on the swing. Gavarai means Goddess Parvati. The idol of Parvati is decorated and worshipped daily by the sisters. This song depicts a situation of Parvati. There is a mango tree (*Mangifera indica* L.) with spreading branches. Goddess Parvati sat on these branches and demanding 'lachchhya' (a silvery chain worn at ankle of a foot).

51. *Pennisetum americanum* L. (Poaceae) Bajara
***Triticum aestivum* L. (Poaceae) Gahu**

"Ghatya re tulhe sangu, male jad nako jau,

Bajarina mise mi dayas valla gahu".

['घट्या रे तुल्हे सांगू, माले जड नको जाऊ,

बाजरीना मिसे मी दयस वल्ला गहू'.]

A lady requests a stone-mill while grinding grains of Bajara (*Pennisetum americanum* L.). She says that in the name of Bajara, I am grinding wet grains of wheat (*Triticum aestium* L.). Please, let me fill easy while grinding them. As a part of preparation of marriage ceremony, maggy is prepared for which wet wheat grains are ground.

52. *Ficus religiosa* L. (Moraceae) Pimpal,Pipay

"Tapi Narmadana mey va may Marmadana may

Tathe kay sonana pipay va may sonana pipay,

Tyale kay chandinya halkadya va may chandinya halkadya

Tyale resamnya dorya va may resamnya dorya

Tahe kay rangit payana va may rangit payana

Tathe kay Mhaisur gadi va may Mhaisur gadi

Mazar basana kanata Hari va may konata Hari

Mazar basana Shankar Hari va may Shankar Hari".

['तापी नर्मदाना मेय व माय नर्मदाना मेय

तठे काय सोनाना पिपय व माय सोनाना पिपय

त्याले काय चांदीन्या हालकड्या व माय चांदीन्या हालकड्या

त्याले रेसमन्या दोऱ्या व माय रेसमन्या दोऱ्या

तठेकाय रंगीत पायना व माय रंगीत पायना

तठे काय म्हैसूर गादी व माय म्हैसूर गादी

मझार बसना कोनता हरी व माय कोनता हरी

मझार बसना शंकर हरी व माय शंकर हरी'.]

This song is sung by young married ladies after returning to their maternal home for 'Aakhaji' or 'Akshaytrutiya' festival. The goddess 'Govrai' (*i.e.* Parvati, wife of Lord Shankar) is worshipped on these days of festival. The song is sung especially on the swings. The song informs that there is a golden 'Pipay' or 'Pipal' tree (*Ficus religiosa* L.) growing near the confluence of the rivers Tapi and Narmada. It has silvery latches to fasten swing with the help of silky ropes. The coloured cradle is also fastened to these latches on which bed bought from Mysore city is placed. In the centre of the cradle is sitting Shankar Hari.

RIDDLES IN MARATHI LANGUAGE AND AHIRANI DIALECT

1. ***Aegle marmelos* L. (Rutaceae) Bel**
   ***Ocimum tenuiflorum* L. (Lamiaceae) Tulashi**

   "Mahadevala bel, vishnula tulas,

   'xyz' ravanche nav ghyayala mala nahi aalas".

'महादेवाला <u>बेल</u>, विष्णूला तुळस,

'xyz' रावांचे नाव घ्यायला मला नाही आळस'.]

Leaves of bel (*Aegle marmelos* L.) are offered to Lord Mahadeo, where as tulas (*Ocimum tenuiflorum* L.) is dear to Lord Vishnu. A married women says that she is not lazy for calling by her husband's name.

2. *Sesamum orientale* L. (Pedaliaceae) Til
***Saccharum officinarum* L.(Poaceae) Oos**

"<u>Tila</u> sarakha aasava sneh, gula sarakhi aasavi godi,

Ishwar sukhi thevo 'XYZ' chi jodi.

['<u>तिळा</u> सारखा असावा स्नेह, गुळासारखी असावी गोडी,

ईश्वर सुखी ठेवो, 'xyz' ची जोडी'.]

Affection should be warm like the seeds of Sesamum (*Sesamum orientale* L.), sweetness should be like the jaggery (*Saccharum officinarum* L.). God should shower happiness for couple of mine.

3. *Jasminum officinale* L. var.*officinale* (Oleaceae) Jai

"<u>Jai-juichya</u> zadakhali harani gheti visava,

'xyz' che nav ghete aashivad asava.

['<u>जाई-जुईच्या</u> झाडाखाली हरणी घेते विसावा

'xyz' चे नांव घेते आशीर्वाद असावा'.]

Deer rests under the plant of Jai (*Jasminum officinale* L. var. *officinale*). I (a married woman) call by the name of my husband. I should be blessed.

4. *Santalum album* L. (Santalaceae) Chandan

"<u>Chandanachy</u> zadala naginicha vedha

'xyz' chya ranila motyacha chuda.

['<u>चंदनाच्या</u> झाडाला नागिणीचा विळखा,

'xyz' च्या राणीचा मोत्याचा चुडा'.]

The tree of Sandal (*Santalum album* L.) is besieged by a female cobra. I (wife of xyz) wear bangles of pearls.

5. *Nymphea pubescens* Willd. (Nympheaceae) Kamal

"Kamalat ubhi Lakshmi, moravar ubhi Saraswati,

'xyz' che nav ghete bhagyavati".

['कमळात उभी लक्ष्मी, मोरावर उभी सरस्वती,

'xyz' चे नांव घेते त्यांची भाग्यवती'.]

Goddess Laxmi is standing on lotus flower (*Nymphaea Pubescens* Willd.) while goddess Saraswati is riding on a peacock. I, the fortunate wife, call by the name of my husband.

6. *Areca catechu* L. (Arecaceae) Supari

"Supbhar suparya mapu kashi, galyat har vaku kashi,

payat painjan chalu kashi, 'xyz' che nav ghete ya divashi".

['सुपभर सुपाऱ्या मापू कशी, ग1यात हार वाकू कशी,

पायात पैंजण चालू कशी, 'xyz' चे नांव घेते या दिवशी'.]

How I can measure painful arecanuts (*Areca catechu* L.). How I can bend down while wearing a necklace around my neck. How I can walk wearing a 'painjan' (silver ornament for ankles). I call by the name of my husband on this auspicious day.

7. *Elettaria cardamomum* Maton. (Zingiberaceae) Weldoda

Syzygium aromaticum L. (Myrtaceae) Lavang

"Lavangine hote usnata, weldodyane jate sardi,

'xyz' che nav ghete soda aata gardi".

['लवंगीने होते उष्णता, वेलदोड्याने जाते सर्दी

'xyz' चे नांव घेते, सोडा आता गर्दी'.]

Clove (*Syzygium aromaticum* L.) raises heat in my body, Cardamom (*Elettraria cardomomum* Maton) cures cold. I call by the name of my husband, now let me lead my way through the crowd.

8. *Ocimum tenuiflorum* L. (Lamiaceae) Tulshi

"Tulashila ghalate pani, Vishnuchi karate shanti,

'xyz' sukhi asavet hich mazi vinanti".

['तुळशीला घालते पाणी, विष्णूची करते शांती,

'xyz' सुखी असावेत हिच माझी विनंती'.]

I water holy basil (*Ocimum tenuiflorum* L.). I thus please God Vishnu. My husband should be happy, that is the only request.

9. *Rosa indica* L. (Rosaceae) Deshi Gulab

"Gulabachye phul dandila hirva

'xyz' ravanche nave ghete mala nani garva"

['गुलाबाचे फुल दांडीला हिरवं,

'xyz' रावांचे नाव घेते मला नाही गर्व'.]

The stalk of rose flower (*Rosa indica* L.) is green. I call by the name of my husband as I am not proud.

10. *Curcuma longa* L. (Zingiberaceae) Halad

"Halad lavati thodi, kunku lavate kinchit,

'xyz' ravansarakhe pati milale,

he maze purvajanmache sanchit".

['हळद लावते थोडी, कुंकू लावते किंचित,

'xyz' रावांसारखे पती मिळाले, हे माझे पूर्वजन्माचे संचित'.]

I apply turmeric powder (*Curcuma longa* L.) and red powder on my forehead. I got husband like xyz. This is a merit of my past life.

11. *Santalum album* L. (Santalaceae) Chandan
***Crocus sativa* Linn. (Iridaceae) Keshar**
***Acacia catechu* (L.*f.*) Willd. (Mimosaceae) Khair**
***Areca catechu* L. (Arecaceae) Supari**
***Myristica fragrans* Houtt. (Myrtaceae) Jaiphal**
***Syzygium aromaticum* L. (Myrtaceae) Lawang**
***Elettaria cinamomum* L. (Zingiberaceae) Jaipatri**

Piper betle **L. (Piperaceae) Pan, Nagin Pan**
Mangifera indica **L. (Anacardiaceae) Amba**
Musa paradisiaca **L.(Musaceae) Kela**
Elettaria cardmomum **Maton. (Zingiberaceae) Weldoda**

"Sonyacha ghangal, Ganga-Bhagirathiche pani aangholila,

Jarikathi dhotarjode nesayala, chandanache pat basayala,

Sahan gevalechi, khod Badodyache, Kesari gandh lawayala

Patra sonyache, pakwanya bhojanche, tarhe tarhechya koshimbirya,

Udbattichi zade, morachya samaya, tabak Dilliche, daba daginyancha,

Adkitta Dharwadcha, chuna Bhokarcha, Katha mumbai shaharacha,

Chikan supari kokanchi, jaypatri Tajapurchi, Lawang Sataryachi,

Weldoda Brindawan, jayaphal killyache, pan Dhulyach,

Alap danila yeto tar, gulabdanila jado bhar,

Gadya Satanichya, ushya Mathurachya,

Parade bhardari, bhardari sarakhe lod, bolan kiti amrutpeksha god,

Kale pat kop amba, najuk kelicha gabha,

Ubhi hoti suwasinichya malyat, hasu yei mala,

Hasu mazya galat, dir-bhavacha thatmat,

Rangmahalachi kay hawa, dir-bhawacha thatmat,

Dari chirbandya aad, var lawali jayphalache zad,

Kalpana kiti bara hazar, hajarat padalya paika,

'xyz' ravanche nave laksha deyun aika".

['सोन्याचा घंगाळ, गंगाभागीरथीचे पाणी अंघोळीला,

जरीकाठी धोतरजोडे नेसायला, चंदनाचे पाट बसायला,

सहान गेवलेची, खोड बडोद्याचे, केसरी गंध लावायला,

पात्र सोन्याचे, पक्वान्न भोजनाचे, तऱ्हेतऱ्हेच्या कोशिंबिऱ्या,

उदबत्तीची झाडे, मोराच्या समया, तबक दिल्लीचे, डबा दागिन्यांचा,

अडकित्ता धारवडचा, चुना भोकरचा, काथ मुंबई शहराचा,

चिकणसुपारी कोकणची, जायपत्री तेजापूरची, लवंग साताऱ्याची,

वेलदोडा ब्रिंदावन, जायफळ किल्याचे, पान धुळ्याचे,

अलाप दानीला येता तार, गुलाबदाणीचा जातो भार,

गाद्या साटणीच्या, उशा मथुराच्या,

पडदे भरदारी, भरदारीसारखे लोड, बोलणं किती अमृतापेक्षा गोड,

काळे पाट कोप आंबा, नाजूक केळीचा गाभा,

उभी होती सुवासिनींच्या मळ्यात, हसू येई मला,

हसू माझ्या गालात, उभी होती रंगमहालात,

रंगमहालाची काय हवा, दीर-भावांचा थाटमाट,

दारी चिरबंद्या आड, वर लावले जायफळचे झाड,

कल्पना किती बारा हजार, हजारात पडल्या पैका

'xyz' रावांचे नाव लक्ष देऊन ऐका'.]

This riddle includes many plant and other useful things to welcome husband. The list is, however, exhaustive. The plants or plant products and other things are especially brought from the cities or places of their repute. It is as follows:

Golden vessel (used to hold water for bath), consecreted water brought from the great river Ganga and Baghirathi for bath, dhoti bordered with gold or silver thread-work to wear, seat for sandalwood to sit, a levigating slab brought from Gevale, stem (of Sandalwood) from Badoda city (of Gujarat), Saffron scent to apply, golden pan and sweetmeats for dinner, different types of salad, scented sticks, a peacock-shaped lamp-stand, a plate from Delhi, a jewel case, nut-cutter from Dharwad city (of Karnataka), lime from Bhokar, Catechu from Mumbai city, areca-nut from Konkan region, leaf of nutmeg from Killa, bettle leaf of Dhule city, alap-dani, rose water sprinker, beds from Satani, Pillows from Mathura (U.P.), beautiful curtains and long bolstar, speak sweeter than 'Amrit', black-green mangoes, delicate pulp of banana fruits, a plant of nut-meg planted near a well etc.

12. *Cucurbita maxima* Duch. ex Lam. (Cucurbitaceae) Chakri

"Mahadevani pindvar chakrini phod,

'xyz' bolata god, pand doya vatari dekhani khod".

['महादेवनी पिंडवर चकरीने फोड,

'xyz' बोलता गोड, पण डोया वटारी देखानी खोड'.]

On the idol of Lord Mahadeva, a part of fruit of red pumkin (*Cucurbita maxima* Duch. ex Lam.) is placed. My huaband is sweet tongued, but he has the habit of enlarging eyes.

PUZZLES IN AHIRANI DIALECT

1. *Allium cepa* L. (Liliaceae) Kanda

"Evadhasa ladu tyale dakhi dakhi ladu

Answer: Onion bulb

['एवढासा लाडू त्याले दखी दखी लडू.

उत्तर : कांदा']

When onion bulb (*Allium cepa* L.) is cut into pieces or sliced, tears appear in the eyes as, if the man is weeping.

2. *Zea mays* L. (Poaceae) Maka

"Battis lugada kese ughada

Answer: Maka Kanis

['बत्तीस लुगडा केसे उघडा

उत्तर : मका कणिस']

The cobe of maize is wrapped by many large bracts, which are compared with 'lugde' *i.e.* Sari (Indian wear of lady). The very long styles are compared with the long hair of Indian women.

3. ***Saccharum officinarum*** **L. (Poaceae) Oos**

"Uchchatad she, deakhama lal she, karama god she.

Answer: Sugar cane.

['उच्चाताड शे, दखामा लाल शे, खावामा गोड शे.

उत्तर : ऊस']

It has great height, it is red to look at and sweet in taste.

4. ***Cicer arientinum*** **L. (Papilionaceae) Harbara**

"Evadhasa zad tyale battis ghunguru.

Answer: Harbara.

['एवढस् झाड त्याले बत्तीस घुंगरु.

उत्तर : हरबरा']

The very small plant has many jingling bells. The dried pods cause singing noise.

5. ***Vigna mungo*** **(L.) Hepper (Papilionaceae) Udid**

"Mana mama gab gaya, gharma udid pasari gaya.

Answer: House-fly.

['मना मामा गाव गया, घरमा उडीत पसरी गया.

उत्तर : घरमाशी.']

My maternal uncle while going outside spread the grains of black gram [*Vigna mungo* (L.) Hepper] on the floor of his home. This is a comparison to house-flies.

6. ***Capsicum annuum*** **L. (Solanaceae) Mirchi**

"Lal pishvi chippat dana.

Answer: Red pepper.

['लाल पिशवी चिप्पट दाना

उत्तर : लाल मिरची']

Red bag contains flat grains.

7. *Syzygium aromaticum* L. (Myrtaceae) Lawang

"Hajar levu, bajar jau, char shingesna bail levu.

Answer: Clove.

['हजार लेवू, बाजार जावू, चार शिंगेसना बैल लेवू.

उत्तर : लवंग']

I will carry with me one thousand rupees and go to market to purchase four-horned bullock.

8. *Cocos nucifera* L. (Arecaceae) Naral, Naray

"Zadavar Zira.

Answer: A spring of water.

['झाडावर झिरा.

उत्तर : नारळ']

A spring of water. Green fruits of coconut contains ample of water which is thought as a spring of water at the top of the coconut tree.

9. *Areca catechu* L. (Arecaceae) Supari

"Agad pheku, dagad pheku, dagad khali goti,

Manha aanna jiki ti sahebani beti.

Answer: Areca-nut.

['आगड फेकू दगड फेकू दगड खाली गोटी,

मन्ह आन्ह जीकी ती साहेबनी बेटी.

उत्तर : सुपारी']

I will throw here and there, you will find a marble below the stone. If you guessed and answered for my puzzle then you will be the daughter of an officer.

10. *Mangifera indica* L. (Anacardiaceae) Amba

"May aanganama, bap jangalma, beta pangatma.

Answer: Mango fruit.

['माय अंगनमा, बाप जंगलमा, बेटा पंगतमा.

उत्तर : आंबा']

Mother is in the fore-ground of a home, father is found in jungle or fields, but son is present in the rows for meals.

11. *Coriandrum sativum* L. (Apiaceae) Dhane

"Vakad tikad babhuy, sarhandi fok,

Sita chalani maher, tina navatina zok.

Answer: Coriander.

['वाकड तिकड बाभूय, सरहंडी फोक,

सिता चालनी माहेर, तीना नवतीना झोक.

उत्तर : कोथंबीर']

Plants of coriander (*Coriandrum sativum* L.) are always bent just like the babul tree (*Acacia nilotica* (L.) Willd. ex Del. subsp. *indica* (Bth.) Brenan. Coriander is always sent to the houses when they are in full youth (for its culinary use). This is compared with the departure of goddess Sita (of Ramayana) for her maternal home.

12. *Saccharum officinarum* L. (Poaceae) Oos

"Evdhas khod, tyanha pani kitla god.

Answer: Sugarcane.

['एवढस् खोड, त्यान्ह पानी कितलं गोड.

उत्तर : ऊस']

There is a small stem or culm, but it contains very sweet water. It is sugarcane (*Saccharum officinarum* L.).

13. *Allium cepa* L. (Liliaceae) Kanda

"Evdhasa gadu, bhuima dadu.

Answer: Onion bulb.

['एवढासा गडू, भूईमा दडू.

उत्तर : कांदा']

A small round drinking vessel is burried.

14. *Cocos nucifera* L. (Arecaceae) Naral

"Evdhasa kutara, devvar mutara.

Answer: Coconut.

['एवढासा कुतरा, देववर मुतरा.

उत्तर : नारळ']

A coconut fruit is split open and its milky water is showered on the idols of gods. This practice is compared with a small dog urinating on the idol of gods.

15. *Daucus carota* L.(Apiaceae) Gajar

"Kaya vavarama thoki mekh, Upadin to rajana lek.

Answer: Carrot.

['काया वावरमा ठोकी मेख, उपाडीन तो राजाना लेक.

उत्तर : गाजर']

There is small peg in the soil of black fields. Those who will root out it, will be very lucky.

16. *Solanum melongena* L. (Solanaceae) Wange

"Kaya kot, hirva uparana, Khale vakana, dakha dhupat}na.

Answer: Brinjal.

['काया कोट, हिरवं उपरनं, खाले वाकन दख धुपाटन.

उत्तर : वांगे']

It has wore black coat (Black pericarp) and a loose green upper garment (Calyx cover). If bent, looks like a censer.

17. *Annona squamosa* L. (Annonaceae) Sitaphal

"Kaye had, dhavya mas,

Jo nahi vayakhe, tyanha satyanas.

Answer: Custard Apple.

['काये हाड, धव्य मास,

जो नही वयखे, त्यान्हा सत्यानास.

उत्तर : सिताफळ']

Black seeds are thought as bones, white pulp is considereed as flesh, one who will not identify this he will be unlucky.

18. *Allium sativum* L. (Liliaceae) Lasun

"Kokanithin uni thaki, mi mari tilhe bukki,

Tinhi gharbhar janyat leki.

Answer: Lasun.

['कोकनीथीन उनी ठकी, मी मारी तिल्हे बुक्की

तिन्ही घरभर जन्यात लेकी.

उत्तर : लसून']

A roughue arrived from Konkan region, she is beaten by a first. She gave birth to many daughters. A garlic is always composed of small bulblets. They are separated by a fist.

19. *Vitis vinifera* (L.) Walp. (Vitaceae) Draksha

"Kachani gotima mungana dana,

Manha aanha jikin to chatur shahana.

Answer: Grapes.

['काचनी गोटीमा मुंगना दाना,

मन्ह आन्ह जिकीन तो चतुर शहाना.

उत्तर : द्राक्ष']

The grapes are compared with marbles containing green seeds like green gram [*Vigna ungiculata* (L.) Walp.] who identifies this, he will be very clever.

20. *Musa paradisiaca* L. (Musaceae) Kela

"Kho kho khokali, man vakali,

Chauda pilasni may ekali.

Answer: Banana.

['खो खो खोकली, मान वाकली,

चौदापलासनी माय एकली.

उत्तर : केळी']

The banana plant is regarded as a mother of many cubs.

21. *Cocos nucifera* L. (Arecaceae) Naral, Naray

"God god gupi, lakudni kupi

Kay thyanha manmha, sange to devpa.

Answer: Coconut fruit.

['गोड गोड गुपी, लाकुडनी कुपी,

काय त्यान्ह मनम्हा, सांगे तो देवपा.

उत्तर : नारळ']

The fruit of coconut (*Cocos nucifera* L.) contains milky sweet water secretly. It is contained in a wooden phial (hard endocarp). This secret is revealed only to the god.

22. *Colocasia esculenta* (L.) Schott. (Araceae) Aluchepan

"Ghudga evdha hatti, tyanha supda evdha kan,

Jik manha aanha, ka upadu tunha kan.

Answer: Arum leaves.

['गुढगा एवढा हात्ती, त्यान्हा सुपडा एवढा कान,

जिक मन्ह आन्हं, का उपाडू तुन्हा कान.

उत्तर : आळूची (पोथीचे) पान']

The small elephant with ear pinnae of a size of winnowing pan is compared with this plant. If you do not identify this, I will pull your ears.

23. *Capsicum annuum* L. (Solanaceae) Mirchi

"Tarun panma zingi zingi, dhakka panama lavu kuku,

Amalnerna bepari sange, khushal sheka kaku.

Answer: Lal mirachi.

['तरुन पनमा झिंगी झिंगी, धक्का पनमा लावा कुकू,

अमळनेरना बेपारी सांगे, खुशाल शेका काकू.

उत्तर : लाल मिरची']

In the prime youth, she was in air. But at old age she married and applied red powder on her forehead. Therefore, the merchant of Amalner city called her 'kaku' (paternal aunt).

24. *Cicer arientinum* L. (Papilionaceae) Harbara

"Don bhau tyasle ekach bandi,

Answer: Gram, Chick pea.

['दोन भाऊ त्यासले एकच बंडी.

उत्तर : हरबरा']

Two uterine brothers wearing a single underwear are compared two green seeds in a pod.

25. *Raphanus sativus*, L. (Brassicaceae) Mula

"Dhawi batali, tyale hirva buchchan.

Answer: Radish.

['धवी बाटली, त्याले हिरवं बुच्चन.

उत्तर : मुळा']

The white root is called white bottle and the green foliage is compared with its stopper.

26. *Zea mays* L. (Poaceae) Maka

"Patil buwa thamb thamb, dadhi misha lamb lamb.

Answer: Ear of maize.

['पाटील बुवा थांब थांब, दाढी मिशा लांब लांब.

उत्तर : मक्याचे कणीस']

The ear of maize has long styles. These are compared with the long beard and mustache of a Patil (head-man of a village).

PROVERBS IN AHIRANI DIALECT

1. *Ricinus communis* L. (Euphorbiaceae) Arand
***Saccharum officinarum* L. (Poaceae) Oos**

"Oos na mise, Arandle pani.

['ऊस ना मिसे एरंडले पानी'.]

While watering sugarcane crop (*Saccharum officinarum* L.), castor plant (*Ricinus communis* L.) is also automatically watered. While completing one purpose, another one is automatically fulfilled.

2. *Saccharum officinarum* L. (Poaceae) Oos

"Oos god lagna mhanisan muisakat khau nahi.

['ऊस गोड लागना म्हणीसन मुईसकट खाऊ नही'.]

One should not eat the sugarcane plant (*Saccharum officinarum* L.) alongwith roots although it is sweet. One should not take undue privilege of a good person.

3. *Ziziphus mauritiana* Lamk. (Rhamnaceae) Bor

Acacia nilotica (L.) Willd. ex Del. subsp.*indica* (Btn.) Brenan. (Mimosaceae) Babhul

"Jathalya bori, thathani bhabhai.

['जठल्या बोरी, तठल्या बाभई'.]

The jujub plant (*Ziziphus mauritiana* Lamk.) and Babul plant [*Acacia nilotica* (L.) Willd.] ex Del. subsp. *indica* (Bth.) Brenan. Two persons from the same region or locality know each other better.

4. *Butea monosperma* (Lamk.) Taub (Papilionaceae) Palas

"Palasale pane tinach.

['पयसले पाने तीनच'.]

Every payas or Palas tree [*Butea monosperma* (Lamk.) Taub.] bears three leaflets. Everywhere circumstances are similar.

5. *Sorghum bicolor* (L.) Moench. (Poaceae) Jawar

"Gharma nahi jwarina dana ni male Bajirao mhana.

['घरमा नही जवारीना दाना नी माले बाजीराव म्हना'.]

There are no grains of Jawar [*Sorghum bicolor* (L.) Moench.] in home, but boasts like a great man Bajirao.

6. *Daucus carota* L. (Apiaceae) Gajar

"Gajarni pungi vajani te vajani, nahi te modisan khadi.

['गाजरनी पुंगी वाजनी ते वाजनी, नही ते मोडीसन खादी'.]

A musical instrument called 'pungi' is carved from the carrot (*Daucus carota* L.) and played. If it works properly, it can be used so, otherwise it can be consumed also. Something can beused with alternative purposes.

7. *Daucus carota* L. (Apiaceae) Gajar
***Raphanus sativus* L.(Apiaceae) Mula**

"Gajarbhar aatha ni muyabhar tatha.

['गाजरभर आथा नी मुयाभर तथा'.]

To tell something roughly.

8. *Cucumis melo* L. var.*utilissimus* Roxb. (Cucurbitaceae) Kakadi

"Nau hat kakadi an daha hat bee.

['नऊ हात काकडी अन दहा हात बी'.]

To tell something without proportion.

9. *Momordica charantia* L.(Cucurbitaceae) Karle

"Kadhu karl tupma ghoya ka sakharma ghoya tari te kadhuch.

['कडू कारल तुपना घोय का साखरमा घोय, तरी ते कडूच'.]

Bitter fruits of Karala (*Momordica charantia* L.) although fried in ghee and sugar, cannot be changed inspite of the best efforts.

10. *Ficus benghalensis* L. (Moraceae) Vad
***Ficus religiosa* L. (Moraceae) Pimpal**

"Vadna pan pimppayale.

['वडना पान पिंपयले'.]

To show connection of banyan (*Ficus benghalensis* L.). leaves with the pipal tree (*Ficus religiosa* L.). It means to explain something without relevant references.

Section III
Vernacular Plant Names

Vernacular Plant Names in Ethnobotanical Perspective

Names Based on Animals

1. *Arisaema murrayi* (Grah.) Hook. *f.* (Araceae) 'Nagphani' (Nag–a cobra; phani-a hood). The inflorescence and the spathe simulate the hood of a cobra.
2. *Uraria picta* (Jacq.) Desv. ex DC. (Papilionaceae): Wandar-sheptyo' (Wandar–a monkey; sheptyo-tailed). The conspicuously bracteate inflorescence resembles the tail of a monkey.
3. *Clitoria ternatea* L. (Papilionaceae):'Gokarn' (Go–a cow; karn-pinnae of an ear). The flower is shaped like the ear pinnae of a cow.

4. *Martynia annua* L. (Martiniaceae): 'Wagh-nakhya (Wagh– a tiger; nakhya-bearing nails). The mature fruits are pronged or hooked like the nails of a tiger.
5. *Helicteres isora* L. (Sterculiaceae): 'Murudsheng' (It is a corruption of 'Mrigashinga'; mriga-a deer; shing-a horn) In allusion to the twisted fruits like deer's horns.
6. *Dolichandrone falcata* (Wall. ex DC.) Seems. (Bignoniaceae): Medhshingi, Medhshing (Mendha, Medha-a ram; shing-horn). The fruits are shaped like the horns of a ram.
7. *Tribulus terrestris* L. (Zygophyllaceae): Gokharu (Go-a cow; kharu-a word torsioned from khur, meaning a hoof). The clefted fruit resembles the hoof of a cow.
8. *Pergularia daemia* (Forssk.) Chiov. (Asclepiadaceae): Popatchonchi (Popat-a parrot; chonch-a beak). The pointed fruit is similar to the beak of a parrot.

Names Based on Colours

1. *Mallotus philippensis* (Lam.) Muell.–Arg. (Euphorbiaceae): 'Shendrya (Shendrya-of red lead). Fruits are referred as they are covered with bright red resinous powder and stellate hairs.
2. *Polycarpaea corymbosa* (L.) Linn. (Caryophyllaceae). 'Ujalifuli', 'Pandharifuli' (Ujal-brighter, shining, Pandhari-white; fuli-flowered). White, shining colour of the flowers is denoted.
3. *Diplocyclos palmatus* (L.) Jeffrey (Cucurbitaceae): Kabrodowo; (Kabro-variegated, stripped; dowo-eyes). The eye-shaped fruits are striped white, hence the name.
4. *Chlorophytum borivilianum* Santapau and Fernandes (Liliaceae): 'Safed-musali' (Safed-white; musali-a pestle). The white, pestle-shaped root tubers find place in coining the name.
5. *Haldina cordifolia* (Roxb.) Ridsdale (Rubiaceae): 'Haldu' (Haldu-turmeric coloured). The yellowish colour of the wood is the basis.
6. *Soymida febrifuga* A. Juss. (Meliaceae): Rangatroda (Rangat-blood; roda-stem). The wood is reddish.

Names Based on Shapes or Appearances

1. *Diplocyclos palmatus* (L.) Jeffrey (Cucurbitaceae). 'Shivlingi'(Shivling-Lord Shivas peculiar idol). The seeds are shaped like the idol of Lord Shiva.
2. *Cyperus rotundus* L. (Cyperaceae): 'Gundyo' (Gundyo-buttonlike). The shape of the root tubers is emphasized.
3. *Morinda pubescens* J.E. Sm. (Rubiaceae): Bartondi (Bar-twelve; tond-mouth). The fluits at maturity are developed into a syncarp. Many marks (about twelve) of the style base after their abscission are left on it which are considered as the mouths of this fruit.

Names After Taste

1. *Oxalis corniculata* L. (Oxalidaceae): 'Ambush' (Ambus-sour, acid). Leaves if chewed, taste sour.
2. *Coccinia grandis* (L.) Voight (Cucurbitaceae): 'Khatgole'(Khat-sour, acid; gole-rounded-elongated body). The ripened fruits are sour and round to elongate.
3. *Canthium parviflorum* Lamk. (Rubiaceae): Ambat (Ambat-sour). The ripe fruit is sour.

Names Based on Latex

1. *Hemidesmus indicus* (L.) Schultes (Asclepiadaceae): 'Dudhwel'(Dudh-milk; wel-a climber). The climber yields white or milky latex.
2. *Euphorbia hirta L., E.geniculata Orteg., E. cyathophora* Murray (Euphorbiaceae): Dudh or Dudhali (Dudh-; dudhali-milk yielder). The white latex is the reason for coining these names.

Names After Miscellaneous Plant Features

1. *Calotropis procera* (Ait.) R.Br. and *C. gigantea* (L.) R. Br.(Asclepiadaceae): Rui (Rui-cotton). The white coma of the seeds is compared with cotton fibres.
2. *Prosopis cineraria* (L.) Druce (Mimosaceae): 'Shemdi' (Shemdi-mucus, viscous). The fruit pulp is viscous in nature.

Names Based on Inconvenicaes

1. *Hygrophila schull* (Buch.-Ham.) M.R. & S.M. Almeida (Acanthaceae): 'Agnikatyo' (Agni-fire, hot, pungent; Katyo-spiny). The pungent sharp spines find place in the narne.
2. *Trichodesma indica* (L.) R. Br.(Boraginaceae): 'Agyokhod' (Agyo-fire, pungent; khod-stem). The characteristic roughness due to hairs of the stem is denoted.
3. *Mucuna pruriens* (L.) DC. (Papilionaceae): 'Khaj-kuiri' (Khaj-itching; kuiri-small "S" shaped box). The fruits are shaped as the'Kuiri'and studded with the stinging hairs which cause itching when touched. This effect is noted in the narne.

Names After Medicinal Usage

1. *Tridax procumbens* L. (Asteraceae): 'Jakham-judi', 'Ghavpala' (Jakham-injury;judi-healer; Ghav-injury; pala-leaves). Leaf juice is applied for healing injuries, cuts or bruises hence the name.
2. *Cayratia trifolia* (L.) Domin (Vitaccae): Bailmal (Bail-a bullock; mal-a garland). The climber is tied around the neck or horns of the fran, bullock, hence the name.
3. *Gardenia gummifera* L.*f.* (Rubiaceae): Kidamari (Kida-worms, germs; mari-killer, destroyer). The gum obtained from this plant is used against intestinal worms or applied for skin diseases.
4. *Blepharis repens* (Vahl) Roth (Acanthaceae): Hadsan (Had-bone; san-abbreviated form of the word 'sandhane', meaning to join). The plant is locally employed to cure bone fracture.
5. *Helicteres isora* L. (Sterculiaceae): Muradsheng (Murad, Murda-intestinal torsion or troubles caused during dysentery; sheng-capsular fruit). During the period of dysentery the intestine is torsioned and serious stomach troubles and pains are experienced by the patient. To treat this condition the fruit paste is taken orally especially for children.

Names After Similar Uses

1. *Pueraria tuberosa* (Roxb. ex Willd.) DC. (Papilionaceae): 'Pithuda' (Pith-flour). The root tubers yield flour which is used as famine food by the aborigines to make bread.
2. *Amaranthus spinosus* L. (Amaranthaceae): 'Katalo-matlo' (Katalo-spiny; *matalo-Amaranthus viridis* L.). The leaves of both species are used as a vegetable, only *A. spinosus* bears spines.

Names After Habitat or Distribution

1. *Ipomoea carnea* Jacq. subsp. *fistulosa* (Mart. ex Choisy) Austin (Convolvulaceae): 'Besharmi' (Besharam-shameless). The name is based on the ability of the species to grow and flourish in any habitat.
2. *Ziziphus glaberrima* (Edgew.) Sant. (Rhamnaceae): 'Ghatbor' (Ghat-hilly region; bor-jujub plant). The occurrence of the species is pointed out in the hilly region.

Names After Habit

1. *Indigofera cordifolia* Heyne ex Roth (Papilionaceae): 'Godhadi' (Godhadi-a quilt). The growing of the plant in dense masses resembles a quilt.
2. *Ziziphus rugosa* Lam. (Rhamnaceae): Toran (Toran–a wreath of leaves hung at entrance or doors). The spreading habit is compared with a wreath of leaves worn on entrances or doors as a decoration or on some special ceremonies.
3. *Solanum virginianum* L. (Solanaceae): 'Bhui-ringani' or 'Bui-ringano' (Bhui-the earth; ringani or ringano-circular spread). The plant trails on the land assuming near about circular shape or appearance.
4. *Millingtonia hortensis* L. *f.* (Bignoniaceae): 'Akash-neem' (Akash-the sky;. *neem-Azadirachta indica* A. Juss.). The tallness of the plant and similarity in foliage with those of Neem plants are recalled in the name.

Name Based on Ecological Condition

1. *Cuscuta chinensis* Lam. (Cuscutaceae): 'Amarwel' (Amar-immortal or eternal; wel-a climber). This climber parasitaze and dhrives best on any host plant for indenfinite period.

Name Based on Phenological Period

1. *Catharanthus roseus* (L.) G. Don (Apocynaceae): 'Sadafuli' (Sada-always, in any season., fuli-bearing flowers). It blooms throughout the year.

Name Based on Plant Response

1. *Biophytum sensitivum* (L.) DC. (Oxalidaceae): 'Lajalu', 'Laaro', 'Saram', 'Sharmi' (All words mean shyness). Closing of leaves on touch is alluded as its shyness.

Names Indicative of Exotic Nature or Country of Origin

1. *Portulaca pilosa* L. subsp. *grandiflora (Hook.)* Greesink (Portulacaceae): 'Chini-gulab' (Chini-of China; gulab-the rose). The plant is a denizen of China and its flowers simulate roses.
2. *Parkinsonia aculeata* L. (Caesalpiniaceae): 'Vilayati-bhabul' (Vilayati-belonging to a foreign country; *babhul-Acacia nilotica* (L.) Willd. ex Del.). It is exotic plant belonging to tropical America and is thorny like the. Babhul plant. These two features are coined in the name.

Name After Method of Propagation

1. *Kalanchoe pinnata* (Lam.) Pers. (Crassulaceae): Ugpan', 'Panphuti' (Lagane or phutane-to sprout out or to grow; pan-leaf). The propagules developed on leaf margin as a means of propagation is incorporated in the name.

Names Suggesting Relationships

1. *Vigna trilobata* (L.) Verd. (Papilionaceae): Ran-mugyo' (Ran-wild; mugyo-*Vigna radiata (L.)* R. Wilczek. var. *radiata).* Either of the species have similar pods and foliage, hence thought wild relative of the cultivated one.

2. *Orthosiphon pallidus* Benth. (Lamiaceae):'Jangli-tulsi (Jangli-wild; *tulsi-Ocimum tenuiflorum* L.). The similarities in foliage and inflorescence render the former as the wild relative of the latter cultivated plant.
3. *Clitoria biflora* Dalz. (Papilionaceae): 'Jangli-tur' (Jangli.wild; tur-pigeon pea, *Cajanus cajan (L.)* Millsp.). The similarity in seeds render the tribals to think it as wild relative of the pigeon pea.
4. *Crotalaria medicaginea* Lam. (Papilionaceae): 'Ran-methi' (Ran-wild; *methi-Trigonelia foenum- graecum* L.). On account of similar, 3-foliolate leaves in both plants, aborigines think it wild relative of the Methi, a vegetable.
5. *Cassia absus* L. (Caesalpiniaceae): 'Tarhota-na- mamal ('Tarhota'–local name for *Cassia tora L.* and *C. obtsifolia* L.; mama-matemal uncle). These taxa are similar to each other w.r.t. leaves, flowers, pods, habit etc. The names suggest their close relationships.
6. *Portulaca quadrifida* L. (Portulacaceae): 'Barik-ghol' (Barik-thin, narrower; 'ghol'–local name for *Portulaca oleracea* L.). These two plants are similar, thiner in case of *P. quadrifida* . This renders them in the eyes of aborigines, related to each other.
7. *Ensete superbum* (Roxb.) Cheesm. (Musaceae): Rankel (Ran-wild, jungle; kel-babana). The fruits, leaves and the overall habit clearly recall the cultivated banana plant *i.e. Musa paradisiaca L.* Hence, it is thought wild relative of cultivated banana plant.

Names After Gods

1. *lpomoea quamoclit* L. (Convolvulaceae),: 'Ganeshpushpa', 'Ganeshwel' (Ganesh-Lord Ganesh, the Hindus god; pushpa-a flower; wel-a climber). The flowers of the climber is an offering to Lord Ganesh, hence the name.
2. *Passiflora edulis* Sims. (Passifloraceae): Krishnakamal' (Krishna-Lord Krishna; Kamal-the Lotus, here it is meant as a flower). The bluish colour of the flower is similar to the colour of Lord Krishna, hence the name.

3. *Cassia fistula* L. (Caesalpiniaceae): Ramdanda (Ram-God Rama or panacea; danda-a stick). The fluits am sticklike. The fruit pulp is an excellent and safe purgative for children. Due to deep faith in its medicinal efficacy it is called Ram considering it a panacea.
4. *Tricholepis indica* (Willd.) Almeida (Asteraceae): Bramhadandi (Bramha–God Bramha, the creator of universe according to Hindu mythology; dandi–a stick). The stem alongwith the leaves and flowers is used as a nutritive tonic and conceived to lengthen one's life span, hence the name.

Names Based on Allusion or Belief

1. *Acacia nilotica* (L.) Wilid. ex Del. subsp. *cupressiformis* (J.L.Stewart) Ali and Faruq (Mimosaceae): 'Ramkathi' (Ram-God Rama of Hindus; kathi-a stick). The erect, tall and cupressiformis habit is idealised as a walking stick of God Rama.
2. *Cassine albens* (Retz.) Kostern. (Celastraceae): Bhutkes (Bhut-ghost; kes-hair). The pendulous branches are compared to the hair of a ghost.

Name Based on Oppositive/Contradictory Character

1. *Tinospora cordifolia* (Willd.) Miers ex Hook. (Menispermaceae): 'Gulwel' (Gul-jaggery, here it is meant sweetness; wel-a climber). Decoction of the stem is used as a general body tonic which is very bitter. However, instead of bitterness, sweetness is given the place ironically in coining the name.

Scientific And Vernacular Names Based on Similar Concepts or Characteristics:

1. *Oxalis corniculata* L. (Oxalidaceae): i) 'Ambushi' (Ambus-sour, acid). Leaves are sour. ii) *Oxalis:* (Greek) oxys-acid; referring to the acid taste of the leaves.
2. *Uraria picta* (Jacq.) Desv. ex DC. (Papilionaceae)

   (*i*) 'Wandar-sheptyo'(Wandar-the monkey; sheptyo-tailed). The conspicuously bracteate inflorescence resembles the tail of a monkey.

(ii) *Uraria:* (Greek) oura-a tail; referred as above.

3. *Mucuna pruriens* (L). DC. (Papilionaceae):

   (*i*) 'Khaj-kuiri'(Khaj-itching; kuiri-small'S'shaped box, here referred to the fruits). The hairs on the fruits cause itching if touched.

   (*ii*) (Latin) pruriens: itching; based on the same reference as above.

4. *Rumex acetosa* L. (Polygonaccae):

   (*i*) 'Ambatchuka' (Ambat-sour, acid; chuka-sorrel *i.e. Rumex acetosa)*. The leafy vegetable has sour taste.

   (*ii*) (a) *Rumex:* (Latin) runex-sorrel (*i.e. R. acetosa).*

   (b) *Acetosa:* sour, acetous or acetose; the same feature or taste is denoted. (However, new name is *R. vesicarius L.).*

Name After Celestial Body

(1) *Jatropha gossypifolia* L. (Euphorbiaceae): 'Chandrajyot' (Chandra-to gladden, pleasing; jyot-a flame). The young leaves or plant tops are reddish and shining, even single glimpse renders one delighted, hence the name.

Names After Numericals

1. *Phyllanthus airy-shawii* Burnel ex Roux. (Euphorbiaceae): Hajardani (Hajar-thousand; dani-grainlike). The fruits resemble grains in appearance and size. The plant bears grainlike fruits in considerably large numbers.
2. *Tridax procumbens* L. (Asteraceae): Ekdandi (Ek-one; dandi-a stick, a peduncle). Each branch terminate in single peduncle of a flowering head.

Names in Relation to Earth and Other Fruits

1. *Pueraria tuberosa* (Roxb. ex Willd.) DC. (Papilionaceae): Bhuikohala [(Bhui-the earth, soil; Kohala–the fruit of pumpion gourd *i.e. Benincasa hispida* (Thunb.) Cogn.] The

underground tuber of this plant attains the fruit size of pumpion gourd.

2. *Phyllanthus amarus* Schumach and Thonn. (Euphorbiaceae): Bhuiawala [Bhui-the earth, soil; awala-fruit of emblic *i.e. Emblica officinalis* Gaertn.]. The fruits resemble to those of *Emblica offlicinalis* Gaertn., and are borne towards the earth or soil.

Discussion

This work embodies the results of planned ethnobotanical explorations of botanically a fairly rich region. It is the outcome of an intensive as well as extensive study. It covered all 13 talukas of Jalgaon district (Maharashtra). A care was taken to tap information from every nook and corner of hilly areas and the plains as well. The tribal and the non-tribal rural populace was consulted for this comprehensive attempt. This work accounts for the plant species used by these people for medicine, food, fodder, timber, fuel wood, oil, fibre, alcohol, gum and some other miscellaneous uses. It also includes plants associated with belief, ceremonies, rituals, festivals, births, deaths, local plant names, proverbs, folk songs, puzzles, riddles etc. It was also attempted to study the ethnology of the major populace with respect to the origin (history), customs, religions, occupations, festivals, abodes, dresses, ornaments etc. The geography, geology, climate, soils and the general vegetation of the district was particularly noticed. It was thought that these natural and social facts have bearings directly or indirectly on the total relationships of mankind with the surrounding plant species. These ancient important discoveries continued to survive by oral tradition over many generations in rural, semi-rural and tribal communities as well. With the rapid intrusion of modern civilization into this region, ancient tribal traditions, culture and gay of rural folks seem to be on the verge of extinction. There is every possibility that valuable data of ethnobotany will be lost in the near future. Attempts, therefore, were made to procure information about the different uses of plants, wild or domesticated, in the life or culture of tribal and rural inhabitants of Jalgaon district.

A perusal of the information included in this work reveals that total of 390 taxa belonging to 310 genera 96 families are in use for various purposes. Exotic floral elements are totally 51. Crop weeds number 16 and mostly useful for foods, medicine, fodder etc. It appears pertinent to invite attention of readers and users to an Indian proverb. "There is not single letter of alphabet which is not a mantra, there is not single person who is not useful, there is not single root which is not medicine". The present efforts could, report 400 total species useful in Jalgaon district. The different utilities noted in the region are expatiated in the following.

Tribal and some of the rural people always lead quite a hard life. They have learned over the centuries to draw as much as possible from the surrounding vegetational wealth. Although their attention is centered about agriculture, they make great use of spontaneous plants in general. They have wide knowledge about plants and a strong interaction with them is deeply rooted in their culture. The wild plants, make an important contribution, particularly to the diet of the local inhabitants, apart from other useful products. They also provide genes for crop improvement. The study of wild plants is significant not only to know the potential sources which could be used as alternative food or in times of scarcity but to select promising types for domestications. In the present contribution, total 61 species have been recorded useful for food of the aborigines and rural people (Appendix-I). Tribals are thoroughly acquainted with the methods of excluding the harmful substances from these wild plants and preparing acceptable recipes. The plant parts mostly used are the tubers, corms, leaves, flowers, fruit, seeds etc. The rural people using these weeds as food naturally lower down the pressure of seeking food and help to resolve our food problems to some extent. They make judicious use of crop weeds also. Following methods are very common for preservation of food materials: (1) Cereals are dried in sun and store in metal vessel for whole year. (2) Cereals are preserved by using leaves of *Azadirachta indica* A.Juss. (3) Some vegetables like *Capsicum annum* L. fruits like *Tamarindus indica* L., *Emblica officinalis* Gaertn., *Ziziphus mauritiana* Lamk. are dried in sun by applying salt as preservative. (4) Some vegetables such as *Coriandrum sativum* L., *Trigonella-foenum graecum* L. are only dried in sun and stored for use.

Leaves are generally used in roasted, fried or boiled forms, few are used raw. Some are cooked fresh, while few others are used after drying and whenever they need, they can make use of them. Individual tastes of few of them are reported not appealing and hence often cooked in combination with other leaves or vegetables or pulses. An intensive survey, however, is needed of these supplementary vegetables and this should form an exclusive topic of research.

Phoenix sylvestris is another boon plant species for the local people. Its ripe fruits are edible. Toddy, a nutritious drink, is tapped from the top of tree trunks. Afterwards, it is also used or sold as 'Neera', a fermented drink. It is sold also in villages and urban areas. The foliage is widely used for broom-making.

The calyces of *Hibiscus sabdariffa*, which is cultivated by the tribals as a mixed crops, are dried in sun. They are sour in taste. They are usually added to flavour vegetable.

The tubers of *Dioscorea oppositifolia* L. and *D.bulbifera* L. are sliced and kept in running water in nallahs or rivulets to remove unwanted chemical contents. Sometimes, they are dipped in water for 2-3 days. Afterwards, they are roasted and the outer part is removed. It is then eaten directly.

Mahu trees [*Madhuca longifolia* (Koen.) Macbride] are the dearest of all the tribals (Appendix-X). Apart from medicine and miscellaneous uses, it is a good source of food as well. Green fruits are used as vegetable. Ripe fruits are edible. Seed oil is edible. Cake after expression of oil, is also used as food for them and feed for cattle. Flour mixed with flowers renders the bread tasty and nutritious. Tribals sell seeds are cake and earn money. Thus this plant is a boon to them.

Tubers of *Pueraria tuberosa* are dried in sun, made in small pieces and ground. This flour is used for making bread. However, this is not practiced often, but useful especially during famine period. Rhizomes of *Ensete superbum* is (Roxb.) Cheesm. also used similarly to prepare bread.

The increasing demands for food with growing human population in a country like ours compels search for new plants as source of food. The aboriginals inhabitating the hilly forest areas and the rural people consume many wild species which are unknown and inacceptable to the urban elite. Some plants enlisted in Appendix-I many have high nutritive value. At least, some of them should been screened phytochemically. Some of these greens are weeds collected in the fields or along the paths in the course of agricultural work or on the way to our from the fields.

Weeds are usually thought as a group of plants, very aggressive, competive and noxious to man. This is the common experience. However, quite a good number of species have significant role in the economy of the local people. The crop-weeds are not only useful as food plants but also as a source of medicine and fodder for pet animals. This subject matter is dilated upon at proper place.

As in other parts of our country, Jalgaon district also has a valuable heritage of herbal remedies. Its rural people and particularly the tribals still depend on the indigenous system of medicine. Alongwith modern system of medicine, homeopathy, Ayurveda and Unani systems are practiced in India. Other than these systems, still there exists a traditional folklore which is being perpetuated verbally from person to person and generations to generations. By and large, this lesser known medicinal plantlore in the absence of official patronization could not receive the necessary research support and therefore remained lesser unknown or entirely unknown to the masses.

Of late, the traditional medicines are being accepted all over the world since the modern systems of medicine have many side effects. The quest for herbal sources is growing all over because these products have milder action in various body functions in preference to the synthetic drugs. There seems a green wave of lay of interest in medicinal plantlore all over. This wave is discernible in India too. Non-availability of satisfactory drugs for the treatment of different diseases or ailments in the modern systems of medicine prompted many of us to study ethnomedicinal claims. It s doubtless that ethnomedicine have provided lead in the development of drugs used in the modern systems of medicine, for example, atropine, ephedrine, reserpiine and more recently gugulip and taxol.

The use of ethnomedicine is widespread especially in inaccessible tribal areas (Appendix-X). The high percentage of population depend on it due to lack of health centers and doctors and even the expenses of such treatments. The different plant species employed by the aborigines to cure various human or animal diseases or ailments even in this age of chemo-and radio-therapies. These should be screened on scientific grounds.

It is pretty clear from foregoing account that the vegetational wealth of this district contains a great variety of useful plants which have been valuable source of medicines to the vast masses of tribal and rural areas. They use various plant parts such as roots, stems, other underground parts, leaves, fruits, flowers and plant products like gum and latex. The different methods of administration of drugs noted during this study are: (a) raw plant part or product, (b) juice, (c) extract, (d) decoction, (f) paste, (g) ash, (h) poultice, (i) fumes or vapours, (j) oil, (k) contact drugs to body part, (l) bath by adding drugs in bath water etc. In majority of cases, they are single drug. However, there also cases of using more than one type of drugs. Few tribal medicine men, Bhagat and Buwas mutter some hymns (mantras). In aboriginal communities, diseases are thought caused by the breach of some taboo or by hostile spirits, the ghost or the dead. Sickness is the usual punishment for every lapse and crime. Alongwith the medicine, the tribal medicine men, Bhagat or Buwas mutter some hymns. Although so, they do not disregard the curative properties of various plants and their parts or products. Many times, it was a common experience of the present investigator that the tribal hardly make any difference between simple fever and the fevers on account of other diseases like malaria, typhoid etc. It is also the case of stomach complaints whether the pains are due to dysentery, constipation or such other intestinal disorders. It was, therefore, very careful enquiries were made while diagnosing the diseases or body complaints. The present investigators noted many diseases or ailments, although some of them are commonly prevailing. These are head-ache, ear-ache, fever, cough and cold, asthma, rheumatism, cholera, diarrhoea, dysentery, urinogenital complaints, burns, cuts, wounds, skin diseases, snake bite, scorpion sting, weakness, dental and eye-complaints, jaundice, abortion, birth, problems, bone fracture and many others.

As far as medicinal plants are concerned, our earlier visits were largely in vain. In subsequent visits to the various localities or hamlets, especially the women were attracted. It was mostly the ladies who revealed the secrets or facts which enquired by a lady-investigator (Dr. Mrs. Pawar). Some diseases are not naturally disclosed to the man-investigator. It is why the quality and authenticity of information appears improved. Due care and repeated verifications also brought out quite precise, specific and reliable information of the present account.

The exact method of drug preparation and doses were recorded and all possible efforts were made to diagnose the disease or ailment on the basis of symptoms or expressions.

Apart from the human medicine, it was also possible to gather information on the plants used especially for domestic animals (Appendix-II). Some tribal people have agricultural land holdings. They practice agriculture, although area available to them is small. They rear few cattle like cows, bullocks, buffaloes and goats. Nature is so kind that it provides practically everything not only to the mankind but also to their pet animals since prehistoric times. The common diseases/ailments like fever, dysentery, reluctantness to eat fodder, tumour, accidental injuries or wounds, stomach complaints etc. have been found treated by various herbs generally available locally.

The above resume clearly suffices that nature has bestowed the region under study with enormous wealth of medicinal plantlore. Systematic efforts to exploit this valuable potential are lacking. The real progress in this realm can be only be achieved through intensive and interdisciplinary work involving ethnobotanists, chemists, pharmacognists and physicians of various systems of medicine, like Allopathy, Homeopathy, Ayurveda, Unani and other indigenous systems of medicine. The scientific search of pharmacological unique principles from existing indigenous remedies continues and compliments the achievements of modern medicine. As said earlier, drugs of today came into use through the examination of folk remedies. Chemists continue to use plant based drugs as prototypes in their attempts to develop more effective and less toxic compounds. There is ample scope to extend scientific efforts to screen the phyctochemical constituents and their pharmacological properties. These properties can be confirmed with biological screening experiments.

Patenting of plant based medicine, and even food and other products has now become necessary. Apart from the tribals, the rural-folk claims should be studied by isolating active principles from different plant species. This treasure house of useful medicinal plants is available in their surroundings. The loss of ethnobotanical diversity and ethnomedicinal in particular, associated with indigenous people is one of the great crisis of our time. The knowledge and culture of indigenous peoples are disappearing even faster than the plant wealth itself. The rapid loss of vegetational cover poses a serious threat to such studies. This subject matter is dilated upon later.

Apart from the indigenous folk medicinal plants, the district is also rich in plant species useful in Ayurvedic system, which is obviously an another part of research.

As in other parts of our country fuel wood need is immense in this region also (Appendix-X). It is the largest non-timber tree product in volume and value. Wood still continues to be the principal energy source in forested tribal area of this region because other options like fossil fuel and electricity are beyond their reach. In the plains and rural parts of this district, the present investigator observed the poor people using stalks to toxic and least valued plants such as *Xanthium strumarium* L., *Ipomoea carnea* Jacq., *Euphorbia neriifolia* L. as fuel. The demand of fuel wood is increasing due to unbridled population growth. Although social forestry programmes are expected to relieve pressure on natural forests by providing fuel wood, it appears that it has failed to meet this goal satisfactorily. Illicit cuttings from forests by head-loders is an usual scene. All these situations warrant raising of plantations on large scale on village common lands, unproductive lands for a agricultural crops, waste lands, hilly terrains, field boundaries, catchment areas near dams and water reservoirs, river and canal banks and degraded forest lands. Selection of tree species has to be based upon characteristics apt for plantation such as deep root system, drought resistance, ability to coppice, rapid growth, adaptability to different edaphic factors of this region etc. As far as possible, these species should have multiple uses. If possible, charcoal yielders should be given priority because charcoal is an ideal smokeless fuel. The present investigator recommends the following tree species which are charcoal yielders and useful multipurposely: *Albizia procera* (Roxb.) Benth., *Anogeissus latifolia* (Roxb. ex DC.) Wall ex Guill., *Butea monosperma* (Lamk.) Taub., *Cordia dichotoma* Forst.*f.*, *Haldina cordifolia* (Roxb.) Ridsd., *Tamarindus indica* L., *Terminalia* species found in this district. The captive plantations would help in improving the rural economy. It is interesting to note that the people of this region avoid to use Neem (*Azadirachta indica* A.Juss.), Banyan (*Ficus benghalensis* L.) and pipal (*Ficus religiosa* L.) trees as fuel. In tribal forested areas, fuel wood collection on state forest lands is illegal or subject to government regulations. Fuel wood collectors are often harassed. Plantation programmes in these areas will certainly help to minimize them.

Non-timber forests products are many which support industries. Such forest products based industries provide opportunities for off-farm employment and income to the forest dwellers and the other people in vicinity. Tannin is one such product which is used in leather industries. No tribals are noted tapping the plants as a source of tannin. Probably they are unware or they have not tried for the same. There are some tree species growing in Jalgaon district which can be used for this purpose, for example, *Acacia nilotica* (L.) Del., *Cassia fistula* L., *Anogeissus latifolia* (Roxb. ex DC.) Wall. ex Guill., *Tamarindus indica* L., *Terminalia chebula*, *Ziziphus xylopyrus*, *Cassia auriculata* L. and few others.

It is to be noted that with the recent awareness regarding the use of ecofriendly materials in leather process industries the demand is growing for vegetable tanning materials which are environmentally safe products.

There is also a need to explore and exploit aromatic plants to feed the essential oil industries. During our visits we could find few taxa worthy of note. Following are some candidature which may help in such a pursuit: *Alangium salvifolium* (L.*f.*) Wang. (scented flowers), *Holarrhena pubescens* (Buch.-Ham.) Wall. ex G.Don (scented flowers), *Morinda pubescens* J.E.Sm.(scented flowers), *Capillipedium filiculme* (Ho.K.*f.*) Stopf. (scented inflorescence), *Clerodendrum multiflorum* (Burm.*f.*) O.Ktze. (scented flowers), *Disosorea oppisitifolia* L. (scented flowers), *Heterophragma quadriloculare* (Roxb.) K.Schum. (scented flowers), *Hymeonodictyon orixense* (Roxb.) Mabberley (scented flowers), *Catunaregam spinosa* (Thunb.) Tirvengadum (scented flowers).

Fodder requirement of the local inhabitants is high. Traditionally livestock is being reared in the district. Apart from agricultural stocks or residues, wild plants meet their need for fodder (Appendix-X). These are derived either from crop weeds or from forests. These are removed manually as a necessity of cultivation practice but are useful for the livestock. These are mostly herbaceous species. Trees are also lopped for this purpose.

The trees which are raised on private lands are lopped systematically but growing on waste land or in forest suffer from ruthless lopping. Certain valuable species may be declared as protected from lopping. Availability and quality of fodder is correlated with the production of milk and meat. Therefore, efforts should be made to increase fodder production. Although efforts are afoot by the agricultural departments, no tangible results are, however, observed. Multicut-fodder crop species requiring less water is the necessity of this region.

Traditionally, in the non-tribal rural area flat-roofed (Dhaba) houses are commonly seen in the district. However, 'Cement Concrete' jungles are perforce being raised in these places. The old houses were commonly constructed using mostly teak wood (*Tectona grandis* L.*f.*). In tribal hilly areas of this district teak wood is still being used. Apart from it, tree species used are as enlisted in Appendix-III.

The other plant materials commonly used for raising huts is *Sorghum controversum* (Steud.) Snowden.

It is a common knowledge that in many Indian states, the Bidi industry is one of the larger employer. Few Tendu plants (*Diospyros melanoxylon* (Roxb.) are observed here and there in the forests of this district (Appendix-X). Trees are not healthy and appear unattended and uncared by the forest department. They appear not prunned from time to time, and as a consequence, they do not produce a healthy flush of young leaves. This has resulted in a poor quality harvest. The overall situation regarding the Tendu plants, the forest department should take due measures for natural regeneration. This tree species is sun-loving and this district is one of the hot district of the state of Maharashtra. Moreover, this species establishes quickly on disturbed and open forest areas. Captive plantations may also help to the tribals, however, it is to be initiated by the government programmes even on private lands. The produce from the captive plantations should not go to the distant markets. The local tribal people should not be deprive of the intended benefits.

The leaves of Beheda [*Terminalia bellirica* (Gaertn.) Roxb.] and Palas or Dhak [*Butea monosperma* (Lamk.) Taub.] are employed to prepare dining plates (Patrawalya) and bowls (Dronas). These leaves are interwoven or stitched by using small sliced thin strips removed from the culms of Boru [*Sorghum controversum* (Steud.) Snowden]. Basket making is practiced at some places especially in the localities nearer to Satpura mountain. The rural people use the stems of Karvi (*Carvia callosa* Nees). They are pliable and elastic. These baskets are sold in village markets and weekly bazaars at Taluka places. The local people, however, made complaint of availability of plant material of this species and harassing by the forest departments. We are aware that *Carvia callosa* Nees. is monocarpic and flowers are fruits only after 7-9 years. Moreover, they are gregarious. As a control measure, the local people should be advised and trained to collect the stands of these plants only after seed setting and their dispersal. This will help regenerate the species and conserve *in situ*. This practice will also help to collect honey from the population of *C.callosa*, if the tribals are trained and persuaded for this programme. The demands for baskets is, however, partly met with in this region as these stalks of some cultivated species of *Gossypium* and *Cajanus cajan* are employed for this purpose.

Sindhi [*Phoenix sylvestris* (L.) Roxb.] is the only source for broom making on commercial basis. Because of over exploitation, this species is threatened. For broom making using the foliage of this species is old one. Hand brooms and long brooms are used for house cleaning. This foliage stay loose and form a homogeneous texture, and hence fits well for sweeping purposes to sweep foregrounds of houses and farmyards, stalks of some cultivated species of *Gossypium* and *Cajanus cajan* (L.) Millsp. are employed for making 'khatta' a tightly tied stack of stalks. These are also used on farmyards while cleaning the food grains.

Stems or branch tips are used as tooth-brushes and chew-sticks. These are chewed until frayed and leathery. They are then rubbed on the teeth and gums. The different species have characteristic, taste and flavour. These also claim responsibility of its medicinal, cleaning and disinfecting properties. This practice strengthens the gums as well. Cutting twigs and young branches of Babul (*Acacia nilotica* (L.) subsp.Del. *indica*) for this purpose is the usual scene especially along the railway lines. This tree species are commonly found growing on waste lands and surrounding areas along the railway tracts. Large bundles of these branches after necessary cleaning of tightly tied and their parceled through railway departments or privately to the cosmopolitan cities, like Mumbai and Pune where they are sold at high costs. This is actually one form of cottage industry. The people, however, observed in this business and not usually the residents of Jalgaon district. If the local poor tribals are persuaded for this business, it will fetch good returns for them. The species which are usually used and enlisted in the Appendix-IV.

There are some gum yielding species naturally found in the forests *e.g. Sterculia urens* Roxb., *Anogeissus latifolia* (Roxb. ex DC.) Wall. ex Guill., *Butea monosperma* (Lamk.) Taub., *Acacia nilotica* (L.) Del. subsp.*indica*, *Lannea coromandelica* (Houtt.) Merr., *Boswellia serrata* Roxb,. ex Coleb., *Soymida febrifuga* (Roxb.) A.Juss., *Albizia lebbeck* (L.) Willd., *Limonia acidissima* L., *Albizia procera* (Roxb.) Benth., *Terminalia bellirica* (Gaertn.) Roxb., *Pterocarpus marsupium* Roxb. Some are valuable as edible, medicinal and as gluing agent. These resources form the main socio-economic potential and fetch some income for annum, provided their utilization as a resource is systematic.

Ritha (*Sapindus emarginatus* Vahl) is only species found wild in this area of which the fruits are used as hair shampoo. The number of trees are very few in the forests. This appears so because all the fruits are gathered and marketed thereby prohibiting natural seed dispersal. This is one such a tree species deserving recommendation as a candidature for afforestation programme of the government or even for captive plantation on private barren lands.

Certain tree species in this area are avoided from using them as a source of fuel wood, *e.g. Ficus benghalensis* L., *Ficus religiosa* L. and *Azadirachta indica* A.Juss. This taboo automatically help conservation of these species. It is also noteworthy that in some economically important trees as *Madhuca longifolia* (Koen.) Macbride, *Buchanania lanzan* Spring., only naturally dried branches and wood from very old trees are used as a fuel. This trend also partly help conserve these species. All these trends clearly indicate wisdom of the tribal/ rural people of the region.

Some of the practices or uses should be avoided to protect ultimately the interests of the tribals in future. Destructive methods of collection of NTFPs should not be practiced. For example, (i) while collecting leaves from Tendu plant (*D.melanoxylon* Roxb.) they are collected as much as they can. This affects the natural growth of these trees, (ii) over collection of seeds of Behada [*Terminalia bellirica* (Gaertn.) Roxb.], fruits of Ritha (*Sapindus emarginatus* Vahl.), naturally reduces the number of saplings in the next season, (iii) Dye is obtained from flowers of Palas [*Butea monosperma* (Lamk.)Taub], which is used by the tribals at the time of 'Holi' festival. This reduces directly the possibility of fruit formation and seed production. Naturally less seed germination is observed in the next rainy season, (iv) lopping of especially timber yielding species for fodder *e.g. Ougeinia oojeinensis* (Roxb.) Hochr. and for thatching *e.g. Tectona grandis* L.*f.* should be checked and the tribals should be convinced. (v) Likewise, while digging out the tubers of *Pueraria tuberosa* (Roxb. ex Willd.) DC., different species of *Dioscorea* the entire plants are thrown. This has resulted in depletion of these species. (vi) Wood of some medicinally important trees *e.g. Holarrhena pubescens* (Buch. Ham.) Wall. is collected on large scale during summer, which is to be used during rainy season, should be obviously stopped.

Identities, economics and culture of local masses are inextricably bounded to their land and resources. The vegetational wealth have sustained their spiritual, cultural and economic necessities.

Forests are being denuded due to illicit cutting, over grazing, fires, shifting cultivation and even permanent cultivation in forest cleared lands. Over-exploitation and modernization have also negative impact on the plant wealth. Thus rapid depletion of forests has obviously become a serious concern, as in the other parts of our country, in this district also. The loss of biodiversity directly culminates in loss of variety of forest products, scarcity of fuel and fodder. Much of the misery of tribals/rurals is because of deforestation which removes the resources on which their lively hood has been based. Moreover, destructive methods of collection of forest products also play a significant role in diminishing the richness of ancient vegetation. It is to be noted that the demand of these plant based products is ever-growing on account of expansion of human population in the region. High demands naturally tend to encourage over- exploitation. Some economic plants or their products are left to go waste in the forest due to ignorance of economic potential and lack of information of economic potential and lack of information of their distribution, availability and their identity. In general, forest departments do not have the exact figure of various economic plants under their control.

The plant resources of Jalgaon district are still quite rich in raw materials needed for various purposes. The ethnomedicinal plants need to be investigated for pharmacological activity on the basis of ethnotherapeutics being practiced by the tribals/rurals. This will bring to light some new sources of drugs of herbal origin. Systematic research at their conservation, is a dire necessity to improve the economy of local inhabitants. The knowledge of proper harvesting, processing and marketing is also necessary.

Large scale propagation techniques for important tree species such as *Madhuca longifolia* (Koen.) Macbride, *Terminalia bellirica* (Gaertn.) Roxb., *Buchanania lanzan* Sprong. and few other medicinal plants are still to be extended.

A grass 'Rosha' or 'Ruicha' [*Cymbopogon martini* (Roxb.) Wats.] yields essential oil which is commonly used in scents and soaps. This grass species is commonly found in this district. It should be exploited paying due attention to its period of collection, quantification, processing, economy and local people's participation.

Balanites aegyptiaca (L.) Del. is another species which produces oil yielding seeds. This oil is used as edible by the tribals. Likewise, *Pongamia pinnata* (L.) Pierre is a commonly found in the forests which also bears oil-yielding seeds. This is can be used as lubricant, for soap making and has insecticidal properties.

It is pretty clear from the foregoing discussion on NTTPs, the forests of Jalgaon district are the home of a large number of host species for such as medicinal, oils, edibles, gums, fibres, charcoal, tannin, bamboos, bidi wrappers, beverage and few other articles of day to day necessities of the local inhabitants. During our visits, ethnobotanical studies were undertaken with a view of prepare resource inventories of plants used by the people of Jalgaon district. This investigation focused especially on life support species providing different plant products. The plant species used for hedge, hut construction, fish killing, as antidote in cases of snake bite, dog bite and scorpion sting and those sometimes sold market for food are enumerated in appendices V to IX respectively. Plant based economic activity throughout a year of the people of this region is depicted month-wise in the Appendix-X.

The precious knowledge of the local people is fast disappearing hand in hand with the loss of plant wealth of this region. In view of this, following recommendations are made: (i) scientific evaluation of ethnomedicines and other plant products, (ii) *in situ* conservation with people's active participation, (iii) afforestation programmes, especially captive plantation of some indigenous useful and important species involving forest dwellers, (iv) patenting of the desirable plants of their parts, if necessary, in view of GATT agreement for protecting intellectual property rights of local inhabitants.

The local tribal communities having intimate relationship with the forests should be accepted as partners in the forestry development efforts in such area. The best protective devices for the existing forests and the new addition is to create an interest among the local communities in the forest wealth and at the same time their dependency upon the forest is required to be minimized to the optimum.

It is commonly observed that people from different religions, castes and races of human being worship one or other form of gods or goddesses. The worship is performed with traditional rituals for well-being of the persons of the respective human society. They preserve traditions and forms of worships through folklore. They offer suitable or available botanical articles, parts, food grains or their products to the deities right from the occasion of birth to mourning a death and even after this period.

Jalgaon district is predominated by Hindu community. Different castes and tribes inhabit this region of the state of Maharashtra. They have their own festivals, deities, ceremonies, rituals and customs. Generally, they perform specific worship with precise offerings. However, there are also worships in which the articles of worship are substituted. For example, in marriage ceremony of Hindus any five kinds of fruits are offered in the 'Oti' of a bridegroom.

Both wild and domesticated plants, whether indigenous or exotic in origin, have been found in use. Some plant species do not occur naturally in this district but their parts or products are available in local markets. These also find place in abstract relationship.

The plants or their parts are usually collected by the worshippers from nearby places or sometimes they are purchased from local markets in a particular season or period of worship. The plant noted in case of abstract relationship have, however, lesser economic significance in comparison to those recorded in concrete relationship.

Majority of plants used are well-known to the users. Still, there are few species appeared unknown and hence substituted by some other species in this district. For example, instead of leaves of *Michelia champaca* L., locally called 'Chafa' leaves of *Plumeria alba* L. or *Plumeria rubra* L. find places in the rituals.

The area is fairly rich as regards the folk songs, folk dances, proverbs, riddles etc. The majority of the people this district speak a local dialect called Ahirani, apart from the state language. Tribals have their own dialects. The songs or proverbs have been collected from the singers and common old ladies or gents. Their meanings were, however, discussed with them. The themes of folk songs, proverbs or riddles etc. have been explained in the text proper. Their subject matter is avoided here for the sake of precision.

Summary

Extensive and intensive field studies were conducted in 13 talukas of Jalgaon district. These include collections and identification of plant species and documentation of information on local uses in tribal, rural and even urban localities. The data base is prepared on ethnomedicines, home remedies as well as the uses for food, fodder, fibre, timber, oil, gum, tooth brushes, fish poisons etc. as obtained from the different informants and actual personal observations. Apart from these aforesaid concrete uses, information is also gathered on the abstract relationship of the people of Jalgaon district with the plants in their surroundings. It was possible to collect data on association of plants with the songs, proverbs riddles, puzzles, marriages, various ceremonies, births, deaths, festivals worship of gods and goddesses, customs, taboos, fasts, games, dances etc. The plants utilities are noted not only for the human beings but also for their pet animals. Information tapped belongs to all tribal, rural and urban societies of the district.

The data collected is first-hand belonging 390 taxa 310 genera 96 families of angiosperms, fungi and pteridophytes. The latter two are, however, have one species each useful. The representation of pteridophytes is poor and rare in this region. The area appeared fairly rich in view of resources of plant-wealth. The plantlore has passed through generations orally. This contribution is the first comprehensive documentation covering every nook and corner of all 13 talukas of Jalgaon District. Nearly 225 species are employed for medicines. The other uses recorded are as: edible (59 species), hut construction (11 species), timber (11 species), veterinary medicines (18 species), hedge plants (14 specdies), fodder (105 species), fish poison (17 species), antidotes (14 species), tooth brushes (17 species), songs, proverbs, riddles, puzzles, games, other religious species including vernacular names are nearly 110 species.

Phytochemical investigations and other potentialities should be extended to identify the active principles, their efficacies and utilities which eventually will lead to divulge new drugs and other miscellaneous substances. This contribution will meet the need for suitable alternative source of medicinal plants. The species having similar therapeutic properties possess same active principles.

The tribal and rural prudence of employing plant species, *in toto*, shows their wisdom on medicine, food, shelter and other economic values. These are obviously the intellectual component of their contribution to contemporary knowledge on food and health security. These alternate sources will widen the base of our food security system. The data base prepared by the present investigator on ethnobotany of this region will serve as a useful tool to prepare development and action plan for herbal drug and other industries. In turn, these will help improve and uplift the life and economy of these local inhabitants and as well as other societies of our country. The modern scientific scrutinisation especially of traditional medicinal plants, however, should obviously include: (a) phytochemical screening, (b) pharmacological analysis, (c) biological activities and (d) clinical trials.

This book also contains information on geography, geology, ethnology of main tribes, crops and plant based industries etc. to shed more light and bring the necessities and utilities in clearer focus. Some photographs of plants, their parts or products and the people using them etc. have been included to authenticate the data accrued.

Some notations on dwindling of vegetation or plant species in this district have been also discussed pertinently. The factors checking the forest resources are illicit cutting for timber and fuel-wood, over-exploitation at some places for miscellaneous purposes, forest fires, overgrazing etc. Certain plant species have been critically and purposefully discussed in view of depletion. Stringent action on collection of some taxa, more encouragement for peoples' participation programmes in afforestation, creation of public awareness regarding the need for protection of plant wealth may help resolve the problem in this region.

This study reveals that there is enormous ethnobotanical plant diversity in Jalgaon district. This data-base is helpful particularly to the botanists, agriculturists, foresters, pharmacists, plant-chemists and research workers working in multidisciplinary projects.

Appendices

Appendix-I
Edible Plants

Plant Names
Vegetable: Leaves
Amaranthus polygonoides L.
Amaranthus viridis L.
Alternanthera tenella Colla
Asparagus racemosus Willd.
Cayratia trifolia (L.) Domin
Chenopodium murale L.
Caesulia axillaries Roxb.
Commelina benghalensis L.
Caralluma adscendens Haw.
Digera muricata (L.) Mart.
Hibiscus sabdariffa L.
Ipomoea aquatica Fors K.
Ipomoea turbinata Lag.

Contd...

Marsilea minuta L.

Pupalia lappacea (L.) Juss.

Rivea hypocrateriformis (Desr.) Choisy

Schrebera swietenioides Roxb.

Vegetable: Flowers

Butea monosperma (Lamk.) Taub.

Capparis decidua (Forsk.) Edgew.

Wattakaka volubilis (L.*f.*) Stapf.

Ensete superbum (Roxb.) Cheesm.

Hibiscus sabdariffa L.

Ougeinia oojeinensis (Roxb.) Hochr.

Vegetable: Fruits

Butea monosperma (Lamk.) Taub.

Benincasa hispida (Thunb.) Cogn.

Capparis deciduas (Forsk.) Edgew.

Capparis zeylanica L.

Vegetable: Seeds

Mucuna pruriens (L.) DC.

Prosopis cineraria (L.) Druce

Edible Fruits

Plant Names

Diospyros melanoxylon Clarke

Dendropthoe falcate (L.*f.*) Etting

Ehretia aspera Willd.

Ficus religiosa f.

Ficus aronottiana (Miq.) Miq.

Flacourtia indica (Burm.f.) Merr.

Grewia tiliifolia Vahl

Kirganellia reticulata (Poir.) Baill.

Lea asiatica (L.) Ridsd.

Manilkara hexandra (Roxb.) Dab.

Pithecellobium dulce (Roxb.) Benth.

Rhus sinuata Thunb.

Contd...

Securinega leucopyrus (Willd.) Muell.-Arg.
Securinega virosa (Roxb. ex Willd.) Baill.
Strychnos potatorum L.f.
Syzygium cumini (L.) Skeels
Tamarindus indica L.
Terminalia bellirica (Gaertn.) Roxb.
Ziziphus nummularia (Burm. f.) Wt. and Arn.
Ziziphus caracutta Roxb.
Ziziphus rugosa Lam.

Edible Seeds

Balanites aegyptiaca (L.) Del.
Garuga pinnata Roxb.
Holoptelea integrifolia (Roxb.) Planch.
Mucuna prunriens (L.) DC.
Pentatropis spiralis (Forsk.) Decne.
Tamarindus indica L.

Plants Used in Preparing Bread

Madhuca longifolia (Koen.) Macbr.
Pueraria tuberosa Roxb.
Bambusa arundinacea Roxb.

Appendix-II
Plants Used in Veterinary Medicine

Plant Names

Agave vera-cruz Mill.

Allium sativum L.

Allium cepa L.

Annona squamosa L.

Ailanthus excelsa Roxb.

Aristolochia bracteata Retz.

Asparagus racemosus Willd.

Cayratia trifolia (L.) Domin

Clerodendrum phlomidis L. f.

Cymbopogon martini (Roxb.) Wats.

Dalbergia sisoo Roxb. ex DC.

Dolichondrone falcata Seem.

Contd...

Echinops echinatus Roxb.
Jatropha curcas L.
Jatropha gossypifolia L.
Solanum anguivi Lamk.
Salmalia malabarica (DC.) Schott. and Endl.
Viscum nepalense Spreng.

Appendix-III
Plants Used as Timber

Plant Names
Anogeissus latifolia Bedd.
Azadirachta indica A. Juss.
Alangium salvifolium (L.f.) Wang.
Buchanania lenzan Spreng.
Cassine albens (Retz.) Kosterm.
Dalbergia sisoo Roxb.
Holoptela integrifolia (Roxb.) Planch.
Haldina cordifolia (Roxb.) Ridsd.
Ougeinia oojeinensis (Roxb.) Hochr.
Pterocarpus marsupium Roxb.
Tectona grandis L.f.

Appendix-IV

Plants Used as Toothbrush and for Cleaning Mouth

Plant Names
Acacia nilotica Willd. ex Del. subsp. *indica* (Bth.) Benan
Azadirachta indica A.Juss.
Alangium salvifolium (L.f.) Wang.
Acacia catechu (L.f.) Willd.
Barleria prionitis Sant.
Baliospermum raziana Keshav. et. Yog.
Cleome viscosa L.
Cassia auriculata L.
Emblica officinalis Gaertn.
Ehretia aspera Willd.
Mimosa hamata Willd.
Meyna laxiflora Robyns.
Spilanthes calva DC.

Contd...

Securinega leucopyrus (Willd.) Muell.-Arg.
Salvadora persica L.
Terminalia bellirica (Gaertn.) Roxb.
Vitex negundo L.

Appendix-V
Hedge Plants

Plant Names
Agave vera-cruz Mill.
Alangium salvifolium (L.f.) Wang.
Capparis decidua (Forsk.) Edgew.
Coccinia grandis (L.) Voigt.
Carrisa carandas L.
Caesalpinia decapetala (Roth) Alst.
Euphorbia nivulia Buch. – Ham.
Euphorbia tirucalli L.
Ipomoea carnea Jacq.
Lawsonia inermis L.
Mimosa hamata Willd.
Maytenus emarginata (Willd.) Ding Hou
Opuntia elatior Mill.
Ziziphus mauritiana Lamk.

Appendix-VI
Plants for Hut Construction

Plant Names

Bridelia airy-shawii P.T.Li.
Cyperus pangorei Rottb.
Combretum albidum G.Don
Grewia tilifolia Vahl
Hardwickia binata Roxb.
Ixora pavetta Andrews
Kydia calycina Roxb.
Securinega virosa (Roxb. ex Willd.) Baill
Sorghum bicolor (L.) Moench.
Tectona grandis L.f.
Vitex negundo L.

Appendix-VII
Plants Used for Fish Killing

Plant Names
Acacia pennata (L.) Willd.
Butea monosperma (Lamk.) Taub.
Cassine albens (Retz.) Kosterm.
Catunaregam spinosa (Thunb.) Tirvengadum
Garuga pinnata Roxb.
Holoptelea integrifolia (Roxb.) Planch.
Lannea coromandelica (Houtt.) Merr.
Milletia extensa Benth.
Ougenia oojeinensis (Roxb.) Hochr.
Lannea coromandelica (Houtt.) Merr.
Milletia extensa Benth.
Securinega leucopyrus (Willd.) Muell.-Arg.
Strychnos potatorum L.f.

Contd...

Terminalia crenulata Roth
Ventilago denticulata Willd.
Veronica anagallis-aquatica L.
Verbascum chinense (L.) Sant.

Appendix-VIII
Plants Used as Antidote

Plant Names

Snake Bite

Butea monosperma (Lamk.) Taub.

Calotropis gigantea (L.) R.Br.

Diplocyclos palmatus (L.) Jeffery

Ensete superbum (Roxb.) Cheesm.

Kedrostis rostrata (Rottl.) Cogn.

Mimosa hamata Willd.

Salmalia malabarica (DC.) Schott. and Endl.

Tamarindus indica L.

Tabernaemontana divaricata (L.) R.Br.ex Roem. and Schult.

Scorpion Sting

Allium cepa L.

Gmelina arborea Roxb.

Martynia annua L.

Strychnos potatorum L.f.

Dog Bite

Ensete superbum (Roxb.) Cheesm.

Appendix-IX
Plants Sold in Market

Plant Names	Part Used
Asparagus racemosus Willd.	Plant
Buchanania lanzan Spreng.	Seed
Carissa carandas L.	Fruit
Chenopodium murale L.	Plant
Cordia gharaf (Forsk.) Ehrenb. and Asch.	Fruit
Ensete superbum (Roxb.) Cheesm.	Fruit
Manilkara hexandra (Roxb.) Dab.	Fruit
Portulaca quadrifida L.	Plant
Tamarindus indica L.	Fruit
Syzygium cumini (L.) Skeels	Fruit
Ziziphus mauritiana Lamk.	Fruit

Appendix-X

Annual Cycle of Forest Products in Jalgaon District (Monthwise)

S.No.	*Product*	*J*	*F*	*M*	*A*	*M*	*J*	*J*	*A*	*S*	*O*	*N*	*D*
1.	**Housing Timber**			*	*	*	*						
2.	**Fire Wood**	*	*	*	*	*						*	*
3.	**Agricultural Implements**					*	*	*	*				
4.	**Fodder**												
	a) Green Grass						*	*	*	*	*		
	b) Lopping			*	*	*			*	*	*		
	c) Tree Leaves	*	*	*			*	*					
5.	**MFPs**												
	a) Flowers (Mahua)				*	*							
	b) Bidi Wrappers					*	*						
	c) Honey	*	*	*							*	*	
	d) Gums			*	*	*							
	e) Thorns					*	*				*	*	
	f) Thatching Materials								*	*	*		

Contd...

S.No.	Product	J	F	M	A	M	J	J	A	S	O	N	D
6.	**Food**												
	a) Leaves (Herbs)							*	*	*	*		
	b) Underground Plants								*	*	*		
	c) Fruits			*	*	*					*	*	*
7.	**Medicinal Materials**	*	*	*	*	*	*	*	*	*	*	*	*

* : Collection.

Bibliography

Almeida M.R. and S.M.Almeida (1981). Folklore and utility of vascular plants from Savantwadi. In: Proc. All India Symposium on Biology and Utility of Wild Plants. Prof. G.L. Shah Commemoration Vol. Department of Biosciences South Gujarat University, Surat, pp.23-39.

Ambasta S.P. (1986). The Useful Plants of India. Publication and Information Directorate, CSIR, New Delhi, India.

Anand Kumar, Kulkarni B.G. and N.P.Singh (1981). Medicinal and economic plants growing at Pune. M.V.M. Patrica **15:64-74.**

Anonymous (1948-1976). The Wealth of India. A Dictionary of Indian Raw Materials and Industrial Products. Vol.1-11. Publication and Information Directorate, CSIR. New Delhi, India.

Anonymous (1982). List of medicinal plants deposited in various herbaria of the Botanical Survey of India. *Bull.Bot.Surv. India* **2**:180-273.

Ansari M.Y. and M.P.Nayar (1978). Some less-known plants from western ghats useful for essential oil industry. *Indian Perfumer* **22(3)**:210-213.

Ansari M.Y. and Rolla S. Rao (1973). *Iphigenia stellata* Blatter (Liliaceae) – Its identity and economic importance. *Bull.Bot. Surv. India.* **15(1-2)**:118-122.

Badhe P.D. and B.N. Sharma (1981). Ethnomedicinal methods of Korku tribes of West Melghat. *B.M.E.B.R.* **3(2-4)**:162-172.

Badhe P.D. and V.K.Pandey (1988). A study of medicinal and economic plants of Amaravati Division, Amaravati Circle, Maharashtra. *B.M.E.B.R.* **11(1-4)**:1-39.

Badhe P.D., Sharma L.K. and Y.K.Pande (1997). Rare folk medicines practiced by Gond tribes of Wadsa forest division. Gadchiroli district (Maharashtra State). *Bull. Medico-ethno-bot. Res.* **18(1-2)**:12-17.

Bhamare P.B. (1995). Some anti-venom medicinal plants from tribals of Dhule district Maharashtra. *JAST* **1(1)**:36-37.

Bhamare P.B. (1998a). Traditional knowledge of plants for skin ailments of Dhule and Nandurbar district. Maharashtra (India). *J.Phytol. Res.* **11(2)**:195-196.

Bhamare P.B. (1998b). Ethnomedicinal plants used by tribals in Dhule district Maharashtra, India. *J.Swamy Bot. Cl.* **15**:18-83.

Bhamare P.B. and D.A.Patil (2005) Ethnobotanical envision and wisdom of tribals in Dhule and Nandurbar Districts (Maharashtra) In: Proceeding of National Conference of 'Plant Diversity and Biotechnology', 10-11 Dec. 2004. P.G. Department of Botany, S.S.V.P.S's L.K.Dr.P.R.Ghogrey Science College, Dhule, Maharashtra (Eds. Nandan, S.N.; D.A.Patil, B.D.Borse and V.B.Salunkhe), pp.205-208.

Bhogaonkar P.Y. and V.D.Devarkar (2001a). Studies in ethnobotany of Korkus of Melghat Amaravati District, Maharashtra. I-Antisterility Drugs. JAST 4(1-2)63-66.

Bhogaonkar P.Y. and V.D.Devarkar (2001b). Studies in Ethnobotany of Korkus of Melghat Amaravati District, Maharashtra. II-Drugs for Urino-Genital Disorders. JAST 4(1-2) 67-68.

Bhuktar A.S. (2002). Traditional use of medicinal plants of Hingoli district In : Proc. Plant Resource Development. (Eds. A.M.Mungikar and A.S.Bhuktar) Dr.B.A.M. University, Aurangabad, pp.146-152

Borse S.C., Bhamare P.B. and D.A.Patil (1990). Medicinal plantlore of tribals of Dhule Maharashtra. *Biojournal* **2(1)**:47-54.

Caius J.F. and K.S. Radha (1939). The gum arabic of bazaars and shops of Bombay. *J.Bombay Nat.Hist. Soc.* **12**:261-271.

Chandras G.S. (1966). Economic forest products of Maharashtra. In:Proc.The Autumn School in Botany Mahabaleshwar (Ed.T.S.Mahable). Department of University of Poona, India, pp.26-36.

Chute G.S. and V.J.Tiwari (1999). Indigenous ethnomedicinal plants used by tribal people of Bhandara and Gadchiroli district of Maharashtra State. *Indian J. Nat. Prod.* **15(1)**:3-8.

Cooke, T. (1958) The Flora of the Presidency of Bombay. Vol. 1-3. B.S.I. Calcutta (Repr.Ed.).

Das P.K. and M.K.Misra (1988). Some medicinal plants among Kondhas around Chandrapur (Koraput). *J.Econ.Tax.Bot.* **12(1)**:103-109.

Datar Rekha and V.D.Vartak (1975). Enumeration of wild edible plants from Karanala bird sanctuary, Maharashtra State. *Biovigyanam* **1**:123-129.

Deokule S.S. and D.K.Magdum (1992). Enumeration of medicinal plants from Baramati area, District Pune, Maharashtra State. *J.Econ. Tax.Bot. Addl. Ser.* **10**:289-299.

Deshmukh V.R. and M.A.Dhore (1994). Nutritive values of some lesser known herbivor food from Melghat tiger reserve Maharashtra. Repr. *Indian Forester* **120(10)** : 920-923.

Deshmukh V.R., Muratkar G.D. and S.P.Rothe (1999). Preliminary observation of the medicinal and economically important leguminous plant species from Amaravati Tehsil. *J.Econ. Tax.Bot.* **23(2)**:283-289.

Elhoven R.E. (1930). Tree and animal worship in Western India. *Proc. Jubilee Congress-1926*. London : 181-187.

Gadgil Madhav and V.D.Vartak (1973). Groves dedicated to the gods. *The Illustrated Weekly of India*. 20-21.

Gadgil Madhav and V.D.Vartak (1974). Sacred groves of India-A plea for continued conservation. *J.Bombay Nat.Hist.Soc.* **72(2)**:314-320.

Gadgil Madhav and V.D.Vartak (1976). The sacred groves of Western Ghats of India. *Econ.Bot.* **30**:152-160.

Gadgil Madhav and V.D.Vartak (1981). Sacred Groves of Maharashtra : An inventory. In : Glimpses of Indian Ethnobotany (Ed. S.K.Jain). Oxford & IBH Publishing Co., New Delhi, India, pp.279-294.

Gaikwad J.S. (1988). Role of minor forest produce in tribal economy. In : A Review of Tribal Research Studies, Silver Jubilee Volume. (Ed.Dr.Govind Gare), Tribal Research and Training Institute Pune (M.S.) India, pp.103-108.

Ghate V., Agate V. and V.D.Vartak (1988). Promising economic potential of Shumul (*Bombax ceiba* L.) as tuber crop. *Ind. J. For.* **11(2)**:158-159.

Ghate V.S., Vartak V.D. and M.S. Kumbhojkar (1990). Indigenous fuel-wood species from western ghats of Maharashtra. *Indian Forester* **116(3)**:233-237.

Ghate Vinaya S. (1998). Plants in patra-pooja : Notes on their identity and utilization. *Ethnobotany* **10**:6-15.

Ghate Vinaya S. and Hemlata D. Sane (1994). Notes on socio-economic status of the flora of Kanakeshwar hills, Raigad District, Maharashtra. *Ind. J. For.* **17(2)**:119-128.

Ghate Vinaya S. and V.D.Vartak (1996). Socio-economic and domestic utilization of forest tree resources by tribal communities in western Maharashtra. In : Ethnobiology in Human Welfare (Ed.S.K.Jain). Deep Publications, New Delhi, India, pp.329-330.

Giri R.B. (1994). Melghatatil Aushadopyogi Vanaspati. Technical Bulletin No.5 Directorate, Project Tiger, Melghat Paratwada, District Amaravati, Maharashtra.

Godbole Archana (1994). Less known uses of three climbers from western Maharashtra. *Ind. J. For.* **17(4)**:351-352.

Godbole Archana (1996). Role of tribals in preservation of sacred forests. In : Ethnobiology in Human Welfare (Ed.S.K.Jain). Deep Publications, New Delhi, India, pp.345-348.

Godbole S.R. (1934). Some forest trees and shrubs from the Bombay Presidency used as fodder for animals. *Poona Agric. Coll. Mag.* **26**:19-25, 70-77.

Gunjatkar Nalini and V.D.Vartak (1982). Enumeration of wild edible legumes from Pune district, Maharashtra State. *J.Econ.Tax.Bot.* **3**:1-9.

Gunjatkar Nalini and V.D.Vartak (1983). Ethnobotanical and floristic studies on Bherli-mad (*Caryota urens* Linn.) from Western India. *J.Econ.Tax. Bot.* **4(2)** :323-329.

Gupta V.C., Hussain S.J. and S.Imam (1997). Important folk medicinal plants and traditional knowledge of tribals of Aurangabad and Nasik Forest divisions of Maharashtra (India). *Hamdard Medicus* **40(2)**:59-61.

*Harshberger J.W. (1885). Some new ideas Philadelphia Evening Telegraph.

Hooker, J.D. (1872-1897) Flora of British India Vol.1-VII. Reeves & Co., London, U.K.

Jain S.K. (1987). A Manual of Ethnobotany, Scientific Publishers Jodhpur, India.

Jain S.K. (1991). Dictionary of Indian Folk Medicine and Ethnobotany. Deep Publications, New Dehli, India.

Jain S.K. and R.R. Rao (1977). Hand Book of Field and Herbarium Methods. Today & Tomorrow, New Delhi, India.

Jain S.K.(1989). Methods and Approaches in Ethnobotany. *Society of Ethnobotanists.* Lucknow, India.

Janardhanan K.P. (1961). Fodder grass and legumes of Khed taluka, Poona District. In: Proc.Second Annual Sess. Acad.Agri. Sci. Coimbtore. pp.106-113.

Janardhanan K.P. (1963). An enumeration of the medicinal plants of Khed taluka, Maharashtra State. *Bull.Bot.Surv. India.* **5(3-4)**:363-374.

Jayananda Tosh (1981). Observations on wild edible plants from hilly regions of Maharashtra State from Hilly regions and future prospects. In : Glimpses of Indian Ethnobotany (Ed.S.K. Jain), New Delhi, India, pp.261-271.

Jayananda Tosh, Vartak V.D. and M.S.Kumbhojkar (1959). Some wild edible plants from the hilly regions of Pune district. *J.Bombay Nat.Hist.Soc.* **65(1)**:9-26.

Jaynanda Tosh (1996). Ethnobotanical study of Western Maharashtra. J.Econ.Tax.Bot.Addl. Ser. 12, Scientific Publishers Jodhpur (India). 169-174.

Kamble S.Y. and S.G. Pradhan (1980). Ethnobotany of the 'Korkus' in Maharashtra. *Bull.Bot.Surv.India* **22(1-4)**:201-202.

Karnik C.R. (1966). Some medicinal plants from Satpura mountains. *Indian Forester* **92**:173-183.

Khandekar V.P. and G.N.Badve (1998). Local medicinal plants from Koyna valley. *Ayurved Update* **1(2)**:7-9.

Kirtikar K.R. (1906). A note on an edible puff-balls from the Thana district. *J.Bombay Nat.Hist.Soc.* **7**:816-817.

Korekar S.L. and P.B.Papadiwal (2002). Studies on forest based cottage industries in Osmanabad and Latur districts. In:Proc. Plant Resource Development. (Eds.A.M.Mungikar and A.S.Bhuktar) Dr.B.A.M. University, Aurangabad, pp.142-144

Kothari M.J. and S. Moorthy (1987). Floristic composition and utility of wild plants in Raigad district, Maharashtra. In : Proc.All India Symposium on Biology and utility of Wild Plants 100-107. Prof.G.L.Shah Commemoration Vol. Department of Biosciences, South Gujarat University, Surat, India.

Kothari M.J. and S.Moorthy (1996). Ethnobotany in human welfare of Raigad district in Maharashtra. In:Ethnobiology in Human Welfare (Ed.S.K.Jain). Deep Publications. New Delhi, India, pp.403-407.

Kshirsagar R.D. and Singh N.P. (2000a). Less known ethnomedicinal uses of plants in Coorg District of K.T. state, Southern Indian *Ethnobotany* **12**:12-14.

Kshirsagar R.D. and Singh N.P. (2000b). Less known ethnomedicinal uses of plants in Jenukuruba tribe Mysore District, Southern Indian *Ethnobotany* **12**:118-122.

Kshirsagar S.R. and D.A.Patil (2002). Ethnobotanical and Ethnomedicinal observations in Jalgaon District of Maharashtra. *Vanyajati* **3**:9-14.

Kshirsagar, S.R. and D.A.Patil (2007). Forest Flora of Jalgaon district (Maharashtra). Bishen Singh Mahendra Pal Singh, Dehra Dun, India.

Kulkarni B.G. (1968). Medicinal p!ants from Vengurla, Sawantwadi, Phonda and Ratnagiri district. Maharashtra. M.V.M. Patrika **3(1)**:17-24.

Kulkarni D.K. and M.S. Kumbhojkar (1992a). Ethnobotanical studies on Mahadeokoli tribe in Western Maharashtra-Part-I. Cordage Plants. *J.Econ. Tax.Bot.* Addl. Ser.**10**:111-115.

Kulkarni D.K. and M.S. Kumbhojkar (1992b). Ethnobotanical studies on Mahadeokoli tribe in western Mahaıashtra-Part-III. Non conventional wild edible fruits. *J.Econ.Tax.Bot.* Addl.Ser.**10**:151-158.

Kulkarni D.K. and M.S. Kumbhojkar (1992c). Ethnobotanical studies on Mahadeokoli tribe in western Maharashtra-Part-II. Fodder plants. *J.Econ.Tax.Bot.* Addl.Ser.**10**:123-128.

Kulkarni D.K. and M.S. Kumbhojkar. (2000). Plant conservation in Maharashtra during 17th Century. *Ethnobotany* **12**:29-38.

Kulkarni D.K. and M.S.Kumbhojkar (1993). Kitchen garden plants of Mahadeokoli tribe in Maharashtra. *Ethnobotany* **5**:119-127.

Kulkarni D.K. and M.S.Kumbhojkar (1999). Dam-disaster to biodiversity in sacred groves. The Deccan Geographer **37**(1):65-71.

Kulkarni D.K. and M.S.Kumhojkar (1996). Pest control in tribal areas of western Maharashtra on ethnobotanical approach. *Ethnobotany* **8**:56-59.

Kulkarni D.K., Agati V.V. and M.S. Kumbhojkar (2003). Leafy vegetables consumed by Mahadeo Koli Tribe in Western Maharashtra with their nutritional potential. *Ethnobotany* **15(1-2)**:34-39.

Kulkarni D.K., Kulkarni M.B. and M.S.Kumhojkar (1996). Quantification of three reosources utilized for medicinal purpose by Mahadeokoli of Western Maharashtra India. In : Ethnobiology in Human Welfare (Ed.S.K.Jain) : 429-432. Deep Publications, New Delhi, India.

Kulkarni D.K., Kumbhojkar M.S. and A.S. Upadhye (1995). Ethnobiological resources used in traditional medicines by Mahadeokoli tribe from western Maharashtra. *Flora and Fauna.* **1(1)**:65-66.

Kulkarni D.K., Kumbhojkar M.S. and D.S.Nipunge (1990). Research notes (IV) Notes on fish stupefying plants from Western Maharashtra. (Rep.) *Indian Forester.* **116(4)**:331-333.

Kulkarni D.K., Kumbhojkar M.S. and Rajeev Khedkar (1998). Rice germplasm collection, conservation and use : A case study in Western Maharashtra. *Ethnobotany* **10**:27-31.

Kulkarni D.K., Kumbhojkar M.S., Agate V.V., Joshi N.S. and V.N.Joshi (1991). Nutritional contents in *Flacourtia* from Western Maharashtra. *J.Fd.Sci. Technol.*, **28(2)**:118-119.

Kumar Anand, Kulkarni B.G. and N.P.Singh. (1980). Medicinal and economic plants growing at Pune. *M.V.M. Patrika* **15(2)**:64-74.

Kumbhojkar M.S. and D.K.Kulkarni (1998). Environmental impact of sacred groves in Western Ghats of Maharashtra. *Sci. & Cult.* **64(9-10)**:205-207.

Kumbhojkar M.S. and V.D.Vartak (1988). Ethnobotanicals studies on wild edible grapes from sacred groves in Western Maharashtra. J.Econ.Tax.Bot. **12(2)**:257-263.

Kumbhojkar M.S., Kulkarni D.K. and A.S. Upadhye (1991). Ethnobotany of *Cissus quadrangularis* L. from India. *Ethnobotany* 21-25.

Kumbhojkar M.S., Upadhye A.S. and D.K.Kulkarni (1996). Religious forest patches among Mahadeokoli tribal localities : Social cultural and environmental relationship. In:Ethnobiology in Human Welfare (Ed. S.K.Jain). Deep Publications, New Delhi, India, pp.349-351.

Kurion J.C. and B.V.Bhanu (1980). Ethnomedicine : The Nomadic Vaidus of Maharashtra. *The Eastern Anthropologist* **33(1)**:71-78.

Kurion, J.C. and R.T.Tribhuwan (1990). The traditional Medical Practitioners of Sahyadri. *The Eastern Anthropologist* **43(3)**:251-258.

Lakshminarasimhan P. and B.D.Sharma (1990). 'Pawari' – an adviasi musical instrument. (Repr.) *Bull. Bot.Surv. India* **30**(1-4):179.

Lakshminarasimhan P. and B.D.Sharma (1991). Flora of Nasik District. Bot. Surv. of India, Calcutta, India.

*Lisboa J.C. (1887). Notes on Mahabaleshwar and other Indian arrow root yielding plants. *J.Bombay Nat.Hist.Soc.*2:140-141.pp.1.

Magdum S.M. and R.M.Gejage (2004). Plants used for Respiratory Disorders by Ethnic Tribes in Chandoli Area of Sangali District. In : Proc. National Conference Human Health and Nutrition : A Biotechnological Approach. Department of Botany, B.N.Bandodkar College of Science, Thane (M.S.). pp.128-130.

Mahajan S.D. and V.V.Divan (1968). Contribution to the flora of Kolhapur district Vegetation and forest resources of Radhanagari area. J. Shivaji Univ. : 69-75.

Majgaonkar C. and S.G. Phadke (1999). Local medicinal plants from Koyna valley-4. Bhumyamlaki Phyllanthus niruri Linn. *Ayurved Update* **2(1)**:8-9.

Malhotra S.K. and S.Moorthy (1973). Some useful and medicinal plants of Chandrapur district, Maharashtra State. *Bull.Bot. Surv. India.* **15(1-2)**:13-21.

Manilal, K.S. (1999). History of plant explorations in Malabar : its relevance in the past and the present. *Rheedea* **9(2)**:183-184.

Mirashi M.V. and J.B.Salpekar (1975). Ornamental plants at Amaravati-I. *Botanique* **6(4)** 195-204.

Mokat D.N. and S.S.Deokule (2004). Plants used as veterinary medicines in Ratnagiri District of Maharashtra. *Ethnobotany* **16(1-2)**:131-135.

Mudaliar S.K., Kulkarni B.G. and B.D. Sharma (1987). Unpublished ethnobotanical information of G.M.Ryan and G.A.Gammie. *J.Econ.Tax. Bot.* **11(2)**:311-320.

Naik V.N. (1998). Flora of Marathwada Vol.1-2. Amrut Prakashan, Aurangabad, India.

Naser Rafiuddin and N.P.Vaikos (2002). Plants used in ethnoveterinary practices in Aurangabad district, Maharashtra-I. In : Proc.Plant Resource Development. Aurangabad. (Eds.A.M.Mungikar and A.S. Bhuktar) Dr.B.A.M. University, Aurangabad, pp.202-210.

Nilegaonkar Smitha, Vartak V.D. and R.G.Chitre (1985). Nutritional evaluation of some wild food plants from Pune and neighbouring districts, Maharashtra State. Part-I. *J.Econ.Tax.Bot.* **6(3)**:629-635.

Nipunge D.S., Kulkarni D.K. and V.D.Vartak (1993). Floristic studies on sacred groves from Sinhagad hills in Pune district, Maharashtra State. In :Higher Plants of Indian Subcontinent, (Additional series of Ind. J. Forestry, No.VII). **4**:153-159. Bishen Singh Mahendrapal Singh, Dehra Dun, U.P. India.

Nipunge D.S., Kumbhojkar M.S. and V.D. Vartak (1988). Studies on sacred groves of Maharashtra, Part-I. Observation of Sagdara grove in Pune district. *Ind. J. Forestry* **11(4)**:282-286.

Nipunge D.S., Vartak V.D. and M.S.Kumbhojkar (1992). Cultural heritage of sacred groves. *Proceedings* South Indian History Congress Pune. 1988. pp.170-176.

Padhye M.D., Deshmukh V.K. and V.J.Tiwari (1992). Ethnobotanical study of Korku tribe of Amaravati District, Maharashtra State (India). *Int. Jl. Pharmacogn.* **30(1)**:17-20.

Patil D.A. (1992) Key To The Angiospermic Families Of North Maharashtra. Pune Vidyarthi Griha Prakashan Pune (M.S.), India.

Patil D.A. (1998a). Vernacular plant names : Their origin and utility in Dhule district, Maharashtra. *Ethnobotany* **10**:130-132.

Patil D.A. (1998b). Census of genus Ipomoea L. in Dhule district Maharashtra. *J.Phytol.Res.* **11(2):**121-126.

Patil D.A. (2002) A few highly exploited species in a dire necessity of conservation from North Maharashtra (India) Journ. Non-Timber Forest Products **8(3-4)**:184-190.

Patil D.A. (2003). Flora of Dhule and Nandurbar Districts (Maharashtra). Bishen Singh Mahendra Pal Singh, Dehra Dun, India.

Patil D.A. (2005). Anthropogemic pressure on some species of forests in north Maharashtra. *Bull. Biol. Sci.* **3(2)**:91-95.

Patil D.A., M.V.Patil and Shubhangi Pawar (2004b). Ethnomedicinal remedies against diabetes In India : A systematic census. In : Ethnomedicine and Human Welfare Vol. III. Ukaaz Publications, Hyderabad, India, pp.90-117.

Patil D.A., M.V.Patil and Shubhangi Pawar (2007). Ethnobotanical aspects of crop weeds of North Maharashtra. In: Indigenous knowledge. Scientific Publishers, Jodhpur, India. (Ed. Prof.T.R. Sahu, Dr.H.S.Gour Univ., Sagar (In Press).

Patil D.A., S.C.Borse and P.B. Bhamare (1990). Medicinal plantlore of the tribals of Dhule (Maharashtra). *Biojounal* **2(l)**:47-54.

Patil D.A., Shubhangi Pawar and M.V.Patil (2002). Mahuwa tree and the aborigines of North Maharashtra (India) *Natural Product Radiance* **3(5)**:356-358

Patil M.V and D.A. Patil (2000a). Bases and utility of vernacular plant names in Nasik District (Maharashtra). *J.Swamy Bot.Cl.* **17(3-4)**:83-85.

Patil M.V. and D.A. Patil (2000b). Some more wild edible plants of Nasik District (Maharashtra). *Ancient Sci. Life* XIX **(3-4)** : 102-104.

Patil M.V. and D.A. Patil (2000c). Folk medicine of Nasik District (Maharashtra), India. *Ancient Sci. Life* XXI : 26-30.

Patil M.V. and D.A. Patil (2002). Some herbal remedies on stomach complaints in Nasik District (Maharashtra). In : Plant Resource Development, Proceeding of National Seminar held at Department of Botany, Dr.B.A.Maharashtra University, Aurangabad. (Eds Prof.A.M.Mungikar and A.S. Bhutkar), pp.123-127.

Patil M.V. and D.A.Patil (2001). Ethnoveterinary herbal medicines from Nasik District (Maharashtra). *Journal Non-timber Forest Products* **8(1-2)**:*19-24.*

Patil M.V. and D.A.Patil (2005a). Ethnomedicinal plantlore of Nasik District (Maharashtra) In : Proceeding of National Conference of 'Plant Diversity and Biotechnology', 10-11 Dec. 2004. P.G. Department of Botany, S.S.V.P.S's L.K.Dr.P.R.Ghogrey Science College, Dhule, Maharashtra (Eds. Nandan, S.N.; D.A.Patil, B.D.Borse and V.B.Salunkhe), pp.209-213.

Patil M.V. and D.A.Patil (2005c). Ethnomedicinal practices of Nasik District (Maharashtra). *Indian J. Traditional Knowledge* **4(3)**:287-290.

Patil M.V. and D.A.Patil (2006a). Certain ethnomedicinal practices amongst aborigines of Nasik District (Maharashtra) For human skin afflictions. In : Advances In Medicinal Plants, Asian Medicinal Plants and Health Care Trust, Jodhpur, India, Vol.II, pp.27-31.

Patil M.V. and D.A.Patil (2006c). Plants Used In Reproductive Ailments By Tribals of Nasik District. *Adv.Bio.Sci.*. **5(1)**:87-89.

Patil M.V., Pawar Shubhangi and D.A.Patil (2004). Native herbal recipes for respiratory afflictions from Jalgaon and Nasik Districts of Maharashtra. In: Proc. National Conference on Human Health and Nutrition : A Biotechnological Approach. Department of Botany, B.N.Bandokar College of Science, Thane (M.S.), pp.183-187.

Patil M.V., Pawar Shubhangi and D.A.Patil (2006). Ethnobotany of *Butea monosperma* in North Maharashtra (India). Natural Product Radiance **5(4)**:323-325.

Patil M.V., Pawar Shubhangi and D.A.Patil (2006d). Ethnomedicinal plants used for dental and oral healthcare in North Maharashtra. *J. Non-Timber Forest Products* **11(3)**: 208-210.

Patil S.L. and D.A. Patil (2007). Unconventional herbal medicine in Shirpur taluka of Dhule District (Maharashtra), India. *Natural Product Radiance* (In Press).

Patil M.V. and D.A.Patil (2005b). Ethnomedicinal practices of Nasik District, Maharashtra. Indian J. Traditional Knowledge **4(3)**:287-290.

Patil M.V. and D.A.Patil (2006b). Traditional Antivenin Practices In Nasik District of Maharashtra. *Flora and Fauna* **11(2)**:50-52.

Pawar Shubhangi and D.A. Patil (2006b). Herbal folk remedies against rheumatic disorders in Jalgaon District (Maharashtra). *Indian J. Traditional Knowledge* **5(3)**:314-316.

Pawar Shubhangi and D.A.Patil (2000a). Certain unconventional food plants of Jalgaon district, Maharashtra. *J. Non-Timber Forest Products* **7(3-4)**:229-232.

Pawar Shubhangi and D.A.Patil (2000b). The origin and utility of vernacular plant names in Khandesh region of Maharashtra. *J.Swamy Bot.Cl.* **17**:43-47.

Pawar Shubhangi and D.A.Patil (2001). Herbal folk medicines of Jalgaon district, Maharashtra. *Ancient Sci. Life.* **21(2)** : 87-91.

Pawar Shubhangi and D.A.Patil (2003). Observations on folkloric medicines of Jalgaon District (Maharashtra). *Indian J. Traditional Knowledge* 3(4):437-441.

Pawar Shubhangi and D.A.Patil (2004). Herbal folk remedies for urinogenital complaints in Jalgaon District of Maharashtra. *Ethnobotany* **16(1-2)**:86-90.

Pawar Shubhangi and D.A.Patil (2005a). Herbal folk medicines in Jalgaon District (Maharashtra) J.Phytol. Res. **18(2)**:251-254.

Pawar Shubhangi and D.A.Patil (2005b). A contribution to the ethnobotany of Jalgaon district (Maharashtra) : Fodder Resources. Adv.Biol.Sci. **5(1)**:79-82.

Pawar Shubhangi and D.A.Patil (2006a). Ethnobotanical aspects of naturalised exotic plants of Jalgaon District, J.Non-Timber Forest Products **13(1)**:55-59.

Pawar Shubhangi and D.A.Patil (2006d). Ethnobotanical aspects of naturalised exotic plants of Jalgaon District. *J.Non-Timber Forest Products* **13(1)**:55-59.

Pawar Shubhangi, M.V. Patil and D.A.Patil (2005) Ethnobotanical diversity of Acacias in North Maharashtra In : Proceeding of National Conference of 'Plant Diversity and Biotechnology', 10-11 Dec. 2004. P.G. Department of Botany, S.S.V.P.S's L.K.Dr.P.R. Ghogrey Science College, Dhule, Maharashtra (Eds. Nandan, S.N.; D.A.Patil, B.D.Borse and V.B.Salunkhe), pp.214-218.

Pawar Shubhangi, M.V. Patil and D.A.Patil (2006). Ethnomedicinal uses of Bel (Aegle marmelos (L.) Corr.) in Khandesh region of Maharashtra, *J. Non-Timber Forest Products* **13(2)**:131-133.

Pawar Shubhangi, Patil M.V. and D.A.Patil (2004). Fish stupefying plants used by tribals of North Maharashtra. *Ethnobotany* **16(1-2)**:136-138.

Pawar, Shubhangi and D.A. Patil (2006c). Native remedies against eye-complaints among the tribals and rural folks of Jalgaon District (Maharashtra) *Flora and Fauna*. **11(2)**:53-55.

Puri G.S. (1959). Survey of medicinal plants in western India. *Indian Journ.Pharm.* **20**:8.

Puri G.S. and S.K.Jain (1960). Survey of some oil-yielding plants of western India. *Bull.Bot.Surv.India.* **2(1-2)**:95-98.

Rao R.S. (1964). The Doum palms in India. *Principes* **8**:49-54.

Rathor O.S., Zare M.K., Kagne R.M. and V.B. Chavan (2002). Ethnobotanical studies in some less known plants of

Marathwada. In : Proc.Plant Resource Development. (Ed.A.M.Mungikar and A.S. Bhuktar) Dr.B.A.M. University, Aurangabad, India, pp.86-89.

Rohidas S.B. and G.J.Bankar (2002). Marathwada : Rich source of medicinal plants In : Plant Resource Development. Proc. of National Seminar held at Department of Botany, Dr.B.A.M. University, Aurangabad, India (Ed.A.M.Mungikar), pp.224-228.

*Roxburgh William (1832). Flora Indica (Descriptions of Indian Plants) Vol. 3 Serampore, Calcutta, India.

Ryan G.M. (1903). Famine foods : *Dioscorea pentaphylla* an important edible. wild yam of the Thane district, Bombay Presidency, *J.Bombay Nat.Hist.Soc.* **14**:772-775.

Santapau H. (1956). The poisonous qualities of *Calotropis gigantea* R.Br. J.Bombay *Nat.Hist.Soc.* **54**:218.

Sensarma Priyadarsan (1984). Ethnobotanical investigation in the Indian *Puranas* V.The *Kurma Purna J.Econ.Tax.Bot.* **5(3)**:634-640.

Shah G.L. (1984). Some economically important plant of Salsette island near Bombay. *J.Econ.Tax.Bot.*5**(4)**:753-566.

Shah G.L., Yadav S.S. and V.Badrinath (1983). Medicinal plants from Dahanu forest divison in Maharashtra State. *J.Econ. Tax.Bot.* **4**:141-151.

Sharma B.D. and B.G. Kulkarni (1980). Floristic composition and peculiarities of Dev Raies (Sacred groves) in Kolhapur District. *J.Econ.Tax.Bot.* **1**:11-32.

Sharma B.D. and P.Lakshminarasimhan (1986). Ethnobotanical studies on the tribals of Nasik district (Maharashtra). *J.Econ.Tax.Bot.* **8(2)**:439-454.

Sharma B.D. and S.K.Malhotra (1984). A contribution to the ethnobotany of tribal areas in Maharashtra. *J.Econ.Tax. Bot.* **5(3)**:533-537.

Sharma B.D., Karthikeyan S. and N.P.Singh (1996). Flora of Maharashtra State : Monocotyledones. Bot.Surv. India, Calcutta, India.

Shekhawat G.S. and Sunil Anand (1984). An ethnobotanical profile of Indian desert. *J.Econ.Tax.Bot.* **5(3)**:591-598.

Shisode S.B. and D.A.Patil (1993). Native medicinal plants of Jalgaon district Maharashtra. *Biojournal* **5(1-2)**:79-82.

Singh N.P. and S.Karthikeyan (2000a). Flora of Maharashtra State : Dicotyledons Vol.I Bot.Surv. India, Calcutta, India.

Singh N.P. and S.Karthikeyan (2000b). Flora of Maharashtra State Dicotyledons Vol.1 (Asstt. Lakshminarasimhan P. and P.V. Prasana) and (2001). Flora of Maharashtra Dicotyledons Vol.2 Bot. Surv. India, Calcutta, India.

Singh N.P. and S.Karthikeyan (2001). Flora of Maharashtra State : Dicotyledons Vol.II. Bot. Surv. India, Calcutta, India.

Tarafder C.R. (1984). Ethnogynaecology in relation to plants Part-III. Plants used to accelerate delivery and Pre & post-natal complaints. *J.Econ.Tax. Bot.* **5(3)**:572-576.

Tayade S.K. and D.A. Patil (2003a). Hitherto untapped plantlore from Nandurbar District (Maharashtra). *Natural Product Radiance* **4(1)**:46-50.

Tayade S.K. and D.A. Patil (2003b). Census of fodder plants used by tribals of Nandurbar District. *J. Swamy Bot. Cl.* **21(3-4)**:81-84.

Tayade S.K. and D.A.Patil (2005). Ethnomedicinal traditions of tribals of Nandurbar District (Maharashtra). J.Phytol. Res. **18(2)**:251-254.

Tayade S.K. and D.A.Patil (2006). Ethnomedicinal wisdom of tribals of Nandurbar District (Maharashtra). *Natural Product Rediance,* **5(1)**:64-69.

Tekale N.S. (2002). Plant resource development for upliptment of Warli tribe in Thane district of Maharashtra. In:Proc.Plant Resource Developments Aurangabad. (Eds. A.M. Mungikar and A.S.Bhuktar) Dr.B.A.M. University, Aurangabad, India, pp.23-33.

Tribhuwan Robin D. and Preeti Peters (1992). Medico-ethnobiology of the Katkaris and Thakurs. *Tribal Res.Bull.* **14**(1):20-24.

Upadhye A.S. and D.K.Kulkarni (2004). Traditional phytotherapy for insect bites among the tribals in Western India *Ethnobotany* **16(1-2)**:59-61.

Upadhye A.S. and M.S. Kumbhojkar (1998). *Schleichera oleosa* (Laur.) Oken. Tree : A boon for tribals. *Ethnobotany* **10**:38-41.

Upadhye A.S., Kumbhojar M.S. and D.K.Kulkarni (1997a). Ethno-medico-botany of some sacred plants of western Maharashtra. *Ethnobotany* **9**:65-68.

Upadhye A.S., Kumbhojkar M.S. and D.K. Kulkarni (1997b). Medico-botany of genus *Mucuna* from western Maharashtra. *B.M.E. B.R.* **8(1-2)**:51-54.

Upadhye Anuradha S. and M.S. Kumbhojkar (1992). Ethnobotany of *Madhuca* from western Maharashtra. *J.Econ.Tax. Bot.* Add. Ser. **10**:70-81.

Upadhye Anuradha S., and M.S.Kumbhojkar (1996). Ethnobotany of genus Sterculia L. from India *J.Econ.Tax. Bot.* Add.Ser.12. Scientific Publishers, Jodhpur (India), pp.40-42.

Upadhye Anuradha S., Kumbhojkar M.S. and V.D.Vartak (1986). Observations on wild plants used in folk medicine in the rural areas of Kolhapur district. *Ancient Sci. Life* **6(2)**:119-121.

Upadhye Anuradha S., S.Kumbhojkar M.S. and V.D.Vartak (1988). Cultivation of medicinal plants along sea-coast in Maharashtra. *B.M.E.B.R.* **7(1-2)**:32-39.

Upadhye Anuradha S., Vartak V.D. and M.S.Kumbhojkar (1994). Ethno-medico-botanical studies in Western Maharashtra, India. *Ethnobotany* **6**:25-34.

Uphadye A.S., Kumbhojkar M.S. and D.K.Kulkarni (1999). Tribal culture and folk medicines from Western Maharashtra. *Sci. & Cult.* **65(9-10)**:288-289.

Vartak V.D. (1957). Aaushadhi vanaspati (Katraj ghat and Parvati). *Ayurvidya* **20(9)**:1-10.

Vartak V.D. (1959a). Medicinal plants from the hilly regions of Poona and Satara districts, Bombay State, A and B. *Ayurvidya* **23**:11-12.

Vartak V.D. (1959b). Some edible wild plants from the hilly regions of Poona district, Bombay State. *J.Bombay Nat.Hist.Soc.* **56**:8-25.

Vartak V.D. (1959c). Medicinal plants from hilly regions of Pune and Satara districts. Bombay state. *Ayurvidya* **22(11)**:455-458.

Vartak V.D. (1980). Note on ambemohar pat (*Pandanus amaryllifolius* Roxb.) from Western India. *J.Bombay Nat.Hist. Soc.* **78**:196-198.

Vartak V.D. (1981a). Observations on wild edible plants from hilly regions of Maharashtra and Goa. In : Glimpses of Indian Ethnobotany (Ed.S.K.Jain), Oxford and IBH Publishing Co., New Delhi, India, pp.231-241.

Vartak V.D. (1981b). Relief forest pockets of Panshet water-catchment area Poona district, Maharashtra State. *Biovigyanum.* **7**:145-148.

Vartak V.D. (1982a). Observations on multifarious uses of sedges from Western India. *J.Econ. Tax. Bot.* **3**:193-196.

Vartak V.D. and Madhav Gadgil (1980). Studies in ethnobotany : A new vistas in botanical sciences. *Biovigyanam* **6**:151-156.

Vartak V.D. and Madhav Gadgil (1981). Studies on sacred groves along the western ghats from Maharashtra and Goa. : Role of Beliefs and Folklores. In : Glimpses of Indian Ethnobotany. (Ed.S.K.Jain), Oxford and IBH Publishing Co., New Delhi, pp.272-278.

Vartak V.D. and Madhav Gadgil (1990). Studies on sacred groves along Western Ghats in Maharashtra and Goa : Role of Belief and folklores. In : Contributions to Ethnobotany of India (Ed.S.K.Jain). Scientific Publishers, Jodhpur, India, pp.329-335.

Vartak V.D. and Rekha Mandavgane (1981). Enumeration of medicinal plants from Karnala tribal area Kolaba district, Maharashtra State. *J.Univ.Poona. (Sci.&Tech.)* **54**:91-99.

Vartak V.D. and V.Ghate (1994). Alu : *Meyna laxiflora* Robyns, a less known but promising wild fruits tree for tribal areas of Western Maharashtra. In : Higher plants of Indian Subcontinent. 3:99-103. (Additional series of Ind. J. Forestry No.4).

Vartak V.D. and Vinaya Ghate (1990). Ethnobotany of Neem. *Boil.Ind.* **1(1)**:55-59.

Vartak V.D.(1982b). Studies on ethnobotany of Maharashtra and Goa. *Tribal Res.Bull.* **4(2)**:9-10.

Vartak V.D., Kumbhojkar M.S. and D.S. Nipunge (1987). Sacred groves of tribal areas along the Western Ghats : Treasure grove of medicinal plants. *B.M.E.B.R.* **8(1-2)**:77-84.

Vartak V.D., Kumbhojkar M.S. and Vinaya S.Ghate (1985). Studies on living hedges from Western Maharashtra and Goa. *J.Econ.Tax.Bot.* **6(2)**:272-282.

Vartak V.D.and D.K.Kulkarni (1987). Monsoon wild leafy vegetables from hilly regions of Pune and neighbouring districts, Maharashtra State. *J.Econ.Tax.Bot.* **11(2)**:231-235.

Ved Prakash and B.N. Mehrotra (1987). Ethnobotanical studies on the flora of Kandala Maharashtra State. *J.Econ. Tax.Bot.* **9(1)**:205-208.

Wangikar B.P. (2002). Study of four edible species of Marchella from Aurangabad district. In : Proc. Plant Resource Development Aurangabad. (Eds.A.M.Mungikar and A.S.Bhuktar) Dr.B.A.M. University, Aurangabad, pp.74-76.

Watt George (1889-1893). A Dictionary of Economic Products of India Vol. 1-6. Periodical Expert, Shahadara, Delhi, India.

Yadav S.S. and P.B.Bhamare (1989). Ethnomedico-botanical studies of Dhule forests in Maharashtra. *J.Econ.Tax. Bot.* **13(2)**:455-460.

* Original not seen.

Index to Plant Names

PART I

PART II

Index to Local Names

PART I

PART II